TOPICS IN MAGNETOHYDRODYNAMICS

Edited by **Linjin Zheng**

Topics in Magnetohydrodynamics
Edited by Linjin Zheng

Published by InTech
Janeza Trdine 9, 51000 Rijeka, Croatia

Publishing Process Manager Silvia Vlase

Technical Editor Teodora Smiljanic

Cover Designer InTech Design Team

Additional hard copies can be obtained from orders@intechopen.com

Topics in Magnetohydrodynamics, Edited by Linjin Zheng
p. cm.
978-953-51-0211-3

For EU product safety concerns:
IN TECH d.o.o., Prolaz Marije Krucifikse Kozulić 3, 51000 Rijeka, Croatia,
info@intechopen.com or visit our website at intechopen.com.

Meet the editor

Dr Linjin Zheng specializes in both analytic theory and large-scale numerical computation of magnetically confined plasmas. He received his M. S. degree from The University of Science and Technology of China and Ph. D. from Institute of Physics – Beijing, Chinese Academy of Sciences. He is currently working at Institute for Fusion Studies, The University of Texas at Austin. He published more than hundred scientific papers for examples in Phys. Rev. Letts, Phys. Letts, Nucl. Fusion, Phys. of Plasmas, and major conferences. His researches cover both ideal/resistive MHD and kinetic theories for MHD modes and drift waves, etc. His major contributions with his colleagues include the discovery of 2nd toroidal Alfven egenmodes, reformulation of gyrokinetic theory, development of physical interpretation of so-called edge localized modes, etc. He also developed the AEGIS and AEGIS-K codes.

Contents

Preface

Magnetohydrodynamics (MHD) is concerned with the dynamic motion of electrically conducting medium (i.e., so-called plasma) in the presence of magnetic field, which is either applied externally or self-generated. MHD is traditionally interpreted as follows: "Magneto" has to do with electro-magnetic fields; "hydro" indicates to deal with fluid medium; while "dynamics" designates to study its time evolution of physical processes. Another kind of interpretation can, however, be as follows: "Magnetohydro" refers to the "magneto" type of fluids, in which particles are stuck together primarily by magnetic field, instead of by collisions. Due to the presentation of strong magnetic field, charged particles gyrate around magnetic field lines, so that the guiding center of charged particle's gyro-motion is primarily frozen in the magnetic field line. Therefore, even in the weak collision case charged particles can move collectively as if fluid in the perpendicular direction. The later interpretation covers a broad scope in which MHD theory applies.

MHD theory has been used to study controlled nuclear fusion, space/astronomic plasma physics, such as solar physics, and low temperature plasma physics, such as plasma processing, MHD generator, etc. The collective motion of charged particles is an intrinsically complicated issue, due to the long-range correlation feature of Coulomb and Lorentz forces. Nevertheless, MHD theory is able to catch the leading behaviors in a clear and tractable manner. This has led to the success of MHD theory.

There are several excellent MHD books in the market. Also, MHD theory is described about in every plasma physics books. Nevertheless, as plasma physics science advances, the gap between cutting-edge researches and textbooks always develops. To fill in this gap, several MHD-related topics are reviewed by eight authors in this book, according to authors' expertise.

In Chapter 1 an overview of MHD theory for toroidal plasma confinement in controlled nuclear fusion is given. Especially, the toroidal theories for four major types of MHD modes: interchange, ballooning, toroidal Alfven egenmodes, and kinetically driven modes (such as energetic particle modes) are reviewed. In Chapter 2 the recent development of sub-fluid models in dissipative MHD is reviewed. Especially, three models to approach "stochastic coherent structures theory" are described. In Chapter 3 the implicit numerical methods for resistive MHD is reviewed. Especially two broad classes of nonlinear methods: Newton-Krylov and nonlinear multigrid are detailed. In Chapter 4 magnetic relaxation in spheromaks is investigated and reviewed. Simulation results show that full relaxation to Taylor state is only achieved when the magnetic fluctuations produce stochastic field line regions of size comparable of that of the whole system. In Chapter 5 MHD activity in an extremely high-beta compact device: field-reversed configuration (FRC) is reviewed. Both the formation methods for FRC plasmas and the MHD behaviors are described. In Chapter 6 MHD theory for solar spicules and X-ray jets is presented, to explain the enduring mystery in solar physics for why Sun's outer atmosphere or corona is much hotter than its surface. In Chapter 7 Hamiltonian representation of MHD for boundary energy controls is described. A passive boundary control formalism for ideal MHD systems is then derived from the distributed port-Hamiltonian (DPH) representations. In Chapter 8 the MHD rotating flow of a fourth grade fluid between two parallel infinite plates is investigated through solving nonlinear momentum and mass equations. The results for Newtonian and non-Newtonian fluids are compared.

These results are published in this book to take the advantage of open-access feature for distribution through InTech.

Dr. Linjin Zheng
The University of Texas at Austin,
Institute for Fusion Studies, Austin, Texas,
USA

1

Overview of Magnetohydrodynamics Theory in Toroidal Plasma Confinement

Linjin Zheng
Institute for Fusion Studies
University of Texas at Austin, Austin, Texas
United States of America

1. Introduction

In this chapter we address magnetohydrodynamics (MHD) theory for magnetically confined fusion plasmas. To be specific we focus on toroidal confinement of fusion plasmas, especially tokamak physics.

The biggest challenges mankind ever faces are falling energy sources and food shortages. If controlled nuclear fusion were achieved with net energy yield, the energy source problem would be solved. If natural photosynthesis were reproduced, food shortage concern would be addressed. Though both nuclear fusion and photosynthesis are universal, the difficulties to achieve them are disproportionally great. Instead, those discoveries harmful to nature, though naturally unpopular, are invented relatively easily. This tendency reminds us of a bible verse (Genesis 3:19): "In the sweat of thy face shalt thou eat bread". This verse basically sketches the dependence of efforts (sweat) demanded for scientific discoveries on the usefulness (bread) of the discoveries to mankind (see Fig. 1). The more the discovery is relevant to mankind, the more the sweat is needed for that discovery. This may explain why controlled nuclear fusion is so difficult and its underlying plasma physics is so complicated.

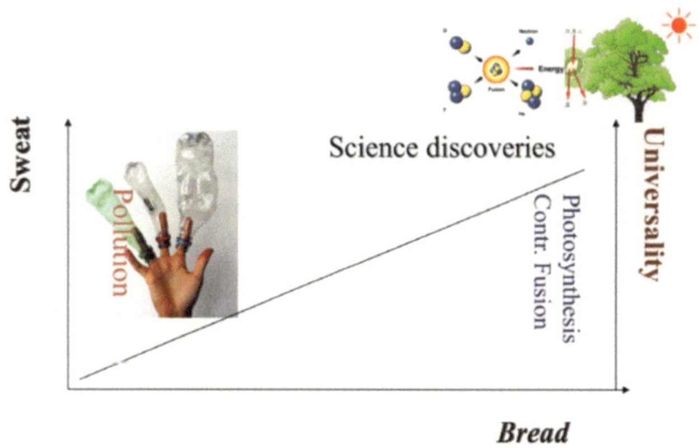

Fig. 1. Schematic interpretation of Genesis 3:19: the dependence of efforts (sweat) demanded for scientific discoveries on the usefulness (bread) of the discoveries to mankind.

When God created universe each day, He always claimed "it was so" and "it was good" (Genesis 1). The hard aspect of fusion plasma physics lies in that we often miss simplicity (it was so) and beauty (it was good) in theoretical formalism. MHD theory seems to be unique in plasma physics. Though many charged particle system, with long mean free path and long range correlation, is intrinsically complicated, MHD theory is relatively simple and, nonetheless, gives rise to rather relevant theoretical predictions for experiments: Tokamaks are designed according to MHD equilibrium theory and nowadays none would expect that a magnetic confinement of fusion plasmas could survive if MHD theory predicted major instabilities. As one will see even with MHD description the theoretical formulation of magnetically confined plasmas in toroidal geometry can still be hard to deal with. Thanks to decades-long efforts many beautiful MHD theoretical formulations for toroidal confinement of fusion plasmas have been laid out. In this chapter we try to give a comprehensive review of these prominent theories. Four key types of modes: interchange/peeling modes Mercier (1962) Greene & Johnson (1962) Glasser et al. (1975) Lortz (1975) Wesson (1978), ballooning modes Connor et al. (1979) Chance et al. (1979), toroidal Alfvén egenmodes (TAEs) Cheng et al. (1985) Rosenbluth et al. (1992) Betti & Freidberg (1992) Zheng & Chen (1998), and kinetically driven modes, such as kinetic ballooning modes (KBMs) Tsai & Chen (1993) Chen (1994) and energetic particle driven modes (EPMs) Zheng et al. (2000), are addressed. Besides, we also describe an advanced numerical method (AEGIS Zheng & Kotschenreuther (2006)) for systematically investigating MHD stability of toroidally confined fusion plasmas. Description of global formulation used for numerical computation can also provide an overall picture of MHD eigen mode structure for toroidal plasmas.

MHD theory is a single fluid description of plasmas. Fluid approach is based on the assumption that particle movements are spatially localized so that a local thermal equilibrium can be established. In the conventional fluid theory particle collision is the ingredient for particle localization. However, for magnetically confined fusion plasmas collision frequency usually is low. One cannot expect particle collisions to play the role for localizing particles spatially. The relevance of partial fluid description of magnetically confined fusion plasmas relies on the presence of strong magnetic field. Charged particles are tied to magnetic field lines due to gyro-motions. Therefore, in the direction perpendicular to magnetic field lines magnetic field can play the role of localization, so that MHD description becomes relevant at least in lowest order. One can expect that perpendicular MHD description needs modification only when finite Larmor radius effects become significant.

In the direction parallel to magnetic field, however, particles can move rather freely. Collision frequency is not strong enough hold charged particles together to establish local thermal equilibrium. One cannot define local thermal parameters, such as fluid density, velocity, temperature, etc. The trapped particle effect, wave-particle resonances, and parallel electric field effects need to be included. Plasma behavior in parallel direction is intrinsically non-fuild and needs kinetic description. Surprisingly, even under this circumstance MHD description still yields valuable and relevant theoretical predictions without major modifications in the concerned low ($\omega \ll \omega_{si}$) and intermediate ($\omega_{si} \ll \omega \ll \omega_{se}$) frequency regimes, where ω is mode frequency, ω_{si} and ω_{se} represent respectively ion and electron acoustic frequencies. In the low frequency regime coupling of parallel motion results only in an enhanced apparent mass effect; while in the intermediate frequency regime kinetic effects only gives rise to a new phenomenological ratio of special heats in leading order.

This chapter is arranged as follows: In Sec. 2 the basic set of ideal MHD equations is described; In Sec. 3 MHD equilibrium is discussed; In Sec. 4 analytical or semi-analytical theories for four types of major MHD modes are presented; In Sec. 5 the formulation of global numerical analyses of MHD modes are given; In the last section the results are summarized. Gyrokinetic and resistive effects are also discussed in this last section.

2. Basic set of ideal MHD equations

The basic set of ideal MHD equations are derived from single fluid and Maxwell's equations. They are given as follows

$$\rho \frac{d\mathbf{v}}{dt} = -\nabla P + \mathbf{J} \times \mathbf{B}, \tag{1}$$

$$\mathbf{E} = -\mathbf{v} \times \mathbf{B}, \tag{2}$$

$$\frac{\partial P}{\partial t} = -\mathbf{v} \cdot \nabla P - \Gamma P \nabla \cdot \mathbf{v}, \tag{3}$$

$$\frac{\partial \rho_m}{\partial t} = -\mathbf{v} \cdot \nabla \rho_m - \rho_m \nabla \cdot \mathbf{v}, \tag{4}$$

$$\mu_0 \mathbf{J} = \nabla \times \mathbf{B}, \tag{5}$$

$$\frac{\partial \mathbf{B}}{\partial t} = \nabla \times \mathbf{E}, \tag{6}$$

where ρ_m is mass density, $\mathbf{v}$ denotes fluid velocity, P is plasma pressure, Γ represents the ratio of specific heats, $\mathbf{E}$ and $\mathbf{B}$ represents respectively electric and magnetic fields, $\mathbf{J}$ is current density, μ_0 is vacuum permeability, and bold faces denote vectors.

The MHD equations (1)-(6) can be linearized. For brevity we will use the same symbols for both full and equilibrium quantities. Perturbed quantities will be tagged with δ, unless specified. Equilibrium equations are

$$\mathbf{J} \times \mathbf{B} = \nabla P, \tag{7}$$

$$\nabla \times \mathbf{B} = \mu_0 \mathbf{J}, \tag{8}$$

$$\nabla \cdot \mathbf{B} = 0. \tag{9}$$

The linearized perturbed MHD equations become

$$-\rho_m \omega^2 \boldsymbol{\xi} = \delta\mathbf{J} \times \mathbf{B} + \mathbf{J} \times \delta\mathbf{B} - \nabla \delta P, \tag{10}$$

$$\delta\mathbf{B} = \nabla \times \boldsymbol{\xi} \times \mathbf{B}, \tag{11}$$

$$\mu_0 \delta\mathbf{J} = \nabla \times \delta\mathbf{B}, \tag{12}$$

$$\delta P = -\boldsymbol{\xi} \cdot \nabla P - \Gamma P \nabla \cdot \boldsymbol{\xi}, \tag{13}$$

where $\boldsymbol{\xi} = \mathbf{v}/(-i\omega)$ represents plasma displacement, and the time dependence of perturbed quantities is assumed to be of exponential type $\exp\{-i\omega t\}$. Inserting Eqs. (11)-(13) into Eq. (10), one obtains a single equation for $\boldsymbol{\xi}$:

$$-\rho_m \omega^2 \boldsymbol{\xi} = \frac{1}{\mu_0} \nabla \times (\nabla \times \boldsymbol{\xi} \times \mathbf{B}) \times \mathbf{B} + \mathbf{J} \times \nabla \times \boldsymbol{\xi} \times \mathbf{B} + \nabla (\boldsymbol{\xi} \cdot \nabla P + \Gamma P \nabla \cdot \boldsymbol{\xi}). \tag{14}$$

We have not included toroidal rotation effects in the linearized equations (10)-(13). For most of tokamak experiments rotation is subsonic, i.e., the rotation speed is much smaller than ion thermal speed. In this case the centrifugal and Coriolis forces from plasma rotation is smaller than the effects from particle thermal motion — plasma pressure effect. Therefore, rotation effect can be taken into account simply by introducing the Doppler frequency shift: $\omega \to \omega + n\Omega_{not}$ in MHD equation (14), where Ω_{rot} is toroidal rotation frequency and n denotes toroidal mode number Waelbroeck & Chen (1991) Zheng et al. (1999).

3. Tokamak MHD equilibrium

In this subsection we discuss tokamak equilibrium theory. MHD equilibrium has been discussed in many MHD books. Here, we focus mainly on how to construct various flux coordinates from numerical solution of MHD equilibrium equations.

We first outline the derivation of Grad-Shafranov equation Grad & Rubin (1958) Shafranov (1966). The cylindrical coordinate system $(X,\ Z,\ \phi)$ is introduced, where X is radius from axi-symmetry axis of plasma torus, Z denotes vertical coordinate, and ϕ is toroidal axi-symmetric angle. We introduce the vector potential $\mathbf{A}$ to represent magnetic field $\mathbf{B} = \nabla \times \mathbf{A}$. Due to toroidal symmetry ϕ is an ignorable coordinate. Using curl expression in cylinder coordinates and noting that $\partial A_X/\partial\phi = \partial A_Z/\partial\phi = 0$, one can prove that the vector potential $\mathbf{A}$ in X and Z directions ($\mathbf{A}_{XZ}$) can be expressed through the single toroidal component: $\mathbf{A}_\phi$. Without losing generality one can express $\mathbf{A}_\phi = -\chi\nabla\phi$. Therefore, total equilibrium magnetic field can be expressed as, by adding (X, Z) components and toroidal component,

$$\mathbf{B} = \nabla \times \mathbf{A}_\phi + XB_\phi\nabla\phi = \nabla\phi \times \nabla\chi + g\nabla\phi, \tag{15}$$

where B_ϕ is toroidal component of magnetic field and $g = XB_\phi$. From Eq. (15) one can prove that $\mathbf{B} \cdot \nabla\chi = 0$ and therefore $\chi = const.$ labels magnetic surfaces. Equation (15) can be used to show that $2\pi\chi$ is poloidal magnetic flux. One can also define the toroidal flux $2\pi\psi_T(\chi)$. The safety factor is then defined as $q = d\psi_T/d\chi$, which characterizes the field line winding on a magnetic surface.

Using Ampere's law in Eq. (8) one can express equilibrium current density as follows

$$\mu_0\mathbf{J} = \nabla g \times \nabla\phi + X^2\nabla \cdot \left(\frac{\nabla\chi}{X^2}\right)\nabla\phi. \tag{16}$$

Here, we have noted that $\nabla\phi \cdot \nabla \times (\nabla\phi \times \nabla\chi) = \nabla \cdot (\nabla\chi/X^2)$ and $\nabla\theta \cdot \nabla \times (\nabla\phi \times \nabla\chi) = \nabla\chi \cdot \nabla \times (\nabla\phi \times \nabla\chi) = 0$.

Inserting Eqs. (15) and (16) into force balance equation (7) one obtains

$$\mu_0\nabla P = -\nabla \cdot \left(\frac{\nabla\chi}{X^2}\right)\nabla\chi - \frac{1}{X^2}g\nabla g + \nabla\phi\nabla g \times \nabla\phi \cdot \nabla\chi. \tag{17}$$

From Eq. (7) one can prove that $\mathbf{B} \cdot \nabla P = 0$. Therefore, one can conclude that plasma pressure is a surface faction, i.e., $P(\chi)$. From Eq. (17) one can further determine that g is a surface function as well, through projecting Eq. (17) on $\nabla\phi$. Therefore, Eq. (17) can be reduced to the so-called Grad-Shafranov equation

$$X^2\nabla \cdot \left(\frac{\nabla\chi}{X^2}\right) = -\mu_0 X^2 P'_\chi - gg'_\chi. \tag{18}$$

Here and later on we use prime to denote derivative with respect to flux coordinate chosen. This is a nonlinear equation for χ for given functions $P(\chi)$ and $g(\chi)$. It generally needs numerical solution. Since it is a two dimensional problem, one needs to introduce a poloidal angle coordinate θ_{eq} around magnetic axis of plasma torus in addition to radial coordinate χ. The solution is usually given in (χ, θ_{eq}) grids for $X(\chi, \theta_{eq})$, $Z(\chi, \theta_{eq})$, or inversely, in (X, Z) grids for $\chi(X, Z)$, $\theta_{eq}(X, Z)$.

Instead of physical cylinder coordinates (X, Z, ϕ) or (χ, θ_{eq}, ϕ), magnetic flux coordinates are often used in theoretical analyses, which is characterized by that the magnetic field line is straight in the covariant representation of coordinate system. Note that the coordinate system $(\psi, \theta_{eq}, \phi)$ usually is not a flux coordinate system. In most equilibrium codes θ_{eq} is just an equal-arc length poloidal coordinate. One of flux coordinate systems is the so-called PEST coordinate system Grimm et al. (1976) (χ, θ_{pest}, ϕ), where θ_{pest} is generalized poloidal coordinate, such that the equilibrium magnetic field can be represented as

$$\mathbf{B} = \chi' \left(\nabla\phi \times \nabla\psi_{pest} + q\nabla\psi_{pest} \times \nabla\theta_{pest} \right). \tag{19}$$

By equating Eqs. (19) and (15) one can find that the Jacobian of PEST coordinates should be

$$\mathcal{J}_{pest} \equiv \frac{1}{\nabla\psi_{pest} \times \nabla\theta_{pest} \cdot \nabla\phi} = \frac{q\chi' X^2}{g}. \tag{20}$$

In PEST coordinate system the flux coordinate is chosen as

$$\psi_{pest} = \frac{2\pi X_0}{c_{pest}} \int_0^{\chi} d\chi \frac{q}{g}, \qquad c_{pest} = \frac{X_0}{2\pi} \int_v d\tau \frac{1}{X^2},$$

where X_0 is major radius at magnetic axis and $\int_v d\tau$ denotes volume integration over entire plasma domain. The PEST poloidal angle θ_{pest} can be related to physical angle coordinate θ_{eq} as follows. Using Eq. (20), one can determine poloidal angle in PEST coordinate

$$\theta_{pest} = \frac{1}{q} \int_0^{\theta_{eq}} d\theta_{eq} \frac{g\mathcal{J}_{eq}}{X^2},$$

where $\mathcal{J}_{eq} = 1/\nabla\chi \times \nabla\theta_{eq} \cdot \nabla\phi$, which can be computed from equilibrium solution. Here, the integration is along the path of constant χ and ϕ.

Next, we discuss construction of general flux coordinates. The covariant type of representation as in Eq. (19) is not unique. It is preserved under the following coordinate transforms

$$\zeta = \phi + \nu(\psi, \theta), \quad \theta = \theta_{pest} + \nu(\psi, \theta)/q, \tag{21}$$

such that

$$\mathbf{B} = \chi' \left(\nabla\zeta \times \nabla\psi + q\nabla\psi \times \nabla\theta \right). \tag{22}$$

Here, θ and ζ are referred to as generalized poloidal and toroidal angles, respectively. PEST coordinates are characterized by its toroidal angle coordinate being axisymmetric toroidal angle. In this general case, by equating Eqs. (22) and (15) in $\nabla\phi$ projection one can find that

$$\left.\frac{\partial\nu}{\partial\theta_{eq}}\right|_{\psi,\phi} \frac{1}{\mathcal{J}_{eq}} + g\frac{1}{X^2} = \chi' q \frac{1}{\mathcal{J}}, \tag{23}$$

where $\mathcal{J} = 1/\nabla\psi \times \nabla\theta \cdot \nabla\zeta$. Using $\mathcal{J}_{eq}$ and $\mathcal{J}$ definitions, one can prove that

$$\left.\frac{\partial\theta}{\partial\theta_{eq}}\right|_{\psi,\phi} = \frac{\chi' \mathcal{J}_{eq}}{\mathcal{J}}. \tag{24}$$

One can solve Eq. (23), yielding

$$\nu(\psi,\theta) = \int_0^{\theta_{eq}} d\theta_{eq}\mathcal{J}_{eq}\left(\chi' q \frac{1}{\mathcal{J}} - g\frac{1}{X^2}\right) = q\theta - \int_0^{\theta_{eq}} d\theta_{eq}\frac{g\mathcal{J}_{eq}}{X^2}, \tag{25}$$

where Eq. (24) has been used.

Equations (21)-(25) can be used to construct various types of flux coordinate systems. There are two classes of them: One is by specifying Jacobian (e.g., Hamada coordinates Hamada (1962) and Boozer coordinates Boozer (1982)) and the other by directly choosing generalized poloidal angle (e.g., equal arc-length coordinate). In the Hamada coordinates the volume inside a magnetic surface is used to label magnetic surfaces, i.e., $\psi = V$, and Jacobian $\mathcal{J}_h = 1/\nabla V \cdot \nabla\theta_h \times \nabla\zeta_h$ is set to be unity. With Jacobian specified, Eq. (24) can be used to solve for θ_h at given (V,ϕ). With ν determined by Eq. (25) the definition Eq. (21) can be used to specify ζ_h. In the Boozer coordinates Jacobian is chosen to be $\mathcal{J}_B = V'\left\langle B^2\right\rangle_s/(4\pi^2 B^2)$, where $\langle\cdot\rangle_s$ represents surface average. The procedure for specifying Boozer poloidal and toroidal coordinates θ_B and ζ_B is similar to that for Hamada coordinates. In the equal-arc-length coordinates poloidal angle is directly specified as equal-arc-length coordinate θ_e. In this case, Jacobian $\mathcal{J}_e$ can be computed through Eq. (24). With ν determined by Eq. (25) the definition Eq. (21) can be used to specify ζ_e.

We can also express current density vector in covariant representation with generalized flux coordinates. Using Ampere's law in Eq. (8) for determining $\mathbf{J} \cdot \nabla\theta$ and Eq. (7) for $\mathbf{J} \cdot \nabla\zeta$, one can also express equilibrium current density in covariant representation

$$\mathbf{J} = -\frac{1}{\mu_0}g'_\psi \nabla\zeta \times \nabla\psi - \left(\frac{q}{\mu_0}g'_\psi + \frac{P'_\psi}{\chi'_\psi}\mathcal{J}\right)\nabla\psi \times \nabla\theta. \tag{26}$$

This general coordinate expression for $\mathbf{J}$ can be alternatively obtained from PEST representation in Eq. (16) and Grad-Shafranov equation (18) through coordinate transform. Equation (26) is significantly simplified in the Hamada coordinates. Due to $\mathcal{J} = 1$, Eq. (26) can be expressed as

$$\mathbf{J} = J'_V \nabla\zeta \times \nabla V + I'_V \nabla V \times \nabla\theta, \tag{27}$$

where $I(V)$ and $J(V)$ are toroidal and poloidal current fluxes, $I' = -g'_V/\mu_0$, and $J' = -qg'_V/\mu_0 - P'_V/\chi'_V$. The force balance equation (7) can be simply expressed as

$$\mu_0 P'_V = J'_V \psi'_V - I'_V \chi'_V. \tag{28}$$

It is also interesting to discuss diamagnetic current and Pfirsch-Schlüter current in plasma torus. Due to the existence of plasma pressure there is diamagnetic current in tokamak system. The diamagnetic current alone is not divergence-free and is always accompanied by a return current in the parallel direction, i.e., the so-called Pfirsch-Schlüter current. The total equilibrium current is therefore can be expressed as

$$\mathbf{J} = \frac{dP}{d\chi}\left(2\lambda\mathbf{B} + \frac{\mathbf{B} \times \nabla\chi}{B^2}\right), \tag{29}$$

where the second term is diamagnetic current and the first term denotes the Pfirsch-Schlüter current. We can determine the Pfirsch-Schlüter current from $\nabla \cdot \mathbf{J} = 0$,

$$\lambda = -\frac{1}{2\chi} \int_0^\theta \nabla \times \frac{\mathbf{B}}{B^2} \cdot \nabla \chi d\theta + \lambda_0, \tag{30}$$

where λ_0 is the integration constant and can be determined by Ohm's law in the parallel direction.

4. Linear MHD instabilities

In this subsection we overview the linear MHD stability theories in toroidal geometry. We will detail major analytical techniques developed in this field in the past decades, such as interchange, ballooning, TAE, and EPM/KBM theories. Due to space limitation, we focus ourselves on ideal MHD theory.

4.1 Decomposition of linearized MHD equations, three basic MHD waves

There are three fundamental waves in magnetic confined plasmas. The compressional Alfvén mode characterizes the oscillation due to compression and restoration of magnetic field. It mainly propagates in the derection perpendicular to magnetic field. Since plasmas are frozen in magnetic field, such a magnetic field compression also induces plasma compression. Note that the ratio of plasma pressure to magnetic pressure (referred to as plasma beta β) usually is low. The compression and restoration forces mainly result from magnetic field energy. The shear Alfvén wave describes the oscillation due to magnetic field line bending and restoration. It mainly propagates along the magnetic field lines. Since long wave length is allowed for shear Alfvén wave, shear Alfvén wave frequency (or restoration force) is usually lower than that of compressional Alfvén wave. Therefore, shear Alfvén wave is often coupled to plasma instabilities. Another fundamental wave in magnetic confined plasmas is parallel acoustic wave (sound wave). Since plasma can move freely along magnetic field lines without being affected by Lorentz's force. Parallel acoustic wave can prevail in plasmas. The various types of electrostatic drift waves are related to it. Due to low beta assumption, the frequency of ion sound wave is lower than that of shear Alfvén wave by oder $\sqrt{\beta}$. The behaviors of these three waves in simplified geometry have been widely studied in many MHD books. Here, we focus on toroidal geometry theories. MHD equation (14) in toroidal geometry can be hard to deal with. One usually needs to separate the time scales for three fundamental waves to reduce the problem. This scale separation is realized through proper projections and reduction of MHD equation (14).

There are three projections for MHD equation, Eq. (14). We introduce three unit vectors: $\mathbf{e}_b = \mathbf{B}/B$, $\mathbf{e}_1 = \nabla\psi/|\nabla\psi|$, and $\mathbf{e}_2 = \mathbf{e}_b \times \mathbf{e}_1$ for projections. The $\mathbf{e}_2$ projection of the MHD equation (14) gives

$$\begin{aligned} \mathbf{e}_1 \cdot \nabla \times \delta\mathbf{B} = & -\frac{gP'}{B^2}\mathbf{e}_1 \cdot \delta\mathbf{B} - g'\mathbf{e}_1 \cdot \delta\mathbf{B} + \frac{1}{B}\mathbf{e}_2 \cdot \nabla\left(P'|\nabla\psi|\mathbf{e}_1 \cdot \xi\right) \\ & +\Gamma P\frac{1}{B}\mathbf{e}_2 \cdot \nabla\left(\nabla \cdot \xi\right) + \frac{\rho_m\omega^2}{B}\mathbf{e}_2 \cdot \xi. \end{aligned} \tag{31}$$

Similarly, the $\mathbf{e}_1$ projection of the MHD equation (14) yields

$$\mathbf{e}_2 \cdot \nabla \times \delta \mathbf{B} = -\frac{gP'}{B^2}\mathbf{e}_2 \cdot \delta \mathbf{B} - g'\mathbf{e}_2 \cdot \delta \mathbf{B} - \frac{P'|\nabla \psi|}{B^2}\mathbf{e}_b \cdot \delta \mathbf{B} - \frac{1}{B}\mathbf{e}_1 \cdot \nabla \left(P'|\nabla\psi|\mathbf{e}_1 \cdot \boldsymbol{\xi}\right)$$
$$-\Gamma P \frac{1}{B}\mathbf{e}_1 \cdot \nabla (\nabla \cdot \boldsymbol{\xi}) - \frac{\rho_m \omega^2}{B}\mathbf{e}_1 \cdot \boldsymbol{\xi}. \quad (32)$$

The $\mathbf{e}_b$ projection of MHD equation (14) can be reduced to, using $\nabla \cdot \boldsymbol{\xi}$ as an independent unknown,

$$\Gamma P \mathbf{B} \cdot \nabla \left(\frac{1}{B^2}\mathbf{B} \cdot \nabla \nabla \cdot \boldsymbol{\xi}\right) + \rho_m \omega^2 \nabla \cdot \boldsymbol{\xi} = \rho_m \omega^2 \nabla \cdot \boldsymbol{\xi}_\perp. \quad (33)$$

Noting that $\delta \mathbf{J}$ and $\delta \mathbf{B}$ are determined completely by $\boldsymbol{\xi}_\perp$, one can see that the set of equations (31) - (33) is complete to determine two components of $\boldsymbol{\xi}_\perp$ and scalar unknown $\nabla \cdot \boldsymbol{\xi}$.

Two perpendicular equations of motion, Eqs. (31) and (32), result from perpendicular projections of MHD equation (10) and therefore contain restoration force due to excitation of compressional Alfvén wave. To suppress compressional Alfvén wave from consideration, one can apply the operator $\nabla \cdot (\mathbf{B}/B^2) \times (\cdots)$ on Eq. (10), yielding

$$\nabla \cdot \frac{\mathbf{B}}{B^2} \times \rho_m \omega^2 \boldsymbol{\xi} = \mathbf{B} \cdot \nabla \frac{\mathbf{B} \cdot \delta \mathbf{J}}{B^2} + \delta \mathbf{B} \cdot \nabla \sigma - \mathbf{J} \cdot \nabla \frac{\mathbf{B} \cdot \delta \mathbf{B}}{B^2} + \nabla \times \frac{\mathbf{B}}{B^2} \cdot \nabla \delta P, \quad (34)$$

where $\sigma = \mathbf{J} \cdot \mathbf{B}/B^2$. Note that compressional Alfvén wave results from the term $\delta \mathbf{J} \times \mathbf{B} + \mathbf{J} \times \delta \mathbf{B} + \nabla \delta P \rightarrow \nabla (\mathbf{B} \cdot \delta \mathbf{B} + \delta P)$ in Eq. (10). Therefore the curl operation in deriving Eq. (34) can suppress compressional Alfvén wave. Equation (34) is often referred to as shear Alfvén law or vorticity equation.

Equations (34), (31), and (33) characterize respectively three fundamental MHD waves: shear Alfvén, compressional Alfvén, and parallel acoustic waves. From newly developed gyrokinetic theory Zheng et al. (2007) two perpendicular equations (31) and (32) are fully recovered from gyrokinetic formulation, expect the plasma compressibility effect.

4.2 Singular layer equation: interchange and peeling modes

Interchange modes are most fundamental phenomena in magnetically confined plasmas. It resembles to the so-called Rayleigh-Taylor instability in conventional fluid theory. Through interchange of plasma flux tubes plasma thermal energy can be released, so that instability develops. Perturbation of magnetic energy from field line bending is minimized for interchange instability. In slab or cylinder configurations such an interchange happens due to the existence of bad curvature region. In toroidal geometry with finite q value, however, the curvature directions with respect to plasma pressure gradient are different on high and low field sides of plasma torus. Good and bad curvature regions appear alternately along magnetic field line. Therefore, one needs to consider toroidal average in evaluating the change of plasma and magnetic energies. This makes interchange mode theory in plasma torus become complicated. The interchange mode theory is the first successful toroidal theory in this field. It includes the derivations of the so-called singular later equation and interchange stability criterion, i.e., the so-called Mercier criterion Mercier (1962) Greene & Johnson (1962).

Early derivation of singular layer equation relies on the assumption that the modes are somewhat localized poloidally. This assumption was released in a later paper by Glasser et al Glasser et al. (1975). However, the details have been omitted in this paper and direct projection method, alternative to the original vorticity equation approach, is used. Here, we detail the derivation of singular layer equation by vorticity equation approach. These derivation can tell analytical techniques to separate the compressional Alfvén wave from low frequency interchange mode and to minimize field line bending effects due to shear Alfvén mode. The singular equation will be used to derive stability criterion for interchange and peeling modes.

In order to investigate the modes which localize around a particular rational (or singular) magnetic surface V_0, we specialize the Hamada coordinates to the neighborhood of mode rational surface V_0 and introduce the localized Hamada coordinates $x,\ u,\ \theta$ as usual, where $x = V - V_0$ and $u = m\theta - n\zeta$. In this coordinate system the parallel derivative becomes $\mathbf{B}\cdot\nabla = \chi'(\partial/\partial\zeta) + (\Lambda x/\Xi)(\partial/\partial u)$, where $\Lambda = \psi'\chi'' - \chi'\psi''$ and $\Xi = \psi'/m = \chi'/n$.

Using the coordinates $(x,\ u,\ \theta)$, we find that, in an axisymmetric torus, equilibrium scalars are independent of u, and therefore perturbations can be assumed to vary as $\exp\{ik_u u\}$ with $k_u = 2\pi n/\chi'$. As in Refs. Johnson & Greene (1967) and Glasser et al. (1975), $\boldsymbol{\xi}$ and $\delta\mathbf{B}$ are projected in three directions as follows:

$$\boldsymbol{\xi} = \xi\frac{\nabla V}{|\nabla V|^2} + \mu\frac{\mathbf{B}\times\nabla V}{B^2} + \nu\frac{\mathbf{B}}{B^2},$$
$$\delta\mathbf{B} = b\frac{\nabla V}{|\nabla V|^2} + v\frac{\mathbf{B}\times\nabla V}{B^2} + \tau\frac{\mathbf{B}}{B^2}.$$

We consider only singular modes whose wavelength across the magnetic surface $\lambda_\perp$ is much smaller than that on the surface and perpendicular to magnetic field line $\lambda_\wedge$. This leads us to choose following ordering scheme as in Ref. Glasser et al. (1975):

$$x \sim \epsilon,\quad \frac{\partial}{\partial V} \sim \epsilon^{-1},\quad \frac{\partial}{\partial u} \sim \frac{\partial}{\partial\theta} \sim 1, \tag{35}$$

where $\epsilon \ll 1$, being a small parameter. Furthermore, we consider only the low-frequency regime

$$|\omega/\omega_{si}| \lesssim 1. \tag{36}$$

where ω_{si} is parallel ion acoustic frequency.

Since the modes vary on a slow time scale, they are decoupled from compressional Alfvén wave. It can be verified *a posteriori* that we can make following ordering assumptions:

$$\xi = \epsilon\xi^{(1)} + \cdots,\quad \mu = \mu^{(0)} + \cdots,\quad \delta P^{(2)} = \epsilon^2\delta P^{(2)} + \cdots,$$
$$b = \epsilon^2 b^{(2)} + \cdots,\quad v = \epsilon v^{(1)} + \cdots,\quad \tau = \epsilon\tau^{(1)} + \cdots,$$

where $\delta P^{(2)} = -\Gamma P\nabla\cdot\boldsymbol{\xi}$. These ordering assumptions are the same as those in Ref. Glasser et al. (1975), except that we use $\delta P^{(2)}$ as unknown to replace ν. With these ordering assumptions we can proceed to analyze the basic set of linearized MHD equations. As usual, perturbed quantities are separated into constant and oscillatory parts along the field lines: $\xi = \bar{\xi} + \tilde{\xi}$, where $\bar{\xi} = \langle\xi\rangle \equiv \oint \xi\, dl/B / \oint dl/B$, l is arc length of magnetic field line, and $\tilde{\xi} = \xi - \langle\xi\rangle$.

The condition that $\delta\mathbf{B}$ be divergence free, as required by Eq. (11), yields

$$\frac{\partial b^{(2)}}{\partial x} + \frac{1}{\Xi}\frac{\partial}{\partial u}v^{(1)} + \frac{\partial v^{(1)}}{\partial \theta}\frac{\mathbf{B}\times\nabla V\cdot\nabla\theta}{B^2} + v^{(1)}\nabla\cdot\frac{\mathbf{B}\times\nabla V}{B^2} + \chi'\frac{\partial}{\partial\theta}\frac{\tau^{(1)}}{B^2} = 0.$$

It can be reduced to

$$\frac{\partial b^{(2)}}{\partial x} + \frac{1}{\Xi}\frac{\partial}{\partial u}v^{(1)} + \frac{J'}{P'}\frac{\partial v^{(1)}}{\partial \theta} - \frac{\chi'}{P'}\frac{\partial \sigma v^{(1)}}{\partial\theta} + \chi'\frac{\partial}{\partial\theta}\frac{\tau^{(1)}}{B^2} = 0. \tag{37}$$

After surface average it gives

$$\frac{\partial \bar{b}^{(2)}}{\partial x} + \frac{1}{\Xi}\frac{\partial \bar{v}^{(1)}}{\partial u} = 0. \tag{38}$$

The two significant orders of induction equation, Eq. (11), in the ∇V–direction are

$$0 = \chi'\frac{\partial \xi^{(1)}}{\partial\theta}, \tag{39}$$

$$b^{(2)} = \chi'\frac{\partial \xi^{(2)}}{\partial\theta} + \frac{\Lambda x}{\Xi}\frac{\partial \xi^{(1)}}{\partial u}. \tag{40}$$

The component of Eq. (11) in the ∇u–direction, in lowest order, yields

$$\chi'\frac{\partial \mu^{(0)}}{\partial\theta} = 0. \tag{41}$$

To satisfy the component of Eq. (11) along the magnetic field line, one must set

$$(\nabla\cdot\boldsymbol{\xi}_\perp)^{(0)} + 2\boldsymbol{\kappa}\cdot\boldsymbol{\xi}^{(0)} = \frac{\partial \xi^{(1)}}{\partial x} + \frac{1}{\Xi}\frac{\partial \mu^{(0)}}{\partial u} = 0. \tag{42}$$

where Eq. (41) and $\nabla\cdot(\mathbf{B}\times\nabla V/B^2) = 2\mathbf{B}\times\boldsymbol{\kappa}\cdot\nabla V/B^2$ have been used, and $\boldsymbol{\kappa} = \mathbf{b}\cdot\nabla\mathbf{b}$ is magnetic field line curvature.

Next, we turn to momentum equation (14). The two components perpendicular to $\mathbf{B}$ of the momentum equation (14) both lead, in lowest order, to

$$\tau^{(1)} - P'\xi^{(1)} = 0. \tag{43}$$

This is consistent to Eq. (42). Since both components yield the same information, we can directly work on the vorticity equation Eq. (34) and obtain

$$\chi'\frac{\partial}{\partial\theta}\left(\frac{|\nabla V|^2}{B^2}\frac{\partial v^{(1)}}{\partial x}\right) + \chi'\frac{\partial\sigma}{\partial\theta}\frac{\partial \xi^{(1)}}{\partial x} = 0, \tag{44}$$

$$\begin{aligned}
&-\omega^2\frac{N_iM_i|\nabla V|^2}{B^2}\frac{\partial \mu^{(0)}}{\partial x} \\
&= -\chi'\frac{\partial}{\partial\theta}\left(\frac{|\nabla V|^2}{B^2}\frac{\partial v^{(2)}}{\partial x} - v\frac{\mathbf{B}}{B^2}\cdot\nabla\times\frac{\mathbf{B}\times\nabla V}{B^2} - \tau\frac{\mathbf{B}}{B^2}\cdot\nabla\times\frac{\mathbf{B}}{B^2} + \frac{J'}{\chi'}\tau^{(1)}\right) \\
&-v^{(1)}\left(\frac{J'}{P'} - \frac{\chi'}{P'}\sigma\right)\frac{\partial\sigma}{\partial\theta} - \tau^{(1)}\frac{\chi'}{B^2}\frac{\partial\sigma}{\partial\theta} - \Lambda x\frac{|\nabla V|^2}{\Xi B^2}\frac{\partial}{\partial u}\frac{\partial v^{(1)}}{\partial x} + \frac{P'}{\Xi B^2}\frac{\partial \tau^{(1)}}{\partial u} \\
&+P'\frac{\nabla V\cdot\nabla(P+B^2)}{\Xi B^2|\nabla V|^2}\frac{\partial \xi^{(1)}}{\partial u} - \chi'\frac{\partial\sigma}{\partial\theta}\Theta\frac{\partial \xi^{(1)}}{\partial u} + \frac{\chi'}{P'}\frac{\partial\sigma}{\partial\theta}\frac{\partial}{\partial x}\left(\delta P^{(2)} - P'\xi^{(2)}\right).
\end{aligned} \tag{45}$$

We will derive the singular layer equation by averaging this equation. Therefore, it is needed to express unknowns in this equation in terms of $\xi^{(1)}$.

It is trivial to get $\mu^{(0)}$ from Eqs. (41) and (42), and $\tau^{(1)}$ from Eq. (43). The rest can be obtained as follows. From Eqs. (39) and (40) one can find that $\bar{b}^{(2)} = (\Lambda x/\Xi)(\partial\xi^{(1)}/\partial u)$. With $\bar{b}^{(2)}$ obtained one can determine $\bar{v}^{(1)}$ from Eq. (38):

$$\bar{v}^{(1)} = -\Lambda \frac{\partial}{\partial x}(x\xi^{(1)}). \tag{46}$$

Using Eq. (46) to determine integration constant, Eq. (44) can be solved, yielding that

$$\frac{\partial v^{(1)}}{\partial x} = -\left(\frac{B^2\sigma}{|\nabla V|^2} - \frac{\langle B^2\sigma/|\nabla V|^2\rangle}{\langle B^2/|\nabla V|^2\rangle}\frac{B^2}{|\nabla V|^2}\right)\frac{\partial\xi^{(1)}}{\partial x} - \Lambda\frac{B^2/|\nabla V|^2}{\langle B^2/|\nabla V|^2\rangle}\frac{\partial^2}{\partial x^2}(x\xi^{(1)}).$$

From Eqs. (40) and (37) one obtains

$$-\chi'\frac{\partial^2\xi^{(2)}}{\partial\theta\partial x} = \frac{1}{\Xi}\frac{\partial\bar{v}^{(1)}}{\partial u} + \frac{J'}{P'}\frac{\partial v^{(1)}}{\partial\theta} - \frac{\chi'}{P'}\frac{\partial\sigma v^{(1)}}{\partial\theta} + \chi'\frac{\partial}{\partial\theta}\frac{\tau^{(1)}}{B^2}.$$

We need also to solve the equation of parallel motion, Eq. (33). Taking into consideration of low frequency assumption in Eq. (36) and the result in Eq. (42), the equation of parallel motion can be reduced to

$$\chi'^2\frac{\partial}{\partial\theta}\left(\frac{1}{B^2}\frac{\partial}{\partial\theta}\delta P^{(2)}\right) = i\frac{\rho_m\omega^2}{k_u\Gamma P}\frac{\mathbf{B}\times\nabla V}{B^2}\cdot\boldsymbol{\kappa}\frac{\partial\xi^{(1)}}{\partial x}. \tag{47}$$

Noting that $\mathbf{B}\times\nabla V\cdot\boldsymbol{\kappa}/B^2 = \chi'(\partial\sigma/\partial\theta)$, equation (47) can be solved to yield

$$\chi'\frac{\partial}{\partial\theta}\delta P^{(2)} = i\frac{\rho_m\omega^2}{k_u\Gamma P}\left(B^2\sigma - \frac{\langle B^2\sigma\rangle}{\langle B^2\rangle}B^2\right)\frac{\partial\xi^{(1)}}{\partial x}.$$

Inserting these results into Eq. (45) and averaging over l, one obtains the singular layer equation

$$\frac{\partial}{\partial x}\left(x^2 - M\omega^2\right)\frac{\partial\xi^{(1)}}{\partial x} + \left(\frac{1}{4} + D_I\right)\xi^{(1)} = 0, \tag{48}$$

where the total mass parameter $M = M_c + M_t$,

$$D_I \equiv E + F + H - \frac{1}{4},$$

$$E \equiv \frac{\langle B^2/|\nabla V|^2\rangle}{\Lambda^2}\left(J'\psi'' - I'\chi'' + \Lambda\frac{\langle\sigma B^2\rangle}{\langle B^2\rangle}\right),$$

$$F \equiv \frac{\langle B^2/|\nabla V|^2\rangle}{\Lambda^2}\left(\left\langle\frac{\sigma^2 B^2}{|\nabla V|^2}\right\rangle - \frac{\langle\sigma B^2/|\nabla V|^2\rangle^2}{\langle B^2/|\nabla V|^2\rangle} + P'^2\left\langle\frac{1}{B^2}\right\rangle\right),$$

$$H \equiv \frac{\langle B^2/|\nabla V|^2\rangle}{\Lambda}\left(\frac{\langle \sigma B^2/|\nabla V|^2\rangle}{\langle B^2/|\nabla V|^2\rangle} - \frac{\langle \sigma B^2\rangle}{\langle B^2\rangle}\right),$$

$$M_c \equiv \frac{N_i M_i}{k_u^2 \Lambda^2}\left\langle \frac{B^2}{|\nabla V|^2}\right\rangle\left\langle \frac{|\nabla V|^2}{B^2}\right\rangle,$$

$$M_t \equiv \frac{N_i M_i}{k_u^2 \Lambda^2 P'^2}\left\langle \frac{B^2}{|\nabla V|^2}\right\rangle\left(\left\langle \sigma^2 B^2\right\rangle - \frac{\langle \sigma B^2\rangle^2}{\langle B^2\rangle}\right).$$

Here, the mass factor M_c results from perpendicular motion and M_t from parallel motion due to toroidal coupling. M_t is often referred to as apparent mass. In the kinetic description the apparent mass is enhanced by the so-called small parallel ion speed effect. In the large aspect ratio configurations this enhancement factor is of order $\sqrt{R/a}$, where R and a are respectively major and minor radii Mikhailovsky (1974) Zheng & Tessarotto (1994b).

From Eq. (48) one can derive the Mercier criterion, i.e., the stability criterion for localized interchange modes in toroidal geometry. In the marginal stability $\omega^2 = 0$, Eq. (48) becomes the Euler differential equation. Its solution is

$$\xi = \xi_0 x^{-\frac{1}{2}\pm\sqrt{-D_I}}. \tag{49}$$

The system stability can be determined by Newcomb's theorem 5 Newcomb (1960): system is unstable, if and only if the solution of Eq. (48) vanishes two or more points. From the solution in Eq. (49) one can see that if $-D_I < 0$ ξ becomes oscillated. Therefore, interchange mode stability criterion is simply $-D_I > 0$.

Interchange modes are internal modes. When internal modes are stable, it is still possible to develop unstable external modes. For external modes one needs to consider the matching condition between plasma and vacuum solutions. As discussed in conventional MHD books, these matching conditions are: (1) the tangential magnetic perturbation ($\delta \mathbf{B}_t$) should be continuous; and (2) total magnetic and thermal force ($\mathbf{B} \cdot \delta\mathbf{B} + \delta P$) should balance across plasma-vacuum interface in the case without plasma surface current. It can be proved that for localized modes the vacuum contribution is of order ϵ^2 and therefore can be neglected Lortz (1975). Consequently, the boundary condition becomes that total magnetic and thermal forces on the plasma side of the plasma-vacuum interface should vanish. This gives the necessary and sufficient stability condition for peeling modes

$$\left[\frac{x^2}{2}\left(\xi^*\frac{d\xi}{dx} + \xi\frac{d\xi^*}{dx}\right) + \left(\Delta + \frac{1}{2}\right)x|\xi|^2\right]_{x=b} > 0, \tag{50}$$

where b is the coordinate of plasma-vacuum interface, relative to the rational surface, and

$$\Delta = \frac{1}{2} + S^{-1}\left\langle \frac{B^2\sigma}{|\nabla V|^2}\right\rangle, \quad S = \chi'\psi'' - \psi'\chi''.$$

Note that the stability condition Eq. (50) can be alternatively obtained by the approach of minimization of plasma energy Lortz (1975) Wesson (1978).

One can derive the peeling mode stability criterion by inserting Eq. (49) into Eq. (50) Wesson (1978). In the derivation of peeling stability criterion we assume system to be Mercier stable, i.e., $-D_I > 0$. For the case with $\Delta < 0$ we assume that rational surface resides inside plasma region, so that $b > 0$. In this case the stability condition becomes

$$\sqrt{-D_I} + \Delta > 0. \tag{51}$$

For the case with $\Delta > 0$ we assume that rational surface resides outside plasma region, so that $b < 0$. In this case the stability condition becomes

$$\sqrt{-D_I} - \Delta > 0. \tag{52}$$

Note that $-D_I \equiv \Delta^2 - \Lambda_p$, where $\Lambda_p = S^{-2}\left\langle \mathbf{J}^2|\nabla V|^2 + I'\psi' - J'\chi''\right\rangle \left\langle B^2|\nabla V|^{-2}\right\rangle$. Therefore, both cases, Eqs. (51) and (52), give rise to the same stability criterion for peeling mode: $\Lambda_P < 0$. This is more stringent than the Mercier criterion.

4.3 Ballooning modes

In this section we review high-n ballooning mode theory. The stability criterion for interchange modes takes into account only average magnetic well effect. As it is well-known tokamak plasmas have good and bad curvature regions, referring to whether pressure gradient and magnetic field line curvature point in same direction or not. Usually bad curvature region lies on low field side of plasma torus; good curvature region on high field side. Although tokamaks are usually designed to have average good curvature, i.e., Mercier stable, the ballooning modes can still develop as soon as the release of plasma thermal energy on bad curvature region is sufficient to counter the magnetic energy resulting from field line bending Connor et al. (1979) Chance et al. (1979). In difference from interchange modes ballooning modes have high toroidal mode number n, while interchange modes can be either low and high n. Also, ballooning modes allow normal and geodesic wave lengths to be of same order $\lambda_\perp \sim \lambda_\wedge$, but both of them are much smaller than parallel wave length $\lambda_\parallel$.

We first derive ballooning mode equation. In high n limit, both components of perpendicular momentum equation, Eqs. (31) and (32), give the same result

$$\mathbf{B}\cdot\delta\mathbf{B} + \delta P = -(B^2 + \Gamma P)\nabla\cdot\boldsymbol{\xi} + \mathbf{B}\cdot\nabla\left(\frac{\mathbf{B}\cdot\boldsymbol{\xi}}{B^2}\right) - 2\boldsymbol{\kappa}\cdot\boldsymbol{\xi} = 0. \tag{53}$$

In lowest order, one has $\nabla\cdot\boldsymbol{\xi}_\perp \sim \xi/R$. This allows to introduce the so-called stream function $\delta\varphi$ Chance et al. (1979): $\boldsymbol{\xi}_\perp = \mathbf{B}\times\nabla\delta\varphi/B^2$. Equation (34) then becomes,

$$\mathbf{B}\cdot\nabla\frac{1}{B^2}\nabla\cdot\left(B^2\nabla_\perp\frac{\mathbf{B}\cdot\nabla\delta\varphi}{B^2}\right) + \nabla\cdot\left(\rho\omega^2\frac{\nabla_\perp\delta\varphi}{B^2}\right)$$
$$+P'_\psi\nabla\times\frac{\mathbf{B}}{B^2}\cdot\nabla\left(\frac{\mathbf{B}\times\nabla\psi}{B^2}\cdot\nabla\delta\varphi\right) + \Gamma P\nabla\times\frac{\mathbf{B}}{B^2}\cdot\nabla\nabla\cdot\boldsymbol{\xi} = 0. \tag{54}$$

Equation (33), meanwhile, can be reduced to

$$\Gamma P\mathbf{B}\cdot\nabla\left(\frac{1}{B^2}\mathbf{B}\cdot\nabla\nabla\cdot\boldsymbol{\xi}\right) + \rho_m\omega^2\nabla\cdot\boldsymbol{\xi} = \rho_m\omega^2\frac{2\mathbf{B}\times\boldsymbol{\kappa}}{B^2}\cdot\nabla\delta\varphi, \tag{55}$$

where Eq. (53) has been used.

The key formalism to ballooning mode theory is the so-called ballooning representation Lee & Van Dam (1977) Connor et al. (1979). Here, we outline its physics basis and derivation, especially to explain the equivalence of two kinds of representations in Refs. Lee & Van Dam (1977) and Connor et al. (1979). In tokamak geometry one can introduce the following Fourier decomposition:

$$\delta\varphi(nq,\theta,\zeta) = \sum_{m=-\infty}^{+\infty} \delta\varphi_m(nq)\exp\{-in\zeta + m\theta\}. \tag{56}$$

For simply to describe ballooning mode representation we have used nq as flux surface label. This is allowed for systems with finite magnetic shear, in which the ballooning representation applies. For high n modes the distance of mode rational surfaces is of order $1/n$, which is much smaller than equilibrium scale length. Therefore, in lowest order we can neglect the spatial variance of equilibrium quantities and require mode Fourier harmonics to have the so-called ballooning invariance:

$$\delta\varphi_m(nq) = \delta\varphi(nq - m), \tag{57}$$

so that the Fourier decomposistion in Eq. (57) can be expressed as

$$\delta\varphi(nq,\theta,\zeta) = \sum_{m=-\infty}^{+\infty} \delta\varphi(nq - m)\exp\{-in\zeta + m\theta\}. \tag{58}$$

We can further introduce the Laplace tranform

$$\delta\varphi(nq) = \frac{1}{2\pi}\int_{-\infty}^{+\infty} \delta\varphi(\eta)\exp\{inq\eta\}d\eta. \tag{59}$$

Using this transform Eq. (58) can be written as

$$\delta\varphi(nq,\theta,\phi) = \frac{1}{2\pi}\exp\{-in\zeta\}\int_{-\infty}^{+\infty} \delta\varphi(\eta)\sum_{m}\exp\{im(\theta-\eta)\}d\eta. \tag{60}$$

Noting that

$$\frac{1}{2\pi}\sum_{m=-\infty}^{+\infty}\exp\{im(\theta-\eta)\} = \sum_{j=-\infty}^{+\infty}\delta(\eta-\theta-j2\pi),$$

Equation (60) is transformed to

$$\delta\varphi(nq,\theta,\zeta) = \sum_{j=-\infty}^{+\infty} \delta\varphi(\theta + j2\pi)\exp\{-in(\zeta - q(\theta + j2\pi))\}. \tag{61}$$

This indicates that we can represent high n modes at a reference surface as

$$\delta\varphi(nq,\theta,\zeta) = \delta\varphi(\theta)\exp\{-in\beta\} \tag{62}$$

without concern of periodicity requirement. Here, $\beta \equiv \zeta - q\theta$. The periodic eigenfunction can always be formed through the summation in Eq. (61). This representation characterizes the most important feature of ballooning modes in a plasma torus that perpendicular wave number is much larger than parallel one: $k_\perp \gg k_\parallel$. This reduction shows the equivalence of two kinds of representations in Eqs. (58) and (61) Lee & Van Dam (1977) Connor et al. (1979).

Uniqueness and inversion of ballooning mode representation were proved in Ref. Hazeltine et al. (1981).

With ballooning mode representation described, we can proceed to derive ballooning mode equation. It is convenient to use the so-called Celbsch coordinates (ψ, β, θ) to construct ballooning mode equations. In this coordinates $\nabla \to -in\nabla\beta$ and $\mathbf{B}\cdot\nabla = \chi'(\partial/\partial\theta)$. Applying Eq. (62) to Eqs. (54) and (55) and employ the high n ordering, one can obtain following coupled ballooning mode equations

$$\chi'\frac{\partial}{\partial\theta}\left(|\nabla\beta|^2\chi'\frac{\partial}{\partial\theta}\delta\varphi\right) + P'\nabla\times\frac{\mathbf{B}}{B^2}\cdot\nabla\beta\delta\varphi + \Gamma P\nabla\times\frac{\mathbf{B}}{B^2}\cdot\nabla\beta\delta\Xi$$
$$+\frac{\omega^2}{\omega_A^2}|\nabla\beta|^2\delta\varphi = 0, \tag{63}$$
$$\Gamma P\chi'\frac{\partial}{\partial\theta}\left(\frac{1}{B^2}\chi'\frac{\partial}{\partial\theta}\delta\Xi\right) + \rho_m\omega^2\delta\Xi = \rho_m\omega^2\frac{2\mathbf{B}\times\boldsymbol{\kappa}}{B^2}\cdot\nabla\beta\delta\varphi, \tag{64}$$

where $\delta\Xi = i\nabla\cdot\boldsymbol{\xi}/n$. These two equations are coupled second order differential equations. The derivatives here are along a reference magnetic field line labeled by ψ and β. The boundary conditions are $\delta\varphi, \delta\Xi \to 0$ at $\theta \to \pm\infty$ to guarantee the convergence of the Laplace transform in Eq. (59).

In studying ballooning stability at finite beta equilibrium, the so called steep-pressure-gradient equilibrium model is often used Connor et al. (1978) Greene & Chance (1981). In this model, finite beta modification is only taken into account for magnetic shear, while others remain to their low beta values. This model has been proved to be successful for ballooning mode studies. Here, we outline the formulation in Ref. Berk et al. (1983). Noting that $\beta = \zeta - q\theta$, one can see that the magnetic shear effect resides at the quantity $\nabla\beta$ in the ballooning mode equations (63) and (64). From Eq. (22) one can prove that

$$\nabla\beta = \Lambda_s\nabla\chi + \frac{\mathbf{B}\times\nabla\chi}{|\nabla\chi|^2}, \tag{65}$$

where Λ_s is the so-called shear parameter and can be obtained by applying operator $\mathbf{B}\times\nabla\chi\cdot\nabla\times\cdots$ on Eq. (65),

$$\chi'\frac{d\Lambda_s}{d\theta} = -\frac{\mathbf{B}\times\nabla\chi\cdot\nabla\times(\mathbf{B}\times\nabla\chi)}{|\nabla\chi|^4}. \tag{66}$$

We need to determine finite beta modification to Λ. We assume that $\chi = \chi_0 + \chi_1$ and $\beta = \beta_0 + \beta_1$, where χ_0 and β_0 are low beta values and χ_1 and β_1 represent finite beta modifications. The linearized Ampere's law can be written as follows:

$$\nabla\times(\nabla\chi_0\times\nabla\beta_1 + \nabla\chi_1\times\nabla\beta_0) = \mathbf{J} = \frac{\partial P}{\partial\chi}\left(2\lambda\nabla\chi_0\times\nabla\beta_0 + \frac{\mathbf{B}_0\times\nabla\chi_0}{B^2}\right). \tag{67}$$

Noting that in the curl operation on left hand side only the gradient component in $\nabla\chi$ direction needs to be taken, i.e., $\nabla\times \to \nabla\chi_0\partial/\partial\chi\times$, equation (67) can be solved

$$2P\lambda\nabla\beta_0 + \mathbf{B}_0 P + \nabla Q = \nabla\chi_0\times\nabla\beta_1 + \nabla\chi_1\times\nabla\beta_0, \tag{68}$$

where ∇Q is integration factor. Taking the divergence of Eq. (68) for only $\partial\{P,Q\}/\partial\chi$ large gives

$$2\lambda\frac{\partial P}{\partial\chi}(\nabla\chi_0\cdot\nabla\beta_0)+\frac{\partial^2 Q}{\partial\chi^2}|\nabla\chi_0|^2=0,$$

and therefore

$$\partial Q/\partial\chi=-2\lambda P\nabla\chi\cdot\nabla\beta/|\nabla\chi|^2. \quad (69)$$

Now taking the dot product of Eq. (68) with $\nabla\beta_0$ gives

$$2P\lambda|\nabla\beta_0|^2+(\nabla\chi_0\cdot\nabla\beta_0)\frac{\partial Q}{\partial\chi}=-\chi'\frac{\partial\beta_1}{\partial\theta},$$

We can remove subscript 0 afterward for brevity. Now substituting Eq. (69) for $\partial Q/\partial\chi$ and noting that $\partial\beta_1/\partial\chi\equiv\Lambda_{s1}$, one finds that

$$\chi'\frac{\partial\Lambda_{s1}}{\partial\theta}=-2\lambda\frac{\partial P}{\partial\chi}\frac{B^2}{|\nabla\chi|^2}.$$

Therefore, the shear parameter can be evaluated as follows

$$\chi'\frac{d\Lambda_s}{d\theta}=-\frac{\mathbf{B}\times\nabla\chi\cdot\nabla\times(\mathbf{B}\times\nabla\chi)}{|\nabla\chi|^4}-2\lambda\frac{\partial P}{\partial\chi}\frac{B^2}{|\nabla\chi|^2}. \quad (70)$$

The second term here gives rise to the finite beta modification to shear parameter Λ_s in steep pressure gradient model. The rest parameters here and in ballooning equations (63) and (64) can be evaluated with low beta values.

We now consider tokamak model equilibrium with circular cross section, low beta, and large aspect ratio (i.e., $1/\epsilon = R/a \gg 1$). The magnetic field in this model can be expressed as $\mathbf{B}=B_\phi(r)/(1+\epsilon\cos\theta)\mathbf{e}_\phi+B_\theta(r)\mathbf{e}_\theta$. The shear parameter can be expressed as $\Lambda_s=s(\theta-\theta_k)-\alpha\sin\theta$. Here, $\alpha=-(2Rq^2/B^2)(dP/dr)$, $s=d\ln q/d\ln r$, and θ_k is integration constant. Therefore, ballooning equations (63) and (64) can be reduced to

$$\frac{d}{d\theta}\left((1+\Lambda_s^2)\frac{d\delta\varphi}{d\theta}\right)+\alpha(\cos\theta+\Lambda_s\sin\theta)\delta\varphi+\frac{2\Gamma RrqP}{B}(\cos\theta+\Lambda_s\sin\theta)\delta\Xi$$
$$+\frac{\omega^2}{\omega_A^2}(1+\Lambda_s^2)\delta\varphi=0, \quad (71)$$

$$\frac{\Gamma P}{R^2q^2}\frac{\partial^2\delta\Xi}{\partial\theta^2}+\rho_m\omega^2\delta\Xi=-\frac{2\rho_m\omega^2}{R^2B_\theta}(\cos\theta+\Lambda_s\sin\theta)\delta\varphi. \quad (72)$$

To further analyze this set of equations it is interesting to consider two limits: the low frequency ($\omega\ll\omega_{si}$) and intermediate frequency limit ($\omega_{si}\ll\omega\ll\omega_{se}$). In the low frequency limit the second term on left hand side of Eq. (72) can be neglected and inertia term is only important in the outer region $\theta\to\infty$. Equation (72) can be solved to yield

$$\delta\Xi=\frac{2\rho_m q^2\omega^2}{\Gamma PB_\theta}s\theta\sin\theta\delta\varphi. \quad (73)$$

Here, it has been considered that in sound wave scale the slow variable $s\theta$ can be regarded as constant. Inserting Eq. (73) into Eq. (71) yields that

$$\frac{d}{d\theta}\left((1+\Lambda_s^2)\frac{d\delta\varphi}{d\theta}\right)+\alpha(\cos\theta+\Lambda_s\sin\theta)\delta\varphi+\frac{\omega^2}{\omega_A^2}(1+2q^2)s^2\theta^2\delta\varphi=0. \tag{74}$$

Here, we see that the sound wave coupling results in the so-called apparent mass effect: i.e., the inertia term is enhanced by a factor $(1+2q^2)$ Greene & Johnson (1962). In the kinetic description the $2q^2$ term is further boosted by the so-called small particle speed effect to become of order $2q^2/\sqrt{r/R}$ for large aspect ratio case Mikhailovsky (1974) Zheng & Tessarotto (1994b). In the marginal stability $\omega^2=0$ the ballooning stability can be determined by Newcomb's theorem 5 Newcomb (1960): system is unstable, if and only if the solution of Eq. (48) vanishes two or more points. Refs. Connor et al. (1978) and Lortz & Nührenberg (1978). have obtained the stability boundaries for ballooning modes.

In the intermediate frequency regime the first term in Eq. (72) can be neglected and therefore one obtains

$$\delta\Xi=-\frac{2}{R^2B_\theta}(\cos\theta+\Lambda_s\sin\theta)\delta\varphi. \tag{75}$$

Inserting Eq. (75) into Eq. (71) yields Tang et al. (1980)

$$\frac{d}{d\theta}\left((1+\Lambda_s^2)\frac{d\delta\varphi}{d\theta}\right)+\alpha(\cos\theta+\Lambda_s\sin\theta)\delta\varphi-\frac{4\Gamma q^2P}{B^2}(\cos\theta+\Lambda_s\sin\theta)^2\delta\varphi$$
$$+\frac{\omega^2}{\omega_A^2}(1+\Lambda_s^2)\delta\varphi=0. \tag{76}$$

The sound wave coupling term (3rd term) results in the so-called second harmonic TAE in the circular cross section case Zheng et al. (1999).

4.4 Toroidal Alfvén eigen modes

In this subsection we review TAE theory. In the last two subsections we see that interchange and ballooning modes are characterized by having only single dominant or resonant mode at resonance surfaces. In particular their resonance surfaces locates at mode rational surface where $m-nq=0$. TAEs are different from them. TAEs involve two mode coupling. In particular, the first TAEs are centered at the surface where $q=(m_0+1/2)/n$. Two neighboring Fourier modes (m_0 and m_0+1) propagate roughly with same speed $v_A/2Rq$ but in opposite directions. They form a standing wave. The toroidal geometry can induce the first frequency gap so that the standing wave becomes an eigen mode, i.e., TAEs Cheng et al. (1985) Rosenbluth et al. (1992). In the second TAE case, although they have same mode resonance surfaces as interchange and ballooning modes, $m_0\pm1$ mode coupling is involved to form standing 2nd TAEs. The frequency gap for second TAEs in circular cross section case is due to plasma compressibility effect Zheng & Chen (1998).

To explain two mode coupling picture, we consider tokamak model equilibrium with circular cross section, low beta, and large aspect ratio (i.e., $1/\epsilon=R/a\gg1$). There is a review paper on TAEs Vlad et al. (1999). Here, we describe the local dispersion relation for even and odd modes and explain the 2nd TAEs together with the 1st TAEs. The magnetic field in this model can simply be expressed as $\mathbf{B} = B_\phi(r)/(1+\epsilon\cos\theta)\mathbf{e}_\phi + B_\theta(r)\mathbf{e}_\theta$. The general case will be addressed in Sec. 5 with AEGIS code formalism. Since their frequency is much larger than shear Alfvén mode frequency, the compressional Alfvén modes are decoupled. Therefore, we can use Eqs. (54) and (55) as starting equations for TAE investigation. Noting that Alfvén frequency is much larger than sound wave frequency, the first term in Eq. (55) can be dropped. Adopting the Fourier decomposition in Eq. (56), the sound wave equation (55) becomes

$$i(\nabla\cdot\xi)_m = \frac{1}{BR}\left(\frac{d\varphi_{m+1}}{dr} - \frac{d\varphi_{m-1}}{dr}\right).$$

Using this solution for $\nabla\cdot\xi$, Eq. (54) can be reduced to Zheng et al. (1999)

$$\frac{d}{dr}\left[r^3\left(\frac{1}{q}-\frac{n}{m}\right)^2\frac{d}{dr}E_m\right] - \frac{d}{dr}\left(r^3\frac{R^2\omega^2}{m^2v_A^2}\frac{d}{dr}E_m\right) - \epsilon\left(r^3\frac{R^2\omega^2}{m^2v_A^2}\frac{d^2}{dr^2}E_{m+1}\right)$$
$$-\epsilon\left(r^3\frac{R^2\omega^2}{m^2v_A^2}\frac{d^2}{dr^2}E_{m-1}\right) + \frac{\Gamma P r^3}{B^2m^2}\frac{d^2E_{m+2}}{dr^2} + \frac{\Gamma P r^3}{B^2m^2}\frac{d^2E_{m-2}}{dr^2} - wE_m$$
$$-\frac{\alpha r^2}{2mq^2}\frac{dE_{m+1}}{dr} + \frac{\alpha r^2}{2mq^2}\frac{dE_{m-1}}{dr} - \frac{\alpha r}{2q^2}E_{m+1} + \frac{\alpha r}{2q^2}E_{m-1} = 0, \quad (77)$$

where $E_m = \varphi_m/r$, $v_A^2 = B_0^2/\rho_m$, B_0 denotes magnetic field at magnetic axis, and w represents the rest magnetic well terms.

We first examine singular layer physics. In this layer only terms contains second order derivative in r need to be taken into consideration. From the first six terms in Eq. (77) one can see that the 2nd TAEs (coupling of E_{m-1} and E_{m+1}) have structure similarity to the 1st TAEs (coupling of E_m and E_{m+1}). The 1st TAE coupling is due to finite value of aspect ratio; while the 2nd TAE coupling is due to finite beta value. For brevity we focus ourselves to discuss the 1st TAE case. Denoting $\omega_0 = \omega_A/2$, $q_0 = (m+1/2)/n$, $\delta\omega = \omega - \omega_0$, and $\delta q = q - q_0$, the singular layer equations describing the coupling of m and $m+1$ modes becomes

$$\frac{\partial}{\partial\delta q}\left[\frac{\delta\omega}{2\omega_0} - \left(1-\frac{1}{2m+1}\right)n\delta q\right]\frac{\partial}{\partial\delta q}\delta\phi_m = -\frac{\epsilon}{4}\frac{\partial^2}{\partial\delta q^2}\delta\phi_{m+1},$$
$$\frac{\partial}{\partial\delta q}\left[\frac{\delta\omega}{2\omega_0} + \left(1+\frac{1}{2m+1}\right)n\delta q\right]\frac{\partial}{\partial\delta q}\delta\phi_{m+1} = -\frac{\epsilon}{4}\frac{\partial^2}{\partial\delta q^2}\delta\phi_m.$$

Introducing even and odd modes: $\delta\phi_\pm = \delta\phi_m \pm \delta\phi_{m+1}$, these two equations become

$$\frac{\partial}{\partial\delta q}\left(\frac{\delta\omega}{2\omega_0} + \frac{1}{2m_0+1}n\delta q\right)\frac{\partial}{\partial\delta q}\delta\phi_+ - \frac{\partial}{\partial\delta q}n\delta q\frac{\partial}{\partial\delta q}\phi_- = -\frac{\epsilon}{4}\frac{\partial^2}{\partial\delta q^2}\delta\phi_+,$$
$$\frac{\partial}{\partial\delta q}\left(\frac{\delta\omega}{2\omega_0} + \frac{1}{2m_0+1}n\delta q\right)\frac{\partial}{\partial\delta q}\delta\phi_- - \frac{\partial}{\partial\delta q}n\delta q\frac{\partial}{\partial\delta q}\phi_+ = \frac{\epsilon}{4}\frac{\partial^2}{\partial\delta q^2}\delta\phi_-.$$

Integrating once one obtains

$$\mathcal{D}\begin{pmatrix}\frac{\partial\delta\phi_+}{\partial\delta q}\\ \frac{\partial\delta\phi_-}{\partial\delta q}\end{pmatrix} \equiv \begin{pmatrix}\frac{\delta\omega}{2\omega_0} + \frac{1}{2m_0+1}n\delta q + \frac{\epsilon}{4} & -n\delta q\\ -n\delta q & \frac{\delta\omega}{2\omega_0} + \frac{1}{2m_0+1}n\delta q - \frac{\epsilon}{4}\end{pmatrix}\begin{pmatrix}\frac{\partial\delta\phi_+}{\partial\delta q}\\ \frac{\partial\delta\phi_-}{\partial\delta q}\end{pmatrix} = \begin{pmatrix}A_+\\ A_-\end{pmatrix}, \quad (78)$$

where $\mathcal{D}$ is 2×2 matrix and $A_\pm$ are integration constants. Integration of Eq. (78) across singular layer (i.e., from δq^- to δq^+) one obtains the dispersion relation

$$\left(\frac{\delta\phi_+|_{\delta q^-}^{\delta q^+}}{A_+} - \int_{\delta q^-}^{\delta q^+} \frac{\mathcal{D}_{22} d\delta q}{\det|\mathcal{D}|}\right)\left(\frac{\delta\phi_-|_{\delta q^-}^{\delta q^+}}{A_-} - \int_{\delta q^-}^{\delta q^+} \frac{\mathcal{D}_{11} d\delta q}{\det|\mathcal{D}|}\right) = \left(\int_{\delta q^-}^{\delta q^+} \frac{\mathcal{D}_{12} d\delta q}{\det|\mathcal{D}|}\right)^2. \tag{79}$$

Here, $\mathcal{D}_{ij}$ are $\mathcal{D}$ matrix elements and two parameters $\Delta_\pm \equiv \delta\phi_\pm|_{\delta q^-}^{\delta q^+} / A_\pm$ are determined by the outer solutions to the left and right of singular layer. As soon as $\Delta_\pm$ are computed from outer regions, Eq. (79) can be used to determine the frequency. In general this frequency can be complex.

The denominators of integrations in Eq. (79) involve $\det|\mathcal{D}|$. The singularity emerges at $\det|\mathcal{D}| = 0$. In this case the Landau integration orbit needs to be used, as in the case for particle-wave resonances, and continuum damping occurs Berk et al. (1992). The so-called 1st TAE frequency gap, in which eigen modes can exit without continuum damping, can be determined by condition $\det|\mathcal{D}| = 0$, i.e.,

$$\left(\frac{\delta\omega}{2\omega_0} + \frac{1}{2m+1} n\delta q + \frac{\epsilon}{4}\right)\left(\frac{\delta\omega}{2\omega_0} + \frac{1}{2m+1} n\delta q - \frac{\epsilon}{4}\right) = n^2 \delta q^2.$$

Its solution is

$$\left[1 - \left(\frac{1}{2m_0+1}\right)^2\right] n^2 \delta q = \frac{\delta\omega}{2\omega_0(2m_0+1)} \pm \sqrt{\left(\frac{\delta\omega}{2\omega_0}\right)^2 - \frac{\epsilon^2}{16}\left[1 - \frac{1}{(2m+1)^2}\right]}. \tag{80}$$

To exclude real δq solution for $\det \mathcal{D} = 0$, mode frequency must fall in the gap between $\delta\omega_\pm$, i.e., $\omega_- < \omega < \omega_+$, where

$$\delta\omega_\pm = \pm\frac{\epsilon}{2}\omega_0\sqrt{1 - 1/(2m+1)^2}.$$

One can obtain the gap width $\Delta\omega = \delta\omega_+ - \delta\omega_- = \epsilon\omega_0\sqrt{1 - 1/(2m+1)^2}$. The 1st TAEs are Alfvén eigen modes with frequency inside this gap. They are marginally stable and tend to be excited by resonances with energetic particles. Note that the gap width is proportional to ϵ. In cylinder limit the gap vanishes. Therefore, existence of TAEs is due to toroidal effects. Also, we note that the dispersion relation, Eq. (79), allows two types of TAEs: even and odd types ($\varphi_\pm$), depending on the values of $\Delta_\pm$.

We have discussed the 1st TEA theory through coupling of neighboring modes. In similar way one can also develop the 2nd TAE theory through coupling of $m \pm 1$ modes Zheng & Chen (1998). If FLR effects are taken into consideration, the Alfvén types of singularities can be resolved, so that discrete modes can emerge in the continuum. This types of modes are referred to as kinetic TAEs (i.e., KTAEs). Due to correction of gyrokinetic theory Zheng et al. (2007), several missing FLR effects are recovered. Consequently, KTAE theories by far need to be reevaluated.

4.5 Kinetically driven modes: KBMs, EPMs, etc.

In this subsection we describe the kinetically driven modes (KDMs), such as KBMs, EPMs, etc. The frequencies of these modes usually reside in continuum spectrum. Therefore, they are generally damped without driving effects. Unlike KTAEs, for which FLR effects are taken into account to resolve singularity, for KDMs strong kinetic effects from wave-particle-resonances are included to overcome continuum damping. That is why they are referred to as kinetically driven modes. Energetic particle drives to marginal stable TAEs can instantly lead unstable TAEs, but the drives to KDMs need to accumulate sufficient energy to overcome continuum damping for unstable KDMs to develop Tsai & Chen (1993) Zheng et al. (2000). In Secs. 4.3 and 4.4 one has seen that there are two types of modes: ballooning and TAEs. Therefore, KDMs also have two types. Those related to ballooning modes are referred to as KBMs, while EPMs are related to TAEs and usually driven by wave-energetic-particle resonances. We employ ballooning representation formalism to discuss them.

We start with the ballooning mode equation in intermediate frequency regime, Eq. (76), with energetic particle effects included. Introducing the transformation $\zeta = \varphi p^{1/2}$, Eq. (76) becomes

$$\frac{\partial^2 \zeta}{\partial \theta^2} + \Omega^2(1+2\epsilon\cos\theta)\zeta + \frac{\alpha\cos\theta}{p}\zeta - \frac{(s-\alpha\cos\theta)^2}{p^2}\zeta$$
$$-\frac{4\Gamma g^2}{p^2} + \frac{1}{p^{1/2}}\int \frac{d\varepsilon d\mu B}{|v_\parallel|}\omega_d \delta g_h = 0, \tag{81}$$

$$\mathbf{v}_\parallel \cdot \nabla \delta g_h - i\omega \delta g_h = i\omega\left(\mu B + v_\parallel^2\right)\frac{\partial F_{0h}}{\partial \varepsilon}(\kappa_r + \kappa_\theta \Lambda)p^{-1/2}\zeta, \tag{82}$$

where $p = 1+\Lambda_s^2$, $g = \cos\theta + \Lambda\sin\theta$, δg_h is perturbed distribution functions for hot ions, κ_r and κ_θ are respectively radial and poloidal components of magnetic field line curvature $\boldsymbol{\kappa}$, $\Omega = \omega/\omega_A$, ω_d is magnetic drift frequency, $\mathbf{v}$ is particle speed, the subscripts $\perp$ and $\parallel$ represent respectively perpendicular and parallel components to the equilibrium magnetic field line, $\varepsilon = v^2/2$ is particle energy, $\mu = v_\perp^2/2B$ is magnetic moment, and F_{0h} is equilibrium distribution function for hot ions. For simplicity we have neglected the finite Larmor radius effects and only take into account the kinetic effects from energetic ions.

To study KDMs one need to investigate singular layer behavior. In ballooning representation space, singular layer corresponds to $\theta \to \infty$ limit. Again, we exclude the 2nd TAE from discussion (i.e., assuming $\Gamma = 0$). Equation (81) in $\theta \to \infty$ limit becomes:

$$\frac{\partial^2 \zeta}{\partial \theta^2} + \Omega^2(1+2\epsilon\cos\theta)\zeta = 0. \tag{83}$$

This is the well-known Mathieu equation. According to Floquet's Theorem, its solution takes following form

$$\zeta(\theta) = P(\theta)\exp\{i\gamma\theta\},$$

where $P(\theta+2\pi) = P(\theta)$. Since modes with longer parallel-to-$\mathbf{B}$ wavelengths tend to be more unstable, we shall examine solutions corresponding to the two lowest periodicities. The first one is related to KBMs Tsai & Chen (1993) and the second one is related to EPMs Zheng et al. (2000).

We first discuss KBMs. The KBM-type solution is given by

$$\zeta_K = \exp\{i\gamma_K\theta\}(A_0 + A_2\cos\theta + \cdots). \tag{84}$$

Inserting Eq. (84) into Eq. (83), one obtains, noting $\epsilon \ll 1$,

$$\gamma_K^2 \approx \Omega^2(1 + 2\epsilon^2\Omega^2), \qquad \frac{A_2}{A_0} \approx 2\epsilon\Omega^2.$$

Therefore, at leading order, one has

$$\zeta_K = \exp\{i\Omega|\theta|\}, \tag{85}$$

where $\Im m\{\Omega\} > 0$ for causality. Note here that Eq. (85) is valid for general Ω, so that frequency at continuum is allowed, as soon as causality condition is satisfied.

Next, we discuss TAE-type KDMs, e.g., EPMs. This type of solutions can be expressed as Zheng et al. (2000)

$$\zeta_T = \exp\{i\gamma_T\theta\}[A_1\cos(\theta/2) + B_1\sin(\theta/2) + \cdots]. \tag{86}$$

Inserting Eq. (86) into Eq. (83), one obtains, for $\left|\Omega^2 - 1/4\right| \sim \mathcal{O}(\epsilon)$ and $\epsilon \ll 1$,

$$\gamma_T = \left[(\Omega^2 - \Omega_+^2)(\Omega^2 - \Omega_-^2)\right]^{1/2}, \qquad \frac{B_1}{A_1} = \left(\frac{\Omega^2 - \Omega_-^2}{\Omega_+^2 - \Omega^2}\right)^{1/2}, \tag{87}$$

and $\Omega_{+,-}^2 = 1/4 \pm \epsilon\Omega^2$. The leading order solution can, therefore, be expressed as

$$\zeta_T = \exp\left\{i\left[(\Omega^2 - \Omega_+^2)(\Omega^2 - \Omega_-^2)\right]^{1/2}|\theta|\right\}\left[\cos(\theta/2) + \left(\frac{\Omega^2 - \Omega_-^2}{\Omega_+^2 - \Omega^2}\right)^{1/2}\sin(\theta/2)\right]. \tag{88}$$

The causality condition is $\Im m\left\{[(\Omega^2 - \Omega_+^2)(\Omega^2 - \Omega_-^2)]^{1/2}\right\} < 0$. Equation (88) can describe both TAEs and KDMs of TAE type (e.g., EPMs). Existence of TAE solution requires mode frequency to fall in the gap: $\Omega_- < \Omega < \Omega_+$, as shown by the TAE theory in configuration space in Sec. 4.4. For KDMs mode frequency can be in the continuum, i.e., outside the gap as soon as causality condition is satisfied. For TAEs ζ_T contains an $\mathcal{O}(1)$ back scattering and, hence, the continuum damping is either suppressed or much reduced. On the other hand, for KDMs ζ_K contains no back scattering from periodic potential in Eq. (83), and, consequently, there is significant mount of continuum damping. Note that in principle both types of solutions can co-exist at $|\Omega| \approx 1/2$. However, the TAE solution tends to be more unstable in this case than KDMs, since its continuum damping is much less or absent while the instability drives are generally comparable.

With outer solutions given by Eq. (85) or Eq. (88), one can obtain the corresponding dispersion relation by matching outer and inner solutions. For KBMs Eq. (81) can be used to construct the following quadratic form in inner region:

$$2\,\zeta^*\frac{d\zeta}{d\theta}\bigg|_{-\infty}^{+\infty} + \delta W_f + \delta W_k = 0, \tag{89}$$

where

$$\delta W_f = \int_{-\infty}^{+\infty} d\theta \left\{ \left| \frac{\partial \zeta}{\partial \theta} \right|^2 - \left[\frac{\alpha \cos\theta}{p} - \frac{(s - \alpha\cos\theta)^2}{p^2} \right] |\zeta|^2 \right\},$$

$$\delta W_k = \int_{-\infty}^{+\infty} d\theta \zeta^* \frac{1}{p^{1/2}} \int \frac{d\varepsilon d\mu B}{|v_\parallel|} \omega_d \delta g_h.$$

Here, the superscript $*$ represents complex conjugate. Matching inner (Eq. (89)) and outer (Eq. (85)) solutions one obtains the dispersion relation Tsai & Chen (1993)

$$-i\Omega + \delta W_f + \delta W_K = 0. \tag{90}$$

Here, we note that kinetic effects from core plasma should also be taken into account in outer region. As proved in Ref. Zheng & Tessarotto (1994a) this results in the so-called apparent mass effect and leads Ω in the first term of Eq. (90) to become complicated function of actual mode frequency.

Similarly, for KDMs of TAE type, for example EPMs, one need to consider even and odd modes. For even modes the dispersion relation is given by Refs. Zheng et al. (2000) and Tsai & Chen (1993)

$$-i\left(\frac{\Omega_-^2 - \Omega^2}{\Omega_+^2 - \Omega^2}\right)^2 + \delta T_f + \delta T_K = 0, \tag{91}$$

where δT_f represents MHD fluid contribution and δT_K is energetic-particle contribution to the quadratic form in inner region.

The dispersion relations in Eqs. (90) and (91) extend respectively MHD ballooning modes in diamagnetic gap and TAEs in Alfvén gap to respective continua. Kinetic drives are the causes to make causality conditions satisfied.

5. Global numerical analyses of MHD modes: AEGIS code formalism

In Sec. 2.3 analytical or semi-analytical theories are presented to describe four types of MHD modes in toroidal geometry. Due to the developments of modern numerical method and computer hardware, conventional asymptotic expansion methods for global modes have become outdated and been substituted by direct numerical computation. Several excellent numerical codes have been developed in the past to study linear MHD stability of toroidally confined plasmas, such as such as PEST Grimm et al. (1976) Chance et al. (1978), GATO Bernard et al. (1981), DCON Glasser (1997), AEGIS Zheng & Kotschenreuther (2006), etc. In this section we focus on description of AEGIS code, in view of that AEGIS is an adaptive MHD shooting code capable to study MHD continuum Zheng et al. (2005). Through describing AEGIS formalism, we can further explain the general features of MHD eigen modes in toroidally confined plasmas.

Let us first describe the toroidal system to be investigated. The core part is plasma torus, which is surround by a resistive wall; Between plasma torus and resistive wall there is inter vacuum region and outside the resistive wall there is outer vacuum region, which extends to infinity. For simplicity, it is assumed that the wall is thin. We denote respectively the interfaces between plasma torus and inner vacuum region, inner vacuum region and wall, and wall and outer vacuum region as ψ_a, ψ_{b-}, and ψ_{b+}.

5.1 MHD equations and numerical solution method for plasma region

In this subsection we describe how to reduce MHD equations for global mode analyses. The starting equation is the single fluid MHD equation (14). This is a vector equation and can be projected onto three directions to get scalar equations. The parallel projection has be derived in Eq. (33). From parallel equation one can solve for $\nabla \cdot \boldsymbol{\xi}$, which is the only unknown needed for two-perpendicular equations to become a complete set of equations. In principle the parallel motion can not be described by MHD model, since particles are not localized along magnetic field line. There are wave-partcle resonance, trapped particle, and so-called small parallel particle speed effects, etc. Nevertheless, from analyses in Sec. 4.3 one can see that in low frequency regime the parallel coupling results only in the so-called apparent mass effect, while in intermediate regime the parallel coupling mainly gives rise to the 2nd TAEs. Note that apparent mass effect can be absorbed by rescaling mode frequency and inclusion of the 2nd TAE effect is straightforward as discussed in Sec. 4.4. For brevity we limit ourselves to treat only two perpendicular components of Eq. (14) with Γ set to zero. AEGIS-K code has been developed to include parallel dynamics using kinetic description Zheng et al. (2010).

Using general flux coordnates in Eq. (22), the magnetic field line displacement is decomposed as follows

$$\boldsymbol{\xi} \times \mathbf{B} = \xi_s \nabla \psi + \xi_\psi \chi' (\nabla \zeta - q \nabla \theta). \tag{92}$$

Since we deal with linear problem, the Fourier transform can be used to decompose perturbed quantities in poloidal and toroidal directions,

$$\xi \exp\{-in\zeta\} = \sum_{m=-\infty}^{\infty} \xi_m \frac{1}{\sqrt{2\pi}} \exp\{i(m\theta - n\zeta)\}, \tag{93}$$

with $\xi_m = \int_{-\pi}^{\pi} d\theta \xi \exp\{-im\theta\}/\sqrt{2\pi}$. With the toroidal symmetry assumed, only a single toroidal Fourier component needs to be considered. As usual, equilibrium quantities can be decomposed as matrices in poloidal Fourier space, for example

$$\mathcal{J}_{mm'} = \frac{1}{2\pi} \int_{-\pi}^{\pi} d\theta J(\theta) e^{i(m'-m)\theta}.$$

In the poloidal Fourier decomposition, the Fourier components are cut off both from lower and upper sides respectively by $m_{\min}$ and $m_{\max}$. Therefore, the total Fourier component under consideration is $M = m_{\max} - m_{\min} + 1$. We also use bold face (or alternatively $[[\cdots]]$) to represent Fourier space vectors, and calligraphic capital letters (or alternatively $\langle \cdots \rangle$) to represent the corresponding equilibrium matrices (*e.g.*, $\mathcal{J}$ for J) in poloidal Fourier space.

To get scalar equations, we project Eq. (14) respectively onto two directions $J^2 \nabla\theta \times \nabla\zeta \cdot \mathbf{B} \times [\cdots \times \mathbf{B}]/B^2$ and $(1/q\chi') J^2 \nabla\zeta \times \nabla\psi \cdot \mathbf{B} \times [\cdots \times \mathbf{B}]/B^2$, and then introduce the Fourier transformation in Eq. (93) to two projected equations. These procedures lead to the following set of differential equations in matrices

$$\left(\mathcal{B}^\dagger \boldsymbol{\xi}_s + \mathcal{D} \boldsymbol{\xi}'_\psi + \mathcal{E} \boldsymbol{\xi}_\psi\right)' - \left(\mathcal{C}^\dagger \boldsymbol{\xi}_s + \mathcal{E}^\dagger \boldsymbol{\xi}'_\psi + \mathcal{H} \boldsymbol{\xi}_\psi\right) = 0, \tag{94}$$

$$\mathcal{A} \boldsymbol{\xi}_s + \mathcal{B} \boldsymbol{\xi}'_\psi + \mathcal{C} \boldsymbol{\xi}_\psi = 0. \tag{95}$$

Here, the equilibrium matrices contain two parts: plasma/field force and inertia contributions, e.g., $\mathcal{A} = \mathcal{A}_p + \gamma_N \mathcal{A}_i$, where

$$\begin{aligned}
\mathcal{A}_p &= n(n\mathcal{G}_{22} + \mathcal{G}_{23}\mathcal{M}) + \mathcal{M}(n\mathcal{G}_{23} + \mathcal{G}_{33}\mathcal{M}),\\
\mathcal{B}_p &= -i\chi' \left[n(\mathcal{G}_{22} + q\mathcal{G}_{23}) + \mathcal{M}(\mathcal{G}_{23} + q\mathcal{G}_{33})\right],\\
\mathcal{C}_p &= -i\left[\chi''(n\mathcal{G}_{22} + \mathcal{M}\mathcal{G}_{23}) + (q\chi')'(n\mathcal{G}_{23} + \mathcal{M}\mathcal{G}_{33})\right]\\
&\quad -\chi'(n\mathcal{G}_{12} + \mathcal{M}\mathcal{G}_{31})\mathcal{Q} + i(g'\mathcal{Q} - \mu_0 nP'\mathcal{J}/\chi'),\\
\mathcal{D}_p &= \chi'^2\left[(\mathcal{G}_{22} + q\mathcal{G}_{23}) + q(\mathcal{G}_{23} + q\mathcal{G}_{33})\right],\\
\mathcal{E}_p &= \chi'\left[\chi''(\mathcal{G}_{22} + q\mathcal{G}_{23}) + (q\chi')'(\mathcal{G}_{23} + q\mathcal{G}_{33})\right] - i\chi'^2(\mathcal{G}_{12} + q\mathcal{G}_{31})\mathcal{Q} + \mu_0 P'\mathcal{J},\\
\mathcal{H}_p &= \chi''\left[\chi''\mathcal{G}_{22} + (q\chi')'\mathcal{G}_{23}\right] + (q\chi')'\left[\chi''\mathcal{G}_{23} + (q\chi')'\mathcal{G}_{33}\right]\\
&\quad +i\chi'\left[\chi''(\mathcal{M}\mathcal{G}_{12} - \mathcal{G}_{12}\mathcal{M}) + (q\chi')'(\mathcal{M}\mathcal{G}_{31} - \mathcal{G}_{31}\mathcal{M})\right]\\
&\quad +\chi'^2\mathcal{Q}\mathcal{G}_{11}\mathcal{Q} + \mu_0 P'\chi''\mathcal{J}/\chi' + \mu_0 P'\mathcal{J}' - g'q'\chi'\mathcal{I},\\
\mathcal{A}_i &= \frac{B_0^2}{X_0^2 q_0^2}\left\langle \mathcal{J}\frac{\rho_N}{B^2}|\nabla\psi|^2\right\rangle,\\
\mathcal{C}_i &= \frac{B_0^2}{X_0^2 q_0^2}\left\langle \chi'\mathcal{J}\frac{\rho_N}{B^2}(\nabla\psi\cdot\nabla\zeta - q\nabla\psi\cdot\nabla\theta)\right\rangle,\\
\mathcal{H}_i &= \frac{B_0^2}{X_0^2 q_0^2}\left\langle \chi'^2\mathcal{J}\frac{\rho_N}{B^2}(|\nabla\zeta|^2 + q^2|\nabla\theta|^2 - 2\nabla\theta\cdot\nabla\zeta)\right\rangle,
\end{aligned}$$

$\mathcal{B}_i = \mathcal{D}_i = \mathcal{E}_i = 0$, $\mathcal{M}_{mm'} = m\mathcal{I}_{mm'}$, $\mathcal{Q}_{mm'} = (m - nq)\mathcal{I}_{mm'}$, γ_N denotes the dimensionless growth rate normalized by the Alfvén frequency at magnetic axis, ρ_N is the dimensionless mass density normalized by the mass density at magnetic axis, subscript "0" refers to quantities at magnetic axis, and

$$\begin{aligned}
\mathcal{G}_{11} &= \langle J(\nabla\theta\times\nabla\zeta)\cdot(\nabla\theta\times\nabla\zeta)\rangle,\\
\mathcal{G}_{22} &= \langle J(\nabla\zeta\times\nabla\psi)\cdot(\nabla\zeta\times\nabla\psi)\rangle,\\
\mathcal{G}_{33} &= \langle J(\nabla\psi\times\nabla\theta)\cdot(\nabla\psi\times\nabla\theta)\rangle,\\
\mathcal{G}_{12} &= \langle J(\nabla\theta\times\nabla\zeta)\cdot(\nabla\zeta\times\nabla\psi)\rangle,\\
\mathcal{G}_{31} &= \langle J(\nabla\psi\times\nabla\theta)\cdot(\nabla\theta\times\nabla\zeta)\rangle,\\
\mathcal{G}_{23} &= \langle J(\nabla\zeta\times\nabla\psi)\cdot(\nabla\psi\times\nabla\theta)\rangle.
\end{aligned}$$

We can reduce the set of equations (94) and (95) into a set of first order differential equations as in the DCON formalism Glasser (1997). By solving Eq. (95), one obtains

$$\xi_s = -\mathcal{A}^{-1}\mathcal{B}\xi'_\psi - \mathcal{A}^{-1}\mathcal{C}\xi_\psi.$$

Inserting this solution into Eq. (94), we get

$$\frac{d}{d\psi}\left(\mathcal{F}\xi' + \mathcal{K}\xi\right) - \left(\mathcal{K}^{+}\xi' + \mathcal{G}\xi\right) = 0, \tag{96}$$

where $\mathcal{F} = \mathcal{D} - \mathcal{B}^{\dagger}\mathcal{A}^{-1}\mathcal{B}$, $\mathcal{K} = \mathcal{E} - \mathcal{B}^{\dagger}\mathcal{A}^{-1}\mathcal{C}$, and $\mathcal{G} = \mathcal{H} - \mathcal{C}^{\dagger}\mathcal{A}^{-1}\mathcal{C}$. These matrices can be further simplified as follows Glasser (1997)

$$\mathcal{F} = \frac{\chi'^2}{n^2}\left\{\mathcal{Q}\mathcal{G}_{33}\mathcal{Q} + \gamma_N^2\mathcal{A}_i - \left[\gamma_N^2\mathcal{A}_i + \mathcal{Q}(n\mathcal{G}_{23} + \mathcal{G}_{33}\mathcal{M})\right] \right.$$

$$\left. \times\mathcal{A}^{-1}\left[\gamma_N^2\mathcal{A}_i + (n\mathcal{G}_{23} + \mathcal{M}\mathcal{G}_{33})\mathcal{Q}\right]\right\}, \tag{97}$$

$$\mathcal{K} = \frac{\chi'}{n}\left\{i\left[\gamma_N^2\mathcal{A}_i + \mathcal{Q}(n\mathcal{G}_{23} + \mathcal{G}_{33}\mathcal{M})\right]\mathcal{A}^{-1}\mathcal{C}\right.$$

$$\left. -\mathcal{Q}\left[\chi''\mathcal{G}_{23} + (q\chi')'\mathcal{G}_{33} - i\chi'\mathcal{G}_{31}\mathcal{Q} - g'\mathcal{I}\right] - i\gamma_N^2\mathcal{C}_i\right\}. \tag{98}$$

Introducing the expanded $2M$ unknowns $\mathbf{u} = \begin{pmatrix}\xi \\ \mathbf{u}_2\end{pmatrix}$, where $\mathbf{u}_2 = \mathcal{F}\xi' + \mathcal{K}\xi$, Eq. (96) is reduced to the set of $2M$ first order equations

$$\mathbf{u}' = \mathcal{L}\mathbf{u}, \tag{99}$$

where $2M \times 2M$ matrix

$$\mathcal{L} = \begin{pmatrix} -\mathcal{F}^{-1}\mathcal{K} & \mathcal{F}^{-1} \\ \mathcal{G} - \mathcal{K}^{\dagger}\mathcal{F}^{-1}\mathcal{K} & \mathcal{K}^{\dagger}\mathcal{F}^{-1}\end{pmatrix}.$$

We note that ξ and $\mathbf{u}_2$ in plasma region are related to the magnetic field and pressure as follows

$$[[\mathcal{J}\nabla\psi \cdot \delta\mathbf{B}]] = i\mathcal{Q}\xi,$$
$$-[[\mathcal{J}(\mathbf{B} \cdot \delta\mathbf{B} - \xi \cdot \nabla P)]] = \mathbf{u}_2.$$

The set of eigen mode equations in Eq. (99) can be solved numerically by independent solution method together with multiple region matching technique as described in Ref. Zheng & Kotschenreuther (2006). With M boundary conditions imposed at magnetic axis, there remain only M independent solutions:

$$\begin{pmatrix}\Xi_p \\ \mathcal{W}_2\end{pmatrix} \equiv \begin{pmatrix}\xi^1, & \cdots, & \xi^M \\ \mathbf{u}_2^1, & \cdots, & \mathbf{u}_2^M\end{pmatrix},$$

where the superscripts are used to label independent solutions. We use the cylinder limit to describe the boundary condition at magnetic axis, *i.e.*, $\xi_{\psi,m} \propto r^m$. The general solution can be then obtained as a combination of the M independent solutions,

$$\begin{pmatrix}\xi \\ \mathbf{u}_2\end{pmatrix} = i\begin{pmatrix}\Xi_p \\ \mathcal{W}_p\end{pmatrix}\mathbf{c}_p, \tag{100}$$

where $\mathbf{c}_p$ is a constant vector with M elements. Without loss of generality (by defining $\mathbf{c}_p = \Xi_p^{-1}\mathbf{c}_p^{new}$ and $\mathcal{W}_p^{new} = \mathcal{W}_p\Xi_p^{-1}$), we can set Ξ_p to be unity $\mathcal{I}$ at plasma edge. Therefore, at plasma-vacuum interface ψ_a we have

$$[[\mathcal{J}\nabla\psi \cdot \delta\mathbf{B}]] = -\mathcal{Q}\mathbf{c}_p, \tag{101}$$
$$-[[\mathcal{J}(\mathbf{B} \cdot \delta\mathbf{B} - \xi \cdot \nabla P)]] = i\mathcal{W}_p\mathbf{c}_p. \tag{102}$$

5.2 The solution of vacuum region

For completeness, in this subsection we briefly review the vacuum solutions in Ref. Zheng & Kotschenreuther (2006). The vacuum regions are described by the Laplace equation

$$\nabla^2 u = 0, \tag{103}$$

where u is the magnetic scalar potential and is related to the perturbed magnetic field by $\delta\mathbf{B} = -\nabla u$. Here, we note that this representation of vacuum magnetic field, although being simple, excludes the consideration of $n = 0$ modes. To study $n = 0$ modes, one more scalar is needed to represent the vacuum magnetic field. For the sake of conciseness, we outline the general solutions for inner and outer vacuum regions simultaneously.

As in the plasma region, Fourier decompositions are introduced for both poloidal and toroidal directions to solve Eq. (103). Then Eq. (103) becomes a set of second-order differential equations of number M for $\mathbf{u}$. This set of second-order differential equations can be transformed into a set of first-order differential equations of number $2M$, by introducing an additional field $\mathbf{v} = -[[\mathcal{J}\nabla\psi \cdot \delta\mathbf{B}]]$, which is related to the magnetic scalar potential in Fourier space as follows:

$$\mathbf{v} = \left\langle \mathcal{J}|\nabla\psi|^2 \right\rangle \frac{\partial \mathbf{u}}{\partial\psi} + \langle i\mathcal{J}\nabla\psi \cdot \nabla\theta\rangle \mathcal{M}\mathbf{u}.$$

There are $2M$ independent solutions for Eq. (103), which can be used to construct the following independent solution matrices:

$$\begin{pmatrix} \mathcal{U}_1 \\ \mathcal{V}_1 \end{pmatrix} \equiv \begin{pmatrix} \mathbf{u}^1, & \cdots, & \mathbf{u}^M \\ \mathbf{v}^1, & \cdots, & \mathbf{v}^M \end{pmatrix},$$
$$\begin{pmatrix} \mathcal{U}_2 \\ \mathcal{V}_2 \end{pmatrix} \equiv \begin{pmatrix} \mathbf{u}^{M+1}, & \cdots, & \mathbf{u}^{2M} \\ \mathbf{v}^{M+1}, & \cdots, & \mathbf{v}^{2M} \end{pmatrix}.$$

The general solutions in the vacuum regions can be expressed as a linear combination of the independent solutions:

$$\begin{pmatrix} \mathbf{u} \\ \mathbf{v} \end{pmatrix} = \begin{pmatrix} \mathcal{U}_1 \\ \mathcal{V}_1 \end{pmatrix} \mathbf{c}_v + \begin{pmatrix} \mathcal{U}_2 \\ \mathcal{V}_2 \end{pmatrix} \mathbf{d}_v, \tag{104}$$

where $\mathbf{c}_v$ and $\mathbf{d}_v$ are constant vectors in the independent solution space. To distinguish the inner and outer vacuum solutions, we let $\mathbf{c}_{v1}$ and $\mathbf{d}_{v1}$ denote the constants for inner vacuum region and $\mathbf{c}_{v2}$ and $\mathbf{d}_{v2}$ for outer vacuum region.

In the outer vacuum region, the scalar potential $\mathbf{u}$ is subjected to M boundary conditions at infinite ψ. With these M boundary conditions imposed, there are only M independent solutions left. Without loss of generality, we can set $\mathbf{c}_{v2}$ to be zero in this case. Consequently, eliminating $\mathbf{d}_{v2}$ in Eq. (104), we obtain

$$\mathbf{u}|_{\psi_{b+}} = \mathcal{T}\mathbf{v}|_{\psi_{b+}},$$

where the $M \times M$ matrix $\mathcal{T}$ is given by $\mathcal{T} = \mathcal{U}_2\mathcal{V}_2^{-1}|_{\psi_{b+}}$. The matrix $\mathcal{T}$ can be computed by means of the Green function method Chance (1997).

In the inner vacuum region, the independent solutions can be constructed, for example, with the use of an inward numerical shooting Zheng & Kotschenreuther (2006), with the following boundary conditions imposed at ψ_{b-}:

$$\begin{pmatrix} \mathcal{U}_1 \\ \mathcal{V}_1 \end{pmatrix}_{\psi_{b-}} = \begin{pmatrix} \mathcal{I} \\ \mathcal{O} \end{pmatrix}, \tag{105}$$

$$\begin{pmatrix} \mathcal{U}_2 \\ \mathcal{V}_2 \end{pmatrix}_{\psi_{b-}} = \begin{pmatrix} \mathcal{T} \\ \mathcal{I} \end{pmatrix}, \tag{106}$$

where $\mathcal{O}$ is $M \times M$ zero matrix. Since the boundary conditions in Eq. (105) give $\delta\mathbf{B} \cdot \nabla\psi = 0$ at wall, these conditions correspond to a set of solutions that corresponds to the perfectly conducting wall type. On the other hand, since the boundary conditions in Eq. (106) guarantee that the independent solutions to be continuous with outer vacuum solutions, these conditions correspond to a set of solutions that corresponds to the no-wall type. Using the general expression for the solutions in Eq. (104), we can express the normal and parallel magnetic fields at the plasma-vacuum interface as follows:

$$[[\mathcal{J}\nabla\psi \cdot \delta\mathbf{B}]] = -\mathcal{V}_1\mathbf{c}_{v1} - \mathcal{V}_2\mathbf{d}_{v1}, \tag{107}$$

$$-[[\mathcal{J}\mathbf{B} \cdot \delta\mathbf{B}]] = i\mathcal{Q}\left(\mathcal{U}_1\mathbf{c}_{v1} + \mathcal{U}_2\mathbf{d}_{v1}\right). \tag{108}$$

5.3 Eigenvalue problem

The solutions in the plasma and vacuum regions described in the last two subsections can be used to construct the eigen value problem Zheng & Kotschenreuther (2006). The normal magnetic field component and the combined magnetic and thermal pressures are required to be continuous at the plasma-vacuum interface. Matching plasma [Eqs. (101) and (102)] and vacuum [Eqs. (107) and (108)] solutions at the interface ψ_a gives

$$\mathbf{d}_{v1} = \mathcal{F}_1^{-1}\delta\mathcal{W}_b\delta\mathcal{W}_\infty^{-1}\mathcal{F}_2\mathbf{c}_{v1}, \tag{109}$$

where $\delta\mathcal{W}_\infty = \mathcal{W}_p - \mathcal{Q}[\mathcal{U}_2\mathcal{V}_2^{-1}]_{\psi_a}\mathcal{Q}$, $\delta\mathcal{W}_b = \mathcal{W}_p - \mathcal{Q}[\mathcal{U}_1\mathcal{V}_1^{-1}]_{\psi_a}\mathcal{Q}$, $\mathcal{F}_1 = \mathcal{Q}\left[\mathcal{U}_2 - \mathcal{U}_1\mathcal{V}_1^{-1}\mathcal{V}_2\right]_{\psi_a}$, and $\mathcal{F}_2 = \mathcal{Q}\left[\mathcal{U}_1 - \mathcal{U}_2\mathcal{V}_2^{-1}\mathcal{V}_1\right]_{\psi_a}$. Note that $\delta\mathcal{W}_\infty$ and $\delta\mathcal{W}_b$ correspond to the energy matrices without a wall and with a perfectly conducting wall at ψ_b, respectively, as can be seen from the boundary conditions in Eqs. (105) and (106).

We now consider the matching across the thin resistive wall. For the radial magnetic field, the Maxwell equation $\nabla \cdot \delta\mathbf{B} = 0$ and the thin wall assumption lead to

$$\mathbf{v}|_{\psi_{b-}} = \mathbf{v}|_{\psi_{b+}} = \mathbf{d}_{v1}.$$

The current in the resistive wall causes a jump in the scalar magnetic potential. This can be obtained from the Ampére law

$$\nabla \times \nabla \times \delta\mathbf{B} = -\gamma\mu_0\sigma\delta\mathbf{B}, \tag{110}$$

where σ is the wall conductivity. Equation (110) can be reduced to

$$\mathcal{V}\left(\mathbf{u}|_{\psi_{b+}} - \mathbf{u}|_{\psi_{b-}}\right) = \tau_w \gamma_N \mathbf{d}_{v1}, \tag{111}$$

where $\tau_w = \mu_0 \sigma d b / \tau_A$, d is the wall thickness, b is the average wall minor radius, and

$$\mathcal{V} = \mathcal{M}\left\langle \mathcal{J}|\nabla\psi||\nabla\theta| - \mathcal{J}|\nabla\psi\cdot\nabla\theta|^2/(|\nabla\psi||\nabla\theta|)\right\rangle \mathcal{M}$$
$$+n^2\left\langle \mathcal{J}|\nabla\phi|^2|\nabla\psi|/|\nabla\theta|\right\rangle.$$

Since $\mathbf{c}_{v2} = 0$, we find that Eqs. (104) - (106) yield

$$\mathbf{u}|_{\psi_{b+}} - \mathbf{u}|_{\psi_{b-}} = -\mathbf{c}_{v1}. \tag{112}$$

From Eqs. (109), (111), and (112) we find the eigen mode equations

$$\mathcal{D}_0(\gamma_N)\mathbf{d}_{v1} \equiv \tau_w \gamma_N \mathbf{d}_{v1} + \mathcal{V}\mathcal{F}_2^{-1}\delta\mathcal{W}_\infty \delta\mathcal{W}_b^{-1}\mathcal{F}_1 \mathbf{d}_{v1} = \mathcal{O}.$$

The dispersion relation for this eigen value problem is given by the determinant equation $\det|\mathcal{D}_0(\gamma_N)| = 0$. In general the Nyquist diagram can be used to determine the roots of this dispersion relation. For RWMs, however, the growth rate is much smaller than the Alfvén frequency. Therefore, the growth rate dependence of $\delta\mathcal{W}_\infty\delta\mathcal{W}_b^{-1}$ can be neglected for determining the stability condition. Consequently, one can use the reduced eigen value problem

$$-\mathcal{V}\mathcal{F}_2^{-1}\delta\mathcal{W}_\infty \delta\mathcal{W}_b^{-1}\mathcal{F}_1 \mathbf{d}_{v1} = \tau_w \gamma_N \mathbf{d}_{v1}, \tag{113}$$

with the RWM mode growth rate γ_N on the right hand side of this equation used as the eigen value to determine the stability.

5.4 Discussion

Now let us discuss the connection of current global theory with localized analytical theories described in Sec. 4. The singular layer equation in Eq. (48) is derived by employing mode localization assumption. Only localized mode coupling is considered. The general eigen mode equation Eq. (96) in plasma region contains all side band couplings. Noting that $\mathcal{Q} \propto x$, one can see from Eqs. (97) and (98) that $\mathcal{F} \propto x^2$ and $\mathcal{K} \propto x$ at marginal stability $\omega = 0$. We can therefore see the root of Eq. (48) in Eq. (96). If ballooning invariance in Eq. (57) is introduced, the set of matrix Eq. (96) can be transformed to a single ballooning equation. The TAE theory in Sec. 4.4 uses just two Fourier components to construct eigen modes. The general Alfvén gap structure can be determined by $\det|\mathcal{F}| = 0$. Note that, if an analytical function is given on a line on complex ω plane, the function can be determined in whole domain through analytical continuation by using the Cauchy-Riemann condition. Note also that one can avoid MHD continuum by scanning the dispersion relation with real frequency $\Re e\{\omega\}$ for a small positive growth rate $\Im m\{\omega\}$. Using the scan by AEGIS one can in principle find damping roots through analytical continuation. Due to its adaptive shooting scheme AEGIS can be used to compute MHD modes with very small growth rate. It has successfully computed Alfvén continuum damping rate by analytical continuation based on AEGIS code Chen et al. (2010).

6. Summary and discussion

In this chapter we have given an overview of MHD theory in toroidal confinement of fusion plasmas. Four types of fundamental MHD modes in toroidal geometry: interchange, ballooning, TAEs, and KDMs, are discussed. In describing these modes we detail some fundamental analytical treatments of MHD modes in toroidal geometry, such as the average technique for singular layer modes, ballooning representation, mode coupling treatment in TAEs/KDMs theories. Note that analytical approach is often limited for toroidal plasma physics. Global numerical treatment of MHD modes is also reviewed in this chapter, especially the AEGIS code formalism. These theories are reviewed in ideal MHD framework. Here, we briefly discuss kinetic and resistive modifications to ideal MHD, as well as the connection of MHD instabilities to transport.

Let us first discuss kinetic effects. Since strong magnetic field is used to contain plasmas in magnetically confined fusion experiments, MHD theory can be rather good to describe fusion plasmas in the direction perpendicular to magnetic field. This is because strong magnetic field can hold plasmas together in perpendicular movement. Therefore, MHD is a very good model to describe perpendicular physics, if FLR effects are insignificant. However, in parallel direction the Lorentz force vanishes and particle collisions are insufficient to keep particles to move collectively. Consequently, kinetic description in parallel direction is generally necessary. Kinetic effect is especially important when wave-particle resonance effect prevails in the comparable frequency regime $\omega \sim \omega_{si}$ Zheng & Tessarotto (1994a). In the low frequency regime $\omega \ll \omega_{si}$, wave-particle resonances can be so small that kinetic description results only in an enhancement of apparent mass effect. Kinetic effect in this case can be included by introducing enhanced apparent mass. Another non-resonance case is the intermediate frequency regime $\omega_{si} \ll \omega \ll \omega_{se}$. In this regime kinetic description results in a modification of ratio of special heats. By introducing proper Γ MHD can still be a good approximation. Recovery of perpendicular MHD from gyrokinetics has been studied in details in Ref. Zheng et al. (2007).

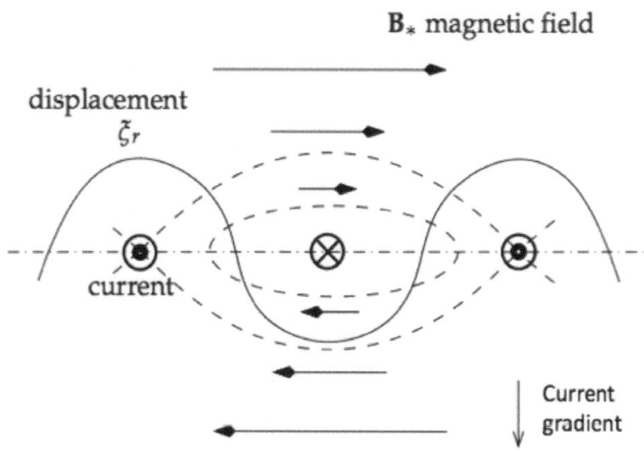

Fig. 2. CITM physics picture. The dot-dashed line represents mode rational surface. Perturbed current at rational surface due to interchange modes leads to field line reconnection and formation of magnetic islands.

Next, let us discuss resistivity effects. Resistivity usually is small in magnetically confined fusion plasmas. Due to its smallness resistivity effects are only important in the singular layer region. With ideal MHD singular layer theory detailed in Sec. 4.2 one can rederive resistive singular layer equations given in Ref. Glasser et al. (1975). However, it should be pointed out that, when kinetic enhancement of apparent mass effect is taken into account, the ratio of resistivity and inertia layer widths changes. This leads kinetic description of resistive MHD modes to become substantially different from fluid description Zheng & Tessarotto (1996) Zheng & Tessarotto (1995). Kinetic analysis of low frequency resistive MHD modes becomes necessary.

The driving force for ideal MHD instabilities is related to pressure gradient. Resistivity can instead cause field line reconnection and induce the so-called tearing modes. It is important to note that if current gradient is taken into account pressure driven modes and tearing modes are coupled to each other. The underlying driving mechanism for pressure driven modes is the release of plasma thermal energy from the interchange of magnetic flux tubes. Actually, interchange-type modes exchange not only thermal and magnetic energies between flux tubes, but also current. In a plasma with a current (or resistivity) gradient, such an interchange can create a current sheet at a mode resonance surface and result in the excitation of current interchange tearing modes (CITMs) as shown in Fig. 2 Zheng & Furukawa (2010).

Instabilities of interchange type have been widely used to explain anomalous transport in tokamaks in terms of the formation of turbulent eddies through nonlinear coupling. However, the explanation for experimental observations that the electron energy transport is much larger than what one would expect from diffusive process due to Coulomb collisions is still unsatisfactory. The electron Larmor radius is much smaller than ion one. Nonetheless, the electron thermal transport often is stronger than ion transport. In Ref. Rechester & Rosenbluth (1978), the broken magnetic surfaces due to formation of magnetic island and stochastic field lines are used to explain the enhanced electron transport. But, how magnetic islands are formed in axisymmetric tokamak plasmas has not been given. CITM theory shows that interchange-type instabilities can directly convert to current interchange tearing modes. This helps to clarify the source of electron transport in tokamaks.

Another transport issue we need to discuss is the so-called flow shear de-correlation of turbulences. This concept has been widely used for explaining suppression of plasma turbulences. In fact, this picture is not right for systems with magnetic shear. We use Fig. 3 to explain it (L. J. Zheng and M. Tessarotto, private communication). In Fig. 3, the dashed long arrow represents a magnetic field line on a given magnetic surface ψ_0, and two solid long arrows denote the magnetic field lines respectively at two time sequences t_0 and $t_0 + \Delta t$ on an adjacent magnetic surface ψ_1. Let us examine the correlation pattern in the local frame moving together with equilibrium velocity of the dashed long arrow on surface ψ_0. The modes are supposed to locate around the point "O" initially at $t = t_0$. After a time interval Δt, the field line on surface ψ_1 moves relatively to the dashed long arrow on the surface ψ_0 due to flow shear. From Fig. 3 one can see that the fixed pattern has not been de-correlated by flow shear, instead the pattern just propagates from point "O" at time $t = t_0$ to point "O'" at subsequent time $t = t_0 + \Delta t$. This indicates that flow shear does not de-correlate turbulence eddies. Only flow curvature can result in the de-correlation. This resembles to ballooning mode behavior in rotating plasmas with Cooper representation Waelbroeck & Chen (1991).

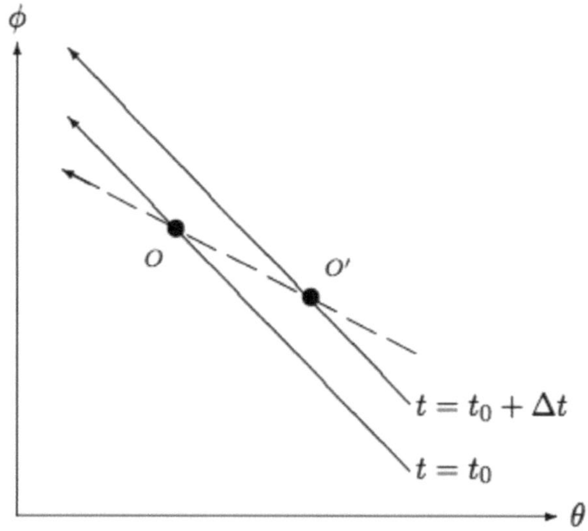

Fig. 3. Schematic explanation for why flow shear does not de-correlate turbulence eddies.

In conclusion significant progresses have been made for linear ideal MHD theories and numerical codes in the past dacades. However, the kinetic effects on MHD remains considerably open. Although correction of gyrokinetics theory has been made recently Zheng et al. (2007), the applications of the new gyrokinetics theory remain to be worked out. The theories for FLR effects on ballooning modes, KTAEs, energetic particle effects, etc. need to be modified with newly corrected gyrokinetics theory. The extension of toroidal resistive MHD theory Glasser et al. (1975) to take into account the small parallel ion speed effect Zheng & Tessarotto (1996) and current interchange effects Zheng & Furukawa (2010) is under consideration.

7. References

Berk, H. L., Rosenbluth, M. N. & Shohet, J. L. (1983). Ballooning mode calculations in stellarators, *Physics of Fluids* 26(9): 2616–2620.
URL: *http://link.aip.org/link/?PFL/26/2616/1*

Berk, H. L., Van Dam, J. W., Guo, Z. & Lindberg, D. M. (1992). Continuum damping of low-n toroidicity-induced shear Alfvén eigenmodes, *Physics of Fluids B: Plasma Physics* 4(7): 1806–1835.
URL: *http://link.aip.org/link/?PFB/4/1806/1*

Bernard, L., Helton, F. & Moore, R. (1981). GATO: an MHD stability code for axisymmetric plasmas with internal separatrices, *Comput. Phys. Commun.* 24: 377.

Betti, R. & Freidberg, J. P. (1992). Stability of Alfvén gap modes in burning plasmas, *Physics of Fluids B: Plasma Physics* 4(6): 1465–1474.
URL: *http://link.aip.org/link/?PFB/4/1465/1*

Boozer, A. H. (1982). Establishment of magnetic coordinates for a given magnetic field, *Physics of Fluids* 25(3): 520–521.
URL: *http://link.aip.org/link/?PFL/25/520/1*

Chance, M., Greene, J., Grimm, R., Johnson, J., Manickam, J., Kerner, W., Berger, D., Bernard, L., Gruber, R. & Troyon, F. (1978). Comparative numerical studies of ideal

magnetohydrodynamic instabilities, *Journal of Computational Physics* 28(1): 1 – 13.
URL: *http://www.sciencedirect.com/science/article/pii/0021999178900438*

Chance, M. S. (1997). Vacuum calculations in azimuthally symmetric geometry, *Physics of Plasmas* 4(6): 2161–2180.
URL: *http://link.aip.org/link/?PHP/4/2161/1*

Chance, M. S., Dewar, R. L., Frieman, E. A., Glasser, A. H., Greene, J. M., Grimm, R. C., Jardin, S. C., Johnson, J. L., Manickam, J., Okabayashi, M., & Todd, A. M. M. (1979). MHD stability limits on high-beta tokamaks, *Plasma Physics and Controlled Fusion Research 1978*, Vol. 1, International Atomic Energy Agency, Vienna, p. 677.

Chen, E., Berk, H., Breizman, B. & Zheng, L. J. (2010). Continuum damping of free-boundary tae with AEGIS, *APS Meeting Abstracts (DPP)* p. 9124P.

Chen, L. (1994). Theory of magnetohydrodynamic instabilities excited by energetic particles in tokamaks, *Physics of Plasmas* 1(5): 1519–1522.
URL: *http://link.aip.org/link/?PHP/1/1519/1*

Cheng, C., Chen, L. & Chance, M. (1985). High-n ideal and resistive shear Alfvén waves in tokamaks, *Annals of Physics* 161(1): 21 – 47.
URL: *http://www.sciencedirect.com/science/article/pii/0003491685903355*

Connor, J. W., Hastie, R. J. & Taylor, J. B. (1978). Shear, periodicity, and plasma ballooning modes, *Phys. Rev. Lett.* 40: 396–399.
URL: *http://link.aps.org/doi/10.1103/PhysRevLett.40.396*

Connor, J. W., Hastie, R. J. & Taylor, J. B. (1979). High mode number stability of an axisymmetric toroidal plasma, *Proceedings of the Royal Society of London. A. Mathematical and Physical Sciences* 365(1720): 1–17.
URL: *http://rspa.royalsocietypublishing.org/content/365/1720/1.abstract*

Glasser, A. (1997). The direct criterion of newcomb for the stability of an axisymmetric toroidal plasma, *Los Alamos Report* LA-UR-95-528.

Glasser, A. H., Greene, J. M. & Johnson, J. L. (1975). Resistive instabilities in general toroidal plasma configurations, *Physics of Fluids* 18(7): 875–888.
URL: *http://link.aip.org/link/?PFL/18/875/1*

Grad, H. & Rubin, H. (1958). Hydromagnetic equilibria and force-free fields, *Proceedings of the 2nd UN Conf. on the Peaceful Uses of Atomic Energy*, Vol. 31, IAEA, Geneva, p. 190.

Greene, J. & Chance, M. (1981). The second region of stability against ballooning modes, *Nuclear Fusion* 21(4): 453.
URL: *http://stacks.iop.org/0029-5515/21/i=4/a=002*

Greene, J. M. & Johnson, J. L. (1962). Stability criterion for arbitrary hydromagnetic equilibria, *Physics of Fluids* 5(5): 510–517.
URL: *http://link.aip.org/link/?PFL/5/510/1*

Grimm, R. C., Greene, J. M. & Johnson, J. L. (1976). *Methods of Computational Physics*, Academic Press, New York, London.

Hamada, S. (1962). Hydromagnetic equilibria and their proper coordinates, *Nuclear Fusion* 2(1-2): 23.
URL: *http://iopscience.iop.org/0029-5515/2/1-2/005*

Hazeltine, R. D., Hitchcock, D. A. & Mahajan, S. M. (1981). Uniqueness and inversion of the ballooning representation, *Physics of Fluids* 24(1): 180–181.
URL: *http://link.aip.org/link/?PFL/24/180/1*

Johnson, J. L. & Greene, J. M. (1967). Resistive interchanges and the negative v" criterion, *Plasma Physics* 9(5): 611.
URL: *http://stacks.iop.org/0032-1028/9/i=5/a=311*

Lee, Y. C. & Van Dam, J. W. (1977). Kinetic theory of ballooning instabilities, *in* B. Coppi & W. Sadowski (eds), *Proceedings of the Finite Beta Theory Workshop*, Vol. CONF-7709167, U.S. Department of Energy, Washington, D.C., p. 93.

Lortz, D. (1975). The general "peeling" instability, *Nuclear Fusion* 15(1): 49.
URL: *http://stacks.iop.org/0029-5515/15/i=1/a=007*

Lortz, D. & Nührenberg, J. (1978). Ballooning stability boundaries for the large-aspect-ratio tokamak, *Physics Letters A* 68(1): 49 – 50.
URL: *http://www.sciencedirect.com/science/article/pii/0375960178907533*

Mercier, C. (1962). Critere de stabilite d'an systeme toroidal hydromagnetique en pression scalaire, *Nucl. Fusion Suppl.* Pt. 2: 801.

Mikhailovsky, A. (1974). *Theory of Plasma Instabilities*, Consultant Bureau, New York.

Newcomb, W. A. (1960). Hydromagnetic stability of a diffuse linear pinch, *Annals of Physics* 10(2): 232 – 267.
URL: *http://www.sciencedirect.com/science/article/pii/0003491660900233*

Rechester, A. B. & Rosenbluth, M. N. (1978). Electron heat transport in a tokamak with destroyed magnetic surfaces, *Phys. Rev. Lett.* 40: 38–41.
URL: *http://link.aps.org/doi/10.1103/PhysRevLett.40.38*

Rosenbluth, M. N., Berk, H. L., Van Dam, J. W. & Lindberg, D. M. (1992). Mode structure and continuum damping of high-n toroidal Alfvén eigenmodes, *Physics of Fluids B: Plasma Physics* 4(7): 2189–2202.
URL: *http://link.aip.org/link/?PFB/4/2189/1*

Shafranov, V. D. (1966). Plasma equilibrium in a magnetic field, *Reviews of Plasma Physics*, Vol. 2, Consultants Bureau, New York, p. 103.

Tang, W., Connor, J. & Hastie, R. (1980). Kinetic-ballooning-mode theory in general geometry, *Nuclear Fusion* 20(11): 1439.
URL: *http://stacks.iop.org/0029-5515/20/i=11/a=011*

Tsai, S.-T. & Chen, L. (1993). Theory of kinetic ballooning modes excited by energetic particles in tokamaks, *Physics of Fluids B: Plasma Physics* 5(9): 3284–3290.
URL: *http://link.aip.org/link/?PFB/5/3284/1*

Vlad, G., Zonca, F. & Briguglio, S. (1999). Dynamics of Alfvén waves in tokamaks, *Nuovo Cimento Rivista Serie* 22: 1–97.

Waelbroeck, F. L. & Chen, L. (1991). Ballooning instabilities in tokamaks with sheared toroidal flows, *Physics of Fluids B: Plasma Physics* 3(3): 601–610.
URL: *http://link.aip.org/link/?PFB/3/601/1*

Wesson, J. (1978). Hydromagnetic stability of tokamaks, *Nuclear Fusion* 18(1): 87.
URL: *http://stacks.iop.org/0029-5515/18/i=1/a=010*

Zheng, L.-J. & Chen, L. (1998). Plasma compressibility induced toroidal Alfvén eigenmode, *Physics of Plasmas* 5(2): 444–449.
URL: *http://link.aip.org/link/?PHP/5/444/1*

Zheng, L.-J., Chen, L. & Santoro, R. A. (2000). Numerical simulations of toroidal Alfvén instabilities excited by trapped energetic ions, *Physics of Plasmas* 7(6): 2469–2476.
URL: *http://link.aip.org/link/?PHP/7/2469/1*

Zheng, L.-J., Chu, M. S. & Chen, L. (1999). Effect of toroidal rotation on the localized modes in low beta circular tokamaks, *Physics of Plasmas* 6(4): 1217–1226.
URL: *http://link.aip.org/link/?PHP/6/1217/1*

Zheng, L. J. & Furukawa, M. (2010). Current-interchange tearing modes: Conversion of interchange-type modes to tearing modes, *Physics of Plasmas* 17(5): 052508.
URL: *http://link.aip.org/link/?PHP/17/052508/1*

Zheng, L.-J. & Kotschenreuther, M. (2006). AEGIS: An adaptive ideal-magnetohydrodynamics shooting code for axisymmetric plasma stability, *Journal of Computational Physics* 211(2): 748 – 766.
URL: *http://www.sciencedirect.com/science/article/pii/S0021999105002950*

Zheng, L.-J., Kotschenreuther, M. & Chu, M. S. (2005). Rotational stabilization of resistive wall modes by the shear Alfvén resonance, *Phys. Rev. Lett.* 95: 255003.
URL: *http://link.aps.org/doi/10.1103/PhysRevLett.95.255003*

Zheng, L. J., Kotschenreuther, M. T. & Van Dam, J. W. (2007). Revisiting linear gyrokinetics to recover ideal magnetohydrodynamics and missing finite Larmor radius effects, *Physics of Plasmas* 14(7): 072505.

Zheng, L. J., Kotschenreuther, M. T. & Van Dam, J. W. (2010). AEGIS-K code for linear kinetic analysis of toroidally axisymmetric plasma stability, *Journal of Computational Physics* 229(10): 3605 – 3622.
URL: *http://www.sciencedirect.com/science/article/pii/S002199911000032X*

Zheng, L.-J. & Tessarotto, M. (1994a). Collisionless kinetic ballooning equations in the comparable frequency regime, *Physics of Plasmas* 1(9): 2956–2962.
URL: *http://link.aip.org/link/?PHP/1/2956/1*

Zheng, L.-J. & Tessarotto, M. (1994b). Collisionless kinetic ballooning mode equation in the low-frequency regime, *Physics of Plasmas* 1(12): 3928–3935.
URL: *http://link.aip.org/link/?PHP/1/3928/1*

Zheng, L.-J. & Tessarotto, M. (1995). Collisional ballooning mode dispersion relation in the banana regime, *Physics of Plasmas* 2(8): 3071–3080.
URL: *http://link.aip.org/link/?PHP/2/3071/1*

Zheng, L.-J. & Tessarotto, M. (1996). Collisional effect on the magnetohydrodynamic modes of low frequency, *Physics of Plasmas* 3(3): 1029–1037.
URL: *http://link.aip.org/link/?PHP/3/1029/1*

2

Sub-Fluid Models in Dissipative Magneto-Hydrodynamics

Massimo Materassi[1], Giuseppe Consolini[2] and Emanuele Tassi[3]
[1]*Istituto dei Sistemi Complessi ISC-CNR, Sesto Fiorentino,*
[2]*INAF-Istituto di Fisica dello Spazio Interplanetario, Roma,*
[3]*Centre de Physique Théorique, CPT-CNRS, Marseille,*
[1,2]*Italy*
[3]*France*

1. Introduction

Magneto-Hydrodynamics (MHD) describes the plasma as a *fluid* coupled with the self-consistent magnetic field. The regime of validity of the MHD description of a plasma system is generally restricted to the temporal and spatial scales much larger than the characteristic plasma temporal scales (such as those associated with the plasma frequency, the ion and electrons cyclotron frequencies and the collision frequency), or the typical spatial scales (as the ion and electron inertial scale, the ion and electron Larmor radii and the Debye length). On the large scale, the plasma can be successfully described in terms of a single magnetized fluid by means of generally differentiable and smooth functions: this description of plasma media has met a wide success. However, the last decade of the 20th Century has brought to scientists' attention a wide amount of experimental and theoretical results suggesting substantial changes in classical magnetized plasma dynamics with respect to the MHD picture. In particular, two fundamental characteristics of the MHD as a dynamical theory have started to appear questionable: *regularity* and *determinism*. The MHD variables are, indeed, analytically smooth functions of space and time coordinates. Physicists refer to this as *regularity*. Moreover, once the initial conditions are assigned (together with some border conditions), the evolution of the MHD variables is unique: hence MHD is strictly *deterministic*. Instead, in in-field and laboratory studies, more and more examples have been brought to evidence, where *irregularity* and *stochastic processes* appear to play a role in magnetized plasma dynamics. This is particularly true when one approaches intermediate and small scales where the validity conditions for the MHD description, although still valid, are no longer valid in a strict sense, or when we are in the presence of topologically relevant structures, whose evolution cannot be described in terms of smooth functions. From now on, the conditions of the MHD variables apparently violating smoothness and/or determinism will be referred to as *irregular stochastic configurations* (ISC). In the following we remind, in some detail, these experimental and theoretical results pointing towards the existence of ISCs, in the context of space plasmas and fusion plasmas.

In the framework of space physics, it has been pointed out that both the global, large scale dynamics and some local processes related to plasma transport could be better explained in terms of stochastic processes, low-dimensional chaos, fractal features, intermittent turbulence, complexity and criticality (see e.g. Chang, 1992; Klimas et al., 1996; Chang, 1999; Consolini, 2002; Uritsky et al., 2002; Zelenyi & Milovanov, 2004 and references therein).

A certain confidence exists in stating that the rate of conversion of the magnetic energy into plasma kinetic one observed in events of *magnetic reconnection* is significantly underestimated by the traditional, smooth and deterministic, MHD (see for example Priest & Forbes, 2000 and also Biskamp, 2000). Lazarian et al. (2004) have found an improvement in the calculation of the magnetic reconnection rate by considering stochastic reconnection in a magnetized, partially ionized medium. This process is stochastic due to the field line probabilistic wandering through the turbulent fluid. In a different context, Consolini et al. (2005) showed that stochastic fluctuations play a crucial role in the current disruption of the geomagnetic tail, a magnetospheric process occurring at the onset of magnetic substorm in the Earth's magnetotail (see, e.g., Kelley, 1989; Lui, 1996). Consistently with the relevance of stochastic processes in space plasmas, tools derived from information theory have been recently applied to describe the near-Earth plasma phenomenology (Materassi et al., 2011; De Michelis et al., 2011). On the other hand, turbulence has been shown to play a relevant role in several different space plasma media as the solar wind (Bruno & Carbone, 2005) or the Earth's magnetotail regions (see, e.g., Borovsky & Funsten, 2003), etc.

In fusion plasmas, phenomena important as anomalous diffusion induced by stochastic magnetic fields (Rechester & Rosenbluth, 1978) have been suggested to be caused by the appearance of irregular modes similar to ISCs: those modes have been documented since a rather long time (Goodall, 1982). In tokamak machines ISCs observed are mesoscopic intermittent and filamentary structures: recently, studies have shown how such structures might be generated by reconnecting tearing modes triggered by a primary interchange instability (Zheng & Furukawa 2010).

The appearance of ISCs should not be expected as an exceptional condition: indeed, time- and space-regular MHD relies on very precise hypotheses, not necessarily holding in real plasmas. As underlined before, it should be considered that MHD is a long time description with respect to the interaction times of particles. In order to expect a smooth deterministic evolution in time, "fast phenomena" should be ignored, and clearly this cannot be done when "fast phenomena" lead to big changes in the MHD variables themselves, on macroscopic scales, as it happens in the *fast magnetic reconnection*.

Space regularity requires the scale at which matter appears as granular to shrink to zero, and this is possible under the hypothesis that such scale is much smaller than the typical scale where the MHD variables do vary. However, in *turbulent regimes* the scale at which the MHD fields vary are so small, that they compare with those scales at which plasma appears as granular.

The phenomenology of plasma ISCs appears to indicate that the role of "fundamental entities" should be played by mesoscopic coherent structures, interacting and stochastically evolving. These *stochastic coherent structures* (SCS) have been observed in several space plasma regions: in solar wind (Bruno et al., 2001) as field-aligned flux tubes, in the Earth's cusp regions (Yordanova et al., 2005), in the geotail plasma sheet as current structures, 2D eddies and so on (see, for instance, Milovanov et al., 2001; Borovsky & Funsten, 2003; Vörös et al., 2004; Kretzschmar & Consolini, 2006). Recent observations of small-scale magnetic field features in the magnetosheath transition region (as described in Retinò et al., 2007; Sundkvist et al., 2007) seem to suggest that the dynamics of such coherent structures can be the origin of a *coherent dissipation mechanism*, a sort of coarse-grained dissipation (Tetrault, 1992a, b; Chang et al., 2003) due to interactions that result non-local in the *k*-space.

A consistent theory of plasmas in ISC should be a consistent theory of SCSs, valid in a suitable "midland" of the "coupling constant space" (Chang et al., 1978). This "midland of SCSs" should be far from the particle scale (because each SCS involves a large amount of correlated particles), but also some steps under the fluid level (because matter should appear granular and fields irregular).

Furthermore, this "midland" is not the usual kinetic-fluid transition as described e.g. in Bălescu (1997). In fact, the kinetic description is sensible under some *weak coupling approximation* allowing for a self-consistent Markovian single particle theory to exist, while if mesoscopic coherent structures appear, the correlation length and inter-particle interaction scale are so big that the single particle evolves only together with a large number of its fellows, excluding such weak coupling. Then, if SCSs exist, the kinetic level of the theory does not.

Well far from trying to give a self-consistent theory of the SCS, here we just discuss some models and scenarios retaining some properties that such a theory should have. The approaches discussed here are exactly the application of the philosophy well described by Bălescu (1997) to dissipative processes in the MHD. Probably, a first principle analytical theory of turbulence is going to be out of reach for decades. However, something useful for applications can be developed in a more advanced framework than "traditional" statistical mechanics by introducing elements of chaos or stochasticity, non-Gaussian or non-Markovian properties, in some "effective" and "sound" models. In this way, one admits a certain "degree of randomness" in the equations, so that the non-Gaussianity of the basic stochastic processes, the role of the non-Markovian equations of evolution, the role of fractal structures and the emergence of "strange transport" are all SCS theoretical features of which one tries to take into account.

The schemes presented here are models with these properties, trying to interpolate between the macroscopic, smooth, deterministic physics of traditional MHD and the mesoscopic, irregular, stochastic physics of "that something else" which has not been formulated yet. Such phenomenological approach is indicated as *sub-fluid.*

In this chapter three sub-fluid models are described, the metriplectic dissipative MHD, the stochastic field theory of resistive MHD and the fractal magnetic reconnection.

In the first model, the metriplectic dissipative MHD (§ 2), we focus on the relationship between the fluid dynamical variables and the microscopic degrees of freedom of the plasma. The thermodynamic entropy of the plasma microscopic degrees of freedom turns out to play an essential role in the metriplectic formalism, a tool developed in the 1980s encompassing dissipation within an algebra of observables, and here adapted to MHD. It is considered that thermodynamics, i.e. statistics, naturally arises for the description of the microscopic degrees of freedom. Fluid degrees of freedom are endowed with energy, linear and angular momenta, while an entropy function, measuring how undetermined their "mechanical" microscopic configuration is, can be attributed to the microscopic degrees of freedom.

In the second model treated, the stochastic field theory (SFT) (§ 3), the dissipation coefficients appearing in the MHD equations of motion are considered as noise, consistently with the fact that, out of its equilibrium, a medium may be treated statistically. In this way, MHD turns into a set of Langevin field equations. These may be treated through the path integral formalism introduced by Phythian (1977), appearing particularly suitable for non equilibrium statistics. Once the resistive MHD theory is turned into a SFT, transition probabilities between arbitrary field configurations may be calculated via a stochastic action formalism, closely resembling what is usually done for quantum fields. This mimics very precisely the idea of an ISC.

A sub-fluid model of *fast magnetic reconnection* (FMR) is dealt with in § 4. FMR clearly belongs to the class of phenomena in which classical fields apparently undergo quantum-like transitions in considerably short times: when magnetic field lines reconnect, the field topology is changed and a big quantity of magnetic energy, associated to the original configuration, is turned into the kinetic energy of fast jets of particles. In order to mimic a reconnection rate high enough, a successful attempt may be done relaxing the assumption that all the local variables of the plasma and the magnetic field are smooth functions. In particular, in a standard 2-dimensional Sweet-Parker scenario (Parker, 1957, 1963; Sweet, 1958), one assumes that the reconnection region, where finite resistivity exists, is a fractal domain of box-counting dimension smaller than 2. This allows for a reconnection rate that varies with the magnetic Reynolds number faster than the traditional one.

2. The dissipation algebrized

Dissipation is a crucial element of the physical mechanism leading to ISCs in plasmas, and dissipative terms already appear in the smooth deterministic MHD. Moreover, the presence of dissipation, together with non-linearity, is a fundamental mechanism in order for coherent structures to form (Courbage & Prigogine, 1983).

Many fundamental phenomena giving rise to plasma ISCs in nature, such as turbulence, magnetic reconnection or dynamo (Biskamp, 1993), are often described by MHD models containing dissipative terms. For instance, this can account for the finite resistivity of the plasma and/or the action of viscous forces.

Where does dissipation come from? Ultimately, MHD is derived from the Klimontovich equations, describing the dynamics of charged particles interacting with electromagnetic fields (Klimontovich, 1967). This is a Hamiltonian, consequently non-dissipative, system. Nevertheless, dissipative terms appear in some versions of MHD equations as a heritage of averaging and approximations carried out along the derivation procedure and which have spoilt the original Hamiltonian structure of the Klimontovich system. The presence of dissipative terms reflects a transfer of energy from the deterministic macroscopic fluid quantities into the microscopic degrees of freedom of the system, to be treated statistically, which lie outside a macroscopic fluid description. Such transfer of energy, in turn, implies an increase of the entropy of the system.

If dissipative terms are omitted, on the other hand, one expects the resulting MHD system to be Hamiltonian, with a conserved energy (the constant value of the Hamiltonian of the system) and a *conserved entropy*. Indeed, the non-dissipative version of MHD, usually referred to as *ideal MHD*, has been shown, long ago, to be a Hamiltonian system (Morrison & Greene, 1980). The elements constituting a Hamiltonian structure are the Poisson bracket, a bilinear operator with algebraic properties, and the Hamiltonian of the system, depending on the dynamical variables: in the case of the MHD, these will be defined in the following (see (7) and (9)). The Hamiltonian formulation of the ideal MHD, apart from facilitating the identification of conserved quantitites, or the stability analysis of the equilibria, renders it evident that the dynamics of the system takes place on *symplectic leaves* that foliate the phase space (Morrison, 1998).

The inclusion of *dissipative terms* invalidates the Hamiltonian representation: this dissipative breakdown matches the fact that, once dissipation is included, the system becomes "less deterministic" in a certain sense, because there is an interaction with microscopic degrees of freedom that are described in a statistical manner (friction forces *are* a statistically effective treatment of microscopic stochastic collisions).

Some dissipative systems possess however an algebraic structure called *metriplectic*, which still permits to formulate the dynamics in terms of a bracket and of an observable, extending the concept of Hamiltonian. Metriplectic structures in general occur in systems which *conserve the energy and increase the entropy*. These are the so called *complete systems*. They are obtained adding friction forces to an originally Hamiltonian system, and then including, in the algebra of observables, the energy and entropy of the microscopic degrees of freedom. The metriplectic formulation permits to reformulate the dynamics of dissipative systems in a geometrical framework, in which information, such as the existence of asymptotically stable equilibria, may be easily retrieved without even trying to solve the equations.

In order to define what a metriplectic structure is, and apply this concept to the case of MHD, it is convenient to start recalling that, very frequently, one deals with the analysis of physical models of the form

$$\partial_t z^i = F_H^i(z) + F_D^i(z)\ , \quad i = 1,\dots,N, \tag{1}$$

where z is the set of the N dynamical variables of the system (N can be infinite; it is actually a continuous real index for field theories or the MHD) evolving under the action of a vector field $F_H(z) + F_D(z)$. Such vector field is the sum of a non-trivial Hamiltonian component $F_H(z)$ and a component $F_D(z)$ accounting for the dissipative terms. If $F_D(z) = 0$, the resulting system is Hamiltonian and consequently can be written as

$$\partial_t z^i = F_H^i(z) = \left[z^i, H(z)\right]\ , \tag{2}$$

where $H(z)$ is the *Hamiltonian* of the system, and [*,*] is the *Poisson bracket*, an antisymmetric bilinear operator, satisfying the Leibniz property and the Jacobi identity (Goldstein, 1980). These properties render the Poisson algebra of group-theoretical nature. An immediate consequence of the antisymmetry of the bracket is that $\partial_t H = [H,H] = 0$, so that H is necessarily a constant of motion.

It is important to point out that, in many circumstances, the Poisson bracket is not of the canonical type. In particular, for Hamiltonian systems describing the motion of continuous media in terms of Eulerian variables, as in the case of ideal MHD, the Poisson bracket is *noncanonical* and no pairs of conjugate variables can be identified. For such brackets, particular invariants, denoted as Casimir invariants, exist. These are quantities $C(z)$ such that $[C,F] = 0$ for every $F(z)$. Consequently $\partial_t C = [C,H] = 0$ in particular, which shows that Casimir functions are indeed conserved quantities.

Energy conservation and entropy increase in metriplectic systems are "algebrized" via a generalized bracket and a generalized energy functional. More precisely, a metriplectic system is a system of the form

$$\partial_t z^i = \left\{ z^i, F(z) \right\} = \left[z^i, F(z) \right] + \left(z^i, F(z) \right), \tag{3}$$

where the metriplectic bracket $\{*,*\} = [*,*] + (*,*)$ is obtained from a Poisson bracket $[*,*]$ and a *metric* bracket $(*,*)$. The latter is a bilinear, symmetric and semidefinite (positive or negative) operation, satisfying also the Leibniz property (strictly speaking, a symmetric semi-definite bracket $(*,*)$ should be referred to as semi-metric). The metric bracket is also required to be such that $(f,H) = 0$, for every function $f(z)$, with H being the Hamiltonian of the system: this means that dissipation does not alter the total energy, since this already includes a part accounting for the energy dissipated.

The function F in (3) is denoted as *free energy*, and is given by

$$F = H + \lambda C, \tag{4}$$

where C is a Casimir of the Poisson bracket, and λ is a constant.

In the cases of interest here, this C is chosen as the entropy of the microscopic degrees of freedom of the plasma, involved in the dissipation.

Let us assume the metric bracket be semi-definite negative (the case in which it is positive is completely analogous). The resulting metriplectic system possesses the following important properties:

- $\partial_t H = 0$, so that the Hamiltonian of the system is still conserved (possibly other quantities such as total linear or angular momenta can also be conserved);
- $\partial_t C = \lambda(C,C)$, so that, due to the semi definiteness of the symmetric bracket one has either $\partial_t C \geq 0$ or $\partial_t C \leq 0$ at all times, depending on whether λ is negative or positive. This candidates C to be an equivalent time coordinate wherever it is strictly monotonic with t (Courbage & Prigogine, 1983);
- isolated minima of F are *stable equilibrium points*.

Metriplectic structures have been identified for different systems as, for instance, Navier-Stokes (Morrison, 1984), free rigid body, Vlasov-Poisson (Morrison, 1986) and, in a looser sense, for Boussinesq fluids (Bihlo, 2008) and constrained mechanical systems (Nguyen & Turski, 2009). An algebraic structure for dissipative systems based on an extension of the Dirac bracket has been proposed by Nguyen and Turski (2001). They have also been used for identifying asymptotic vortex states (Flierl and Morrison, 2011).

Also the visco-resistive plasma falls into the category of complete systems. Indeed, the following version of the visco-resistive MHD equations

$$
\begin{cases}
\partial_t V^i = -V^k \partial_k V^i - \dfrac{1}{2\rho}\partial^i B^2 + \dfrac{B_k \partial^k B^i}{\rho} - \dfrac{\partial^i p}{\rho} + \dfrac{\partial_k \sigma^{ik}}{\rho}, \\
\partial_t B^i = B^j \partial_j V^i - B^i \partial_j V^j - V^j \partial_j B^i + \zeta \partial^2 B^i, \\
\partial_t \rho = -\partial_j \left(\rho V^j \right), \\
\partial_t s = -V^j \partial_j s + \dfrac{\sigma_{ik}}{\rho T} \partial^k V^i + \dfrac{\zeta}{\rho T} \varepsilon_{ikh} \varepsilon^h{}_{mn} \partial^i B^k \partial^m B^n + \dfrac{\kappa}{\rho T} \partial^2 T
\end{cases}
\tag{5}
$$

can be shown to possess a metriplectic formulation. In (5) we adopted a notation with $SO(3)$-indices, which turns out to be practical in this context. We specify that

$$
\sigma_{ik} = \left[\eta \left(\delta_{ni}\delta_{mk} + \delta_{nk}\delta_{mi} - \frac{2}{3}\delta_{ik}\delta_{mn} \right) + \nu \delta_{ik}\delta_{mn} \right] \partial^m V^n
$$

is the stress tensor, with η and ν indicate the viscosity coefficients, κ is the thermal conductivity, T the plasma temperature and s the entropy density per unit mass. In the limit $\kappa = \zeta = \nu = \sigma_{ik} = 0$, one recovers the ideal MHD system treated by Morrison and Greene (1980), reading:

$$
\begin{cases}
\partial_t V^i = -V^k \partial_k V^i - \dfrac{1}{2\rho}\partial^i B^2 + \dfrac{B_k \partial^k B^i}{\rho} - \dfrac{\partial^i p}{\rho}, \\
\partial_t B^i = B^j \partial_j V^i - B^i \partial_j V^j - V^j \partial_j B^i, \\
\partial_t \rho = -\partial_j \left(\rho V^j \right), \\
\partial_t s = -V^j \partial_j s.
\end{cases}
\tag{6}
$$

Morrison and Greene (1980) showed that the system (6) can indeed be cast in the form (2). This is accomplished first, by identifying the dynamical variables z^i with the fields $(\mathbf{B}(\mathbf{x},t), \mathbf{V}(\mathbf{x},t), \rho(\mathbf{x},t), s(\mathbf{x},t))$ (here, the space coordinate $\mathbf{x}$ labels the dynamical variables and plays the role of a continuous 3-index). The Hamiltonian for ideal MHD is then

$$
H[\mathbf{B}, \mathbf{V}, \rho, s] = \int d^3x \left[\frac{\rho V^2}{2} + \frac{B^2}{2} + \rho U(\rho, s) \right]. \tag{7}
$$

The three addenda in the integrand correspond to the kinetic, magnetic and internal energy of the system, respectively. $U(\rho,s)$ is related to the plasma pressure and the temperature as:

$$
p = \rho^2 \frac{\partial U}{\partial \rho}, \quad T = \frac{\partial U}{\partial s}. \tag{8}
$$

The Poisson bracket giving rise to the frictionless (6) through the Hamiltonian (7) is given by:

$$\begin{aligned}[f,g]=-\int d^3x\Bigg[&\frac{1}{\rho}\partial_i s\left(\frac{\delta f}{\delta s}\frac{\delta g}{\delta V_i}-\frac{\delta g}{\delta s}\frac{\delta f}{\delta V_i}\right)+\\&+\frac{1}{\rho}\frac{\delta f}{\delta V_i}\varepsilon_{ijk}\varepsilon^{kmn}B^j\partial_m\left(\frac{\delta g}{\delta B^n}\right)+\frac{\delta f}{\delta B_i}\varepsilon_{ijk}\partial^j\left(\frac{1}{\rho}\varepsilon^{kmn}B_m\frac{\delta g}{\delta V^n}\right)+\\&+\frac{\delta f}{\delta \rho}\partial_i\left(\frac{\delta g}{\delta V_i}\right)+\frac{\delta g}{\delta \rho}\partial_i\left(\frac{\delta f}{\delta V_i}\right)-\frac{1}{\rho}\frac{\delta f}{\delta V_i}\varepsilon_{ikj}\varepsilon^{jmn}\frac{\delta g}{\delta V_k}\partial_m V_n\Bigg].\end{aligned} \tag{9}$$

This bracket possesses Casimir invariants (e.g. Morrison, 1982, Holm et al., 1985), such as the magnetic helicity; particularly relevant in our context, the total *entropy* is defined as:

$$S=\int \rho s d^3x\,. \tag{10}$$

S is conserved along the motion of the non-dissipative system (6).

Some observation should be made here about the role of the plasma entropy as a Casimir. Casimir are invariants that a theory shows because of the singularity of its Poisson bracket, which is not full-rank. Typically this can happen when a Hamiltonian system is obtained by reducing some larger parent one, which possesses some symmetry (see, e.g., Marsden & Ratiu, 1999, Thiffeault & Morrison, 2000). In the case of ideal MHD, the reduction which leads to the Poisson bracket (9), is the map leading from the Lagrangian to the Eulerian representation of the fluid (Morrison, 2009a). When the system of microscopic parcels is approximated as a continuum, its (Lagrangian or Eulerian) fluid variables (as the velocity $\mathbf{V}(\mathbf{x},t)$) pertain to the centre-of-mass of the fluid parcels of size d^3x within which they may be approximated as constants. However, fluids are equipped with some thermodynamic variable, as the entropy s per unit mass here, which represent statistically the degrees of freedom relative-to-the-centre-of-mass of the parcels in d^3x. In the Lagrangian description, the value of the entropy per unit mass is attributed to each parcel at the initial time, and remains constant, for each parcel, during the motion. In the Eulerian description, the total entropy appears as a Casimir, after the reduction, and the symmetry involved in this case is the relabelling symmetry, which is related to the freedom in choosing the label of each parcel at the initial time. In this respect, it is worth recalling that this reduction process implies a loss of information (e.g. Morrison, 1986) in the sense that, through the Eulerian description, one can observe properties of the fluid at a given point in space, but cannot identify which parcel is passing at a given point at a given time.

In a sense, this observation renders the metriplectic a sub-fluid description, because those microscopic degrees of freedom interact with the continuum variables through the role of S in (10) in the metric part of the evolution.

If the dissipative terms are re-introduced into Eq. (6) and one goes back to Eq. (5), a *complete* system is obtained, in the sense that H in (7) doesn't change along the motion (5), while entropy S in (10) is increased (Morrison, 2009b).

Let us illustrate the metriplectic formulation for the system (5). The non-dissipative part of the dynamics is algebrized through the Hamiltonian (7) and the Poisson bracket (9). As far as the construction of the free energy F in (4) is concerned, the entropy S is taken as the Casimir C, whereas the *metric bracket* reads:

$$\begin{aligned}(f,g)=\frac{1}{\lambda}\int d^3x\Bigg\{&\kappa T^2\partial^k\left(\frac{1}{\rho T}\frac{\delta f}{\delta s}\right)\partial_k\left(\frac{1}{\rho T}\frac{\delta g}{\delta s}\right)+\\ &+T\Lambda_{ikmn}\left[\partial^i\left(\frac{1}{\rho}\frac{\delta f}{\delta V_k}\right)-\frac{1}{\rho T}\partial^i V^k\frac{\delta f}{\delta s}\right]\left[\partial^m\left(\frac{1}{\rho}\frac{\delta g}{\delta V_n}\right)-\frac{1}{\rho T}\partial^m V^n\frac{\delta f}{\delta s}\right]+\\ &+T\Theta_{ikmn}\left[\partial^i\left(\frac{\delta f}{\delta B_k}\right)-\frac{1}{\rho T}\partial^i B^k\frac{\delta f}{\delta s}\right]\left[\partial^m\left(\frac{\delta g}{\delta B_n}\right)-\frac{1}{\rho T}\partial^m B^n\frac{\delta f}{\delta s}\right]\Bigg\}.\end{aligned} \tag{11}$$

This metric bracket can be decomposed into two parts. A "fluid" part, corresponding to its first two terms, which was shown to produce the viscous terms of the Navier-Stokes equations (Morrison, 1984), and a "magnetic" part, which accounts for the resistive terms. The proof that the above metric bracket satisfies the properties required by the metriplectic formulation has been given in Materassi & Tassi (2011). The $SO(3)$-tensors needed are defined as:

$$\Lambda_{ikmn}=\eta\left(\delta_{ni}\delta_{mk}+\delta_{nk}\delta_{mi}-\frac{2}{3}\delta_{ik}\delta_{mn}\right)+\nu\delta_{ik}\delta_{mn},$$

$$\Theta_{ikmn}=\zeta\varepsilon_{ikh}\varepsilon^{h}{}_{mn}\ .$$

The bracket (11) together with the free energy functional

$$F[\mathbf{B},\mathbf{V},\rho,s]=\int d^3x\left[\frac{\rho V^2}{2}+\frac{B^2}{2}+\rho U(\rho,s)+\lambda\rho s\right] \tag{12}$$

produces the dissipative terms of the system (5).

Thanks to the metriplectic formulation, it appears evident that the dynamics of the complete visco-resistive MHD takes place on surfaces of constant energy but, unlike Hamiltonian systems, it crosses different surfaces of constant Casimirs. Choosing $C = S$, it becomes evident that the fact that the dynamics does not take place at a surface of constant Casimir reflects of course the presence of dissipation in the system, and in particular the increase in entropy.

Free extremal points of F in (12) (i.e., configurations at which one has $\delta F = 0$ regardless other conditions) correspond to equilibria of the system (5) (even if other equilibria are possible). These can be found by setting to zero the first variation of F and solving the resulting equation in terms of the field variables. These equilibrium solutions are given by

$$\begin{aligned}&\mathbf{V}_{eq}=0,\quad \mathbf{B}_{eq}=0,\quad T_{eq}=-\lambda,\\ &p_{eq}=\rho_{eq}\left(Ts-U\right)_{eq}=\text{constant}\end{aligned} \tag{13}$$

(since it has been obtained as extremal of the free energy functional, this solution is also an equilibrium for ideal MHD). The equilibrium (13) is rather peculiar because it corresponds to a situation in which all the kinetic and magnetic energy have been dissipated and converted into heat. It ascribes a physical meaning to the constant λ, that corresponds to the opposite of the homogeneous temperature the plasma reaches at the equilibrium. Other equilibria with non trivial magnetic or velocity fields can in principle be obtained by considering Casimir constants other than the entropy, and a different metric bracket, or simply by constraining the condition $\delta F = 0$ onto some manifold of constant value for suitable physical quantities. Moreover, the boundary conditions for the system to work in this way must be such that all the fields behave "suitably" at the space infinity. All the results are obtained for a *visco-resistive isolated plasma*: indeed, all the algebraic relationships invoked hold if **V**, **B**, ρ and s show suitable boundary conditions, rendering visco-resistive MHD a "complete system".

Such metriplectic formulation conserves, in addition to the energy H, also the total linear momentum **P**, the total angular momentum **L** and the generator of Galileo's boosts **G**, which are defined by:

$$\mathbf{P} = \int \rho \mathbf{V} d^3x, \quad \mathbf{L} = \int \rho(\mathbf{x} \times \mathbf{V}) d^3x, \quad \mathbf{G} = \int \rho(\mathbf{x} - \mathbf{V}t) d^3x.$$

About these quantities **P**, **L** and **G**, it should be stressed that, besides modifying the scheme with other quantities conserved in the ideal limit, more interesting equilibria than (13) may be identified by conditioning the extremization of F to the initial finite values of the Galilean transformation generators.

3. Sub-fluid physics as noise: A stochastic field theory for the MHD

The metriplectic theory of the MHD discussed in § 2 clarifies how the dissipative part of the dynamics must be attributed to the presence of *statistically treated* degrees of freedom, through their entropy. On the one hand, the metriplectic MHD gives a role to the statistics of the medium properties; on the other hand, local equilibrium and space-time-smoothness of field variables are still assumed. In the sub-fluid model presented in this paragraph, the statistical nature of the microscopic degrees of freedom is cast into a form going beyond the local equilibrium condition. In particular, strong reference to plasma ISCs is made.

Plasma ISC dynamics resembles more closely a quantum transition than a classical evolution: the idea presented here is that localized occurrence of big fluctuations in the medium probably initiate and determine those quantum-like transitions of the variables **B** and **V**. If the fluctuations of the medium are treated as *probabilistic stirring forces*, or *noises*, a totally new scenario appears.

The formalism turning those considerations into a mathematical theory was introduced in Materassi & Consolini (2008); then, an application of it to the visco-resistive reduced MHD in 2 dimensions was obtained in Materassi (2009).

Let's consider the resistive incompressible MHD equations:

$$\begin{cases} \partial_t B^i = B^j \partial_j V^i - V^j \partial_j B^i - \varepsilon^{ijk} \partial_j \left(\zeta_{kh} J^h \right) , \\ \partial_t V^i = -V^j \partial_j V^i + \dfrac{J_j}{\rho} B_k \varepsilon^{jki} - \dfrac{\partial^i p}{\rho} \end{cases} \tag{14}$$

(the choice of incompressible plasma is done for reasons to be clarified later). ζ is the resistivity tensor and p is the plasma pressure. The dynamical variables are the fields **V** and **B**. The viscosity ν is assumed to be zero. The form of ζ and p, and of the mathematical relationships among them (necessary to close the system (14)), depend on the micro-dynamics of the medium. Usually, constitutive hypotheses provide the information on the microscopic nature of the medium (Kelley, 1989). When the (at least local) thermodynamic equilibrium is assumed, the constitutive hypotheses read something like:

$$\zeta = \zeta(T,...) \ , \quad \Phi(p,T) = 0, \tag{15}$$

being T the local temperature field. Then, some heat equation is invoked for T, requiring other constitutive hypotheses about the specific heat of the plasma.

The aforementioned procedure will only give ζ and p regular quasi-everywhere. Instead, in the sub-fluid approach presented here, irregularities of ζ and p are explicitly considered by stating that these local quantities are *stochastic fields*, and by assigning their probability density functions (PDF). The probabilistic nature of the terms ζ and p will be naturally transferred to **B** and **V** through a suitable SFT. The following vector quantities are defined

$$\Xi^i = -\varepsilon^{ijk}\partial_j\left(\zeta_{kh}J^h\right) \ , \quad \Delta^i = \frac{J^i}{\rho}, \quad \Theta^i = -\frac{\partial^i p}{\rho}: \tag{16}$$

these Ξ, $\mathbf{\Delta}$, and $\mathbf{\Theta}$ are considered as *stochastic stirring forces*, and their probability density functional is assigned as some $Q[\Xi,\mathbf{\Delta},\mathbf{\Theta}]$. The resistive MHD equations are then re-written as the following *Langevin field equations*:

$$\begin{cases} \partial_t B^i = B^j\partial_j V^i - V^j\partial_j B^i + \Xi^i, \\ \partial_t V^i = -V^j\partial_j V^i + \Delta_j B_k \varepsilon^{jki} + \Theta^i, \\ (\Xi,\mathbf{\Delta},\mathbf{\Theta}) \overset{\text{iid}}{\approx} Q[\Xi,\mathbf{\Delta},\mathbf{\Theta}]. \end{cases} \tag{17}$$

This scheme, clearly, is not self-consistent because the PDF of the noise terms must be assigned *a priori*, as the outcome of a microscopic dynamics not included in this treatment and not predictable by it. Plasma microscopic physics will enter through some PDF $P_{\text{dyn}}[\zeta,p]$: as far as $P_{\text{dyn}}[\zeta,p]$ keeps trace of the plasma complex dynamics, this represents a (rather general) way to provide constitutive hypotheses. Then, the positions (16) are used to construct mathematically the passage:

$$(\zeta,p) \overset{\text{iid}}{\approx} P_{\text{dyn}}[\zeta,p] \ \Rightarrow \ (\Xi,\mathbf{\Delta},\mathbf{\Theta}) \overset{\text{iid}}{\approx} Q[\Xi,\mathbf{\Delta},\mathbf{\Theta}].$$

A closed form for $Q[\Xi,\mathbf{\Delta},\mathbf{\Theta}]$ should be obtained consistently with any microscopic dynamical theory of the ISC plasma, from the very traditional equilibrium statistical mechanics to the fractional kinetics reviewed in Zaslavsky (2002).

Due to the presence of the stochastic terms Ξ, $\mathbf{\Delta}$, and $\mathbf{\Theta}$ two important things happen: first of all, from each set of initial conditions, *many possible evolutions* of **B** and **V** develop according to

(17), each corresponding to a particular realization of Ξ, $\boldsymbol{\Delta}$, and $\boldsymbol{\Theta}$ (Haken, 1983); then **B** and **V** can be arbitrarily irregular, because they inherit stochasticity from noises; they will possibly show sudden changes in time or non-differentiable behaviours in space, as it happens in ISCs. The description of such a system may be given in terms of *path integrals* (Feynman & Hibbs, 1965). The positions (16) and their consequence (17) are chosen because they reproduce exactly the Langevin equations treated in Phythian (1977), on which this model is based.

The construction introduced in the just mentioned work is the definition of a path integral scheme out of a suitable set of Langevin equations. One starts with a dynamical variable ψ, with any number of components, undergoing a certain equation with noises. Then, another variable χ is defined, referred to as *stochastic momentum conjugated to* ψ. In this way, it is possible to define a kernel

$$A[\psi,\chi;t_0,t) = N\left(t_0,t\right)e^{-i\int_{t_0}^{t} L(\psi,\chi)d\tau}, \tag{18}$$

so that any statistical outcome of the history of the system between t_0 and t is calculated as:

$$\langle F\rangle = \int[d\psi]\int[d\chi]A[\psi,\chi;t_0,t)F(\psi).$$

In the kernel in (18) the quantity $L(\psi,\chi)$ is referred to as *stochastic Lagrangian of the system.* In Phythian (1977) the key result is a closed "recipe" to build up $L(\psi,\chi)$ out of the Langevin equation of motion.

The same procedure may be applied to the system governed by the Langevin equations (17); these may be turned into a SFT by identifying the dynamical variables ψ of the system as **B** and **V**, and introducing as many stochastic momenta χ as the components of ψ (Materassi & Consolini, 2008):

$$\psi = \mathbf{B}\oplus\mathbf{V}, \quad \chi = \boldsymbol{\Omega}\oplus\boldsymbol{\Pi}.$$

The variables $\boldsymbol{\Omega}$ and $\boldsymbol{\Pi}$ are two vector quantities representing the stochastic momenta of **B** and **V** respectively. A stochastic kernel $A[\boldsymbol{\Omega},\boldsymbol{\Pi},\mathbf{B},\mathbf{V};t_0,t)$ is constructed by involving a noise factor

$$\begin{aligned} &\mathrm{C}[\boldsymbol{\Omega},\boldsymbol{\Pi},\mathbf{B},\mathbf{V};t_0,t) = \\ &= \iiint[d\Xi][d\boldsymbol{\Delta}][d\boldsymbol{\Theta}]Q[\Xi,\boldsymbol{\Delta},\boldsymbol{\Theta}]e^{i\int_{t_0}^{t} d\tau\int d^3x\left[\Xi\cdot\boldsymbol{\Omega}+\boldsymbol{\Theta}\cdot\boldsymbol{\Pi}+\boldsymbol{\Delta}\cdot(\boldsymbol{\Pi}\times\mathbf{B})\right]} : \end{aligned} \tag{19}$$

all the statistical dynamics of the resistive MHD interpreted as a SFT is then encoded in the kernel

$$\begin{aligned} &A[\boldsymbol{\Omega},\boldsymbol{\Pi},\mathbf{B},\mathbf{V};t_0,t) = N\left(t_0,t\right)\mathrm{C}[\boldsymbol{\Omega},\boldsymbol{\Pi},\mathbf{B},\mathbf{V};t_0,t)e^{-i\int_{t_0}^{t} d\tau\int L_0(\boldsymbol{\Omega},\boldsymbol{\Pi},\mathbf{B},\mathbf{V})d^3x}, \\ &L_0\left(\boldsymbol{\Omega},\boldsymbol{\Pi},\mathbf{B},\mathbf{V}\right) = \boldsymbol{\Omega}\cdot\dot{\mathbf{B}} + \boldsymbol{\Pi}\cdot\dot{\mathbf{V}} + \\ &+\boldsymbol{\Omega}\cdot\left(\left(\mathbf{V}\cdot\partial\right)\mathbf{B} - \left(\mathbf{B}\cdot\partial\right)\mathbf{V}\right) + \boldsymbol{\Pi}\cdot\left(\left(\mathbf{V}\cdot\partial\right)\mathbf{V}\right). \end{aligned} \tag{20}$$

The quantity $L_0(\mathbf{\Omega},\mathbf{\Pi},\mathbf{B},\mathbf{V})$ is interpreted as the part of the Lagrangian of the SFT not containing noise terms. L_0 shows only space- and time-local terms, always: as it is stressed in Chang (1999), the integration of the noise term $C[\mathbf{\Omega},\mathbf{\Pi},\mathbf{B},\mathbf{V};t_0,t)$ brings terms in L that are non-local in space and in time, due to the self- and mutual correlations of noises. Those terms will be collected in a noise-Lagrangian L_C, so that all in all one has:

$$C[\mathbf{\Omega},\mathbf{\Pi},\mathbf{B},\mathbf{V};t_0,t) = e^{-i\int_{t_0}^{t} d\tau \int L_C(\mathbf{\Omega},\mathbf{\Pi},\mathbf{B},\mathbf{V}) d^3x},$$

$$L = L_C + L_0, \quad A[\mathbf{\Omega},\mathbf{\Pi},\mathbf{B},\mathbf{V};t_0,t) = N(t_0,t)\, e^{-i\int_{t_0}^{t} d\tau \int L(\mathbf{\Omega},\mathbf{\Pi},\mathbf{B},\mathbf{V}) d^3x}.$$

The form of $Q[\mathbf{\Xi},\mathbf{\Delta},\mathbf{\Theta}]$, hence of $C[\mathbf{\Omega},\mathbf{\Pi},\mathbf{B},\mathbf{V};t_0,t)$, may render the SFT long-range correlated and with a finite memory: these conditions of the ISC plasmas described by such a SFT is what encourages people to work through the techniques of *dynamical renormalization group* (Chang et al., 1978). Possibly, the stochastic momenta may be eliminated, so that one obtains a kernel W involving only physical fields

$$W[\mathbf{B},\mathbf{V};t_0,t) = \int[d\mathbf{\Omega}]\int[d\mathbf{\Pi}]A[\mathbf{\Omega},\mathbf{\Pi},\mathbf{B},\mathbf{V};t_0,t)\,. \tag{21}$$

Once $W[\mathbf{B},\mathbf{V};t_0,t)$ has been obtained, the calculation of processes in which the magnetized plasma changes arbitrarily, from an initial configuration $(\mathbf{B}(t_0),\mathbf{V}(t_0)) = (\mathbf{B}_i,\mathbf{V}_i)$ to a final one $(\mathbf{B}(t),\mathbf{V}(t)) = (\mathbf{B}_f,\mathbf{V}_f)$, may be done, for any time interval (t_0,t): the rate of such transitions should be calculated as

$$P_{\mathbf{B}_i,\mathbf{V}_i \to \mathbf{B}_f,\mathbf{V}_f} = \int[d\mathbf{B}]\int[d\mathbf{V}]W[\mathbf{B},\mathbf{V};t_0,t)\Big|_{\substack{\mathbf{B}(t_0)=\mathbf{B}_i,\mathbf{V}(t_0)=\mathbf{V}_i \\ \mathbf{B}(t)=\mathbf{B}_f,\mathbf{V}(t)=\mathbf{V}_f}}\,. \tag{22}$$

As a further development of Materassi & Consolini (2008), a complete representation *à la Feynman* of such processes is to be derived from the SFT, with a suitable perturbative theory of graphs.

In order to arrive to a closed expression for a stochastic action at least in one example case, hereafter a toy model is reported, in which $\mathbf{\Xi}$, $\mathbf{\Delta}$ and $\mathbf{\Theta}$ are assumed to be *Gaussian processes without any memory, and δ-correlated in space*. This hypothesis is surely over-simplifying for a plasma in ISC, since there are experimental results stating the presence of non-Gaussian distributions (Yordanova et al., 2005), and also of memory effects (Consolini et al., 2005). Nevertheless, the Gaussian example is of some use in illustrating the SFT at hand, because a Gaussian shape for $Q[\mathbf{\Xi},\mathbf{\Delta},\mathbf{\Theta}]$ allows for the full integration of $C[\mathbf{\Omega},\mathbf{\Pi},\mathbf{B},\mathbf{V};t_0,t)$, and the explicit calculation of $W[\mathbf{B},\mathbf{V};t_0,t)$ from $A[\mathbf{\Omega},\mathbf{\Pi},\mathbf{B},\mathbf{V};t_0,t)$ in (21). The probability density functional $Q[\mathbf{\Xi},\mathbf{\Delta},\mathbf{\Theta}]$ is obtained via a *continuous product* out of distributions of the local values of the fields $\mathbf{\Xi}$, $\mathbf{\Delta}$ and $\mathbf{\Theta}$ of Gaussian nature; for instance, the PDF of the local variable $\mathbf{\Xi}(\mathbf{x},t)$ reads:

$$q_{\Xi}(\mathbf{\Xi}(\mathbf{x},t)) = \sqrt{\frac{a_{\Xi}^3(\mathbf{x},t)}{\pi^3}}\, e^{-a_{\Xi}(\mathbf{x},t)(\mathbf{\Xi}(\mathbf{x},t)-\mathbf{\Xi}_0(\mathbf{x},t))^2}\,. \tag{23}$$

The quantity $a_{\Xi}(\mathbf{x},t)$ indicates how peaked the distribution $q_{\Xi}(\mathbf{\Xi}(\mathbf{x},t))$ is, i.e. how deterministic are the terms in (16) describing the medium: the larger $a_{\Xi}(\mathbf{x},t)$ is, the less stochastic is the plasma. Formally equal distributions $q_\Delta(\mathbf{\Delta}(\mathbf{x},t))$ and $q_\Theta(\mathbf{\Theta}(\mathbf{x},t))$ describe the local occurrence of the values of $\mathbf{\Delta}$ and $\mathbf{\Theta}$. From (23), the expression of the noise kernel $C[\mathbf{\Omega},\mathbf{\Pi},\mathbf{B},\mathbf{V};t_0,t)$ defined in (19) can be calculated explicitly (Materassi & Consolini, 2008), and the noise Lagrangian $L_C(\mathbf{\Omega},\mathbf{\Pi},\mathbf{B},\mathbf{V})$ determined in a closed form:

$$L_C(\mathbf{\Omega},\mathbf{\Pi},\mathbf{B},\mathbf{V}) = -\frac{i\mathbf{\Omega}^2}{4a_\Xi} - \Xi_0\cdot\mathbf{\Omega} + -\frac{i}{4}\left(\frac{\mathbf{\Pi}^2}{a_\Theta} + \frac{\mathbf{\Pi}^2\mathbf{B}^2 - (\mathbf{B}\cdot\mathbf{\Pi})^2}{a_\Delta}\right) - \left(\mathbf{\Theta}_0\cdot\mathbf{\Pi} + (\mathbf{B}\times\mathbf{\Delta})\cdot\mathbf{\Pi}\right) \tag{24}$$

This noise Lagrangian is space-local and does not contain any memory term, because the PDF $Q[\Xi,\mathbf{\Delta},\mathbf{\Theta}]$ was constructed as the continuous products of infinite terms, each of which representing the independent probability $q_\Xi(\Xi(\mathbf{x},t))q_\Delta(\mathbf{\Delta}(\mathbf{x},t))q_\Theta(\mathbf{\Theta}(\mathbf{x},t))$. The total Lagrangian is the sum of the noise term $L_C(\mathbf{\Omega},\mathbf{\Pi},\mathbf{B},\mathbf{V})$ and of the "deterministic" addendum $L_0(\mathbf{\Omega},\mathbf{\Pi},\mathbf{B},\mathbf{V})$ presented in (20). The sum $L_0 + L_C$ gives rise to a perfectly local theory. The total Lagrangian $L_0 + L_C$ gives a kernel $A[\mathbf{\Omega},\mathbf{\Pi},\mathbf{B},\mathbf{V};t_0,t)$ that is the continuous product of the exponentiation of quadratic terms in $\mathbf{\Omega}$ and $\mathbf{\Pi}$, so that the calculation (21) is an infinite-dimensional Gaussian path integral, which is again feasible. This means that, under the hypothesis (23) on Ξ, and similar assumptions on the two other noises $\mathbf{\Delta}$ and $\mathbf{\Theta}$, the calculation of the stochastic evolution kernel can be done in terms of pure "physical fields" $\mathbf{B}$ and $\mathbf{V}$, obtaining $W[\mathbf{B},\mathbf{V};t_0,t)$. If the calculation is performed to the end, the expression of $W[\mathbf{B},\mathbf{V};t_0,t)$ reads:

$$S[\mathbf{B},\mathbf{V};t_0,t) = N'\left[a_\Xi,a_\Delta,a_\Theta,\zeta_0,p_0;t_0,t\right)\, e^{-i\int_{t_0}^{t} d\tau \int L'(\mathbf{B},\mathbf{V})d^3x},$$

$$\begin{aligned} L'(\mathbf{B},\mathbf{V}) = & -i\ln\left(1+\tfrac{a_\Theta}{a_\Delta}\mathbf{B}^2\right) + \\ & -ia_\Xi\left(\dot{\mathbf{B}} + (\mathbf{V}\cdot\partial)\mathbf{B} - (\mathbf{B}\cdot\partial)\mathbf{V} + \tfrac{\zeta_0}{\mu_0}\partial\times(\partial\times\mathbf{B})\right)^2 + \\ & -\frac{ia_\Theta}{1+\frac{a_\Theta}{a_\Delta}\mathbf{B}^2}\left\{\left(\dot{\mathbf{V}} + (\mathbf{V}\cdot\partial)\mathbf{V} + \tfrac{1}{\rho}\left(\partial p_0 + \tfrac{\partial\mathbf{B}^2}{2\mu_0}\right) - \tfrac{(\mathbf{B}\cdot\partial)\mathbf{B}}{\rho\mu_0}\right)^2 + \right. \\ & \left. + \tfrac{a_\Theta}{a_\Delta}\left[\left(\dot{\mathbf{V}} + (\mathbf{V}\cdot\partial)\mathbf{V} + \tfrac{1}{\rho}\left(\partial p_0 + \tfrac{\partial\mathbf{B}^2}{2\mu_0}\right) - \tfrac{(\mathbf{B}\cdot\partial)\mathbf{B}}{\rho\mu_0}\right)\cdot\mathbf{B}\right]^2\right\}. \end{aligned} \tag{25}$$

The functions ζ_0 and p_0 are defined as the ensemble expectation value of the homonymous stochastic variables. The expression (25) is ready to be used in (22) to calculate the transition probabilities between arbitrary field configurations. The quantity N' in (25), whatever it looks like, will not enter the calculations of processes like (22), since it doesn't depend on $\mathbf{V}$ and B, and will be cancelled out. Last but not least, consider that the functions defining noise statistics, i.e. a_Ξ, a_Δ, a_Θ, ζ_0 and p_0, do enter the Lagrangian as "coupling constants".

Intrinsic limitations of the proposed scheme can be recognized.

First of all, no discussion has been even initiated yet about the convergence of all the quantities defined.

There is an apparent "necessity" of making the choice (16) in order to follow the scheme traced in Phythian (1977). It could be useful to extend the reasoning presented here to other forms of the Langevin equations so to avoid the positions (16) and work directly with ζ and p as stirring forces in (14).

It is also to mention that the problem of defining a good functional measure is still to be examined, by studying the consistency condition of a Fokker-Planck equation for the SFT, starting for example with the Lagrangian density (25), obtained under drastically simplifying hypotheses.

A comment is deserved by the choice of the *incompressible plasma hypothesis*. The MHD as a dynamical system is given by (5): in the absence of incompressibility, the mass density ρ is a distinct variable on its own, with a proper independent dynamics. In the stochastic theory *à la Phythian* each dynamical variable should satisfy a Langevin equation, in which noise is in principle involved. Now, altering the equation for ρ with noise could invalidate the mass conservation, which is a big fact one would like to avoid. Hence, the "sacred principle" of non-relativistic mass conservation $\partial_t\rho + \partial\cdot(\rho\mathbf{V}) = 0$ is saved excluding ρ from dynamics, rendering it a pure parameter of the theory, via incompressibility. The compressible case could be studied considering the local mass conservation a constrain to be imposed to the path integrals as it happens in quantum gauge field theories (Hennaux & Teitelboim, 1992).

Last but not least, the fourth equation in (5) has not been considered at all in this scheme: in Phythian's scheme plasma thermodynamics must be discussed in some deeper way before enlarging the configuration space of stochastic fields to the entropy s.

4. Fractal model of fast reconnection

Among the many interesting fast and irreversible processes occurring in plasmas, *magnetic reconnection* is surely one of the most important (see e.g. Biskamp, 2000; Birn and Priest, 2007). The name "magnetic reconnection", originally introduced by Dungey (1953), refers to a process in which a particle acceleration is observed consequently to a change of the magnetic field line topology (*connectivity*). Being associated to a change in the magnetic field line topology, the magnetic reconnection process involves the occurrence of magnetic field line diffusion, disconnection and reconnection and it is also accompanied by plasma heating and particle acceleration, sometimes termed as *dissipation* (actually, in this case dissipation means transfer of energy from the magnetic field to the particle energy, both bulk motion energy, the term $\rho V^2/2$ in the integrand in (7), and thermal energy, the term $U(\rho,s)$ in the same expression of H; in the context of metriplectic dynamics, dissipation is simply the transfer of energy into the addendum $U(\rho,s)$).

The traditional approach to magnetic reconnection is based on resistive MHD theory. In this framework one of the most famous and first scenarios of magnetic reconnection, able to make some quantitative predictions, was proposed by Parker (1957) and Sweet (1958). The Sweet-Parker model provides a simple 2-dimensional description of steady magnetic reconnection in a non-compressible plasmas (see Figure 1). In this model there are two relevant scales: the global scale L of the magnetic field and the thickness Δ of the current sheet (or of the diffusion region). The main result of such a model may be resumed in the very-well known expression for the *Alfvèn Mach number* M_A,

$$M_A = R_m^{-1/2} \quad / \quad R_m = \mu_0 L V_A / \eta , \tag{26}$$

where R_m is the *Lundquist number* (often referred as *magnetic Reynolds number*), V_A is the Alfvén velocity and η is the resistivity.

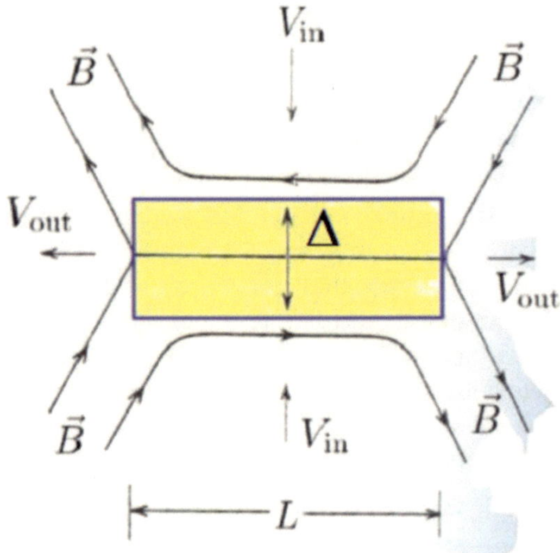

Fig. 1. A schematic view of 2-dimensional geometry for the Sweet-Parker reconnection scenario

Indeed, being a measure of the electric field normalized by the global electric field, i.e.

$$M_A = V / V_A = E / V_A B , \tag{27}$$

the Alfvén Mach number M_A, reported in Eq. (26), provides an estimate of the *reconnection rate*, which is generally expressed in terms of the electric field at the reconnection site.

The typical Lundquist number R_m in astrophysical and space plasmas is $R_m >> 10^6$, implying reconnection rates $M_A << 10^{-4}$. These reconnection rates are *too slow to explain the explosive nature of several space processes* associated with the occurrence of reconnection, so that the Sweet-Parker model is considered not suitable to explain reconnection in space plasmas.

In the course of the time, to overcome such a limitation of the Sweet-Parker model several other models have been proposed. Among these models one of the most successful is the *Petscheck model* (Petscheck, 1964), where the diffusion region (associated with the current sheet) is greatly reduced in length and the energy conversion is associated with the presence of two pairs of standing slow-modes. As a result, the reconnection rate in terms of Alfvénic Mach number is

$$M_A = \frac{\pi}{8 \ln R_m} \tag{28}$$

which is for most of the space and laboratory plasma situations of the order of $M_A \approx 10^{-1}$ to 10^{-2}.

Although several other models have been proposed (see e.g.: Birn and Priest, 2007), some recent MHD simulation have shown that, when the Hall effect is included, it is possible to obtain fast magnetic reconnection rates, which are independent on the current sheet or reconnection region size. For instance, Huba & Rudakov (2004) obtained a reconnection rate $M_A \leq 0.1$ in the case of Hall magnetic reconnection.

All the above approaches to magnetic reconnection move from the assumption that plasma media can be viewed as *noncollisional fluid*. This assumption is clearly valid when the inherent local fluctuations δx of any local field X are negligible with respect to the large scale means,

$$\frac{\left\langle \delta x^2 \right\rangle^{\frac{1}{2}}}{\left\langle X \right\rangle} << 10^{-1}, \tag{29}$$

being $\delta x = X - \langle X \rangle$. Conversely, recent observations evidenced that space plasmas are characterized by an intrinsic stochastic character, and that in many situations *turbulence* is present. This is for instance the case of interplanetary space plasmas, such as the solar wind, and the Earth's magnetotail current sheet, characterized by stochastic and turbulent fluctuations of the same order of magnitude of the average fields.

Several attempts have been done to include the *stochastic and turbulent nature of the plasma media* and to discuss its effects on the magnetic reconnection process (see e.g. Yankov, 1997; Lazarian & Vishniac, 1999). The common point of such models is the idea that as a consequence of the inherent stochasticity and/or turbulent nature of plasma media, the current sheet and the diffusion region topology cannot be associated with a simple continuous regular medium. Conversely, the current sheet could be imagined like a filamentary, complex and not space-filling region.

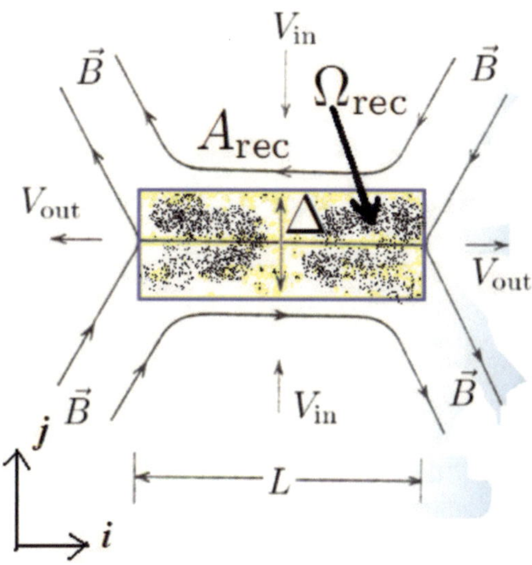

Fig. 2. A schematic view of 2-dimensional geometry for the fractal reconnection model, of size L and Δ, with reconnection area A_{rec} and reconnection active area Ω_{rec}.

In 2007 Materassi & Consolini proposed a revised version of the historical Sweet-Parker model, in which the diffusion/current sheet region, where the magnetic reconnection takes place, is imagined like a fractal object in the plane. The very basic assumption of such a fractal reconnection model is that the reconnection active sites form a *not space-filling domain* Ω_{rec} contained in the diffusion region of measure A_{rec}, and that such a non-space-filling domain is characterized by a Hausdorff dimension $D_H < E$, being E the embedding dimension (here $E = 2$). Figure 2 shows a schematic view of the 2-dimensional geometry of the diffusion region.

Due to the fractal nature of the diffusion region, the constraint of flux conservation can be written as

$$\Phi^{eff}_{S_{out}}[\mathbf{V}] = \Phi^{eff}_{S_{in}}[\mathbf{V}], \tag{30}$$

where S_{in} (S_{out}) is the entrance (exit) surface for the plasma passing through the fractal domain Ω_{rec}. Here, the flux over the entrance and exit surfaces is given by the following expression,

$$\Phi^{eff}_{S}[\mathbf{V}] = \int_{\Omega_S} \mathbf{V} \cdot \hat{\mathbf{n}} d\mu_{\Omega_S}, \tag{31}$$

where $d\mu_{\Omega_S}$ is a proper elemental measure for the fractal domain Ω_S. Thus, the evaluation of such fluxes requires *an integration over a fractal domain*, which can be performed using the definitions by Tarasov (2005, 2006) involving irregular integrals.

According to the results shown in Tarasov (2006), if f is a regular function defined in $\mathbf{R}^n$ to be integrated over a fractal domain Ω characterized by a Harsdorff dimension $D < n$, then the integration can be performed by introducing a proper weight function ξ_D, i.e.

$$\int_{\Omega} f d\mu_{\Omega} = \int_{A_\Omega} f \xi_D dA, \tag{32}$$

where A_Ω is the regular set of dimension n embedding the considered fractal set Ω.

When the above integration technique is applied to the condition of flux conservation (30), one gets for the *fractal reconnection rate*

$$M_A^{FRM} = k\left(\frac{\sqrt{\pi}}{\ell_0}\right)^{\delta} \frac{D_{in}}{D_{out}} \frac{\Gamma\left(\frac{D_{in}}{2}\right)}{\Gamma\left(\frac{D_{out}}{2}\right)} \frac{\Delta^{D_{out}}}{L^{D_{in}}}, \tag{33}$$

where k is a positive constant such that $V_{out} = kV_A$, Δ and L are the thickness (typically of the order of the ion-inertial length) and the length of the diffusion region, respectively, $\delta = D_{out} - D_{in}$ is the difference of the Hausdorff dimensions of the projection of the fractal domain in the direction of the entrance (D_{in}) and exit (D_{out}) directions, and finally ℓ_0 is a *reference microscopic length scale*. Such a reference length scale has to be much smaller than the typical scales at which the medium displays fractal features (Tarasov, 2005).

Moving from the above result and assuming $k = 1$ and $D_{in} = D_{out} = D$, Eq. (33) for the *fractal reconnection rate* can be reduced to a more simple expression in terms of the *Lundquist number* R_m:

$$M_A^{FRM} = R_m^{-\left(\frac{D}{D+1}\right)}. \tag{34}$$

We note that this expression reduces to the standard Sweet-Parker solution of the reconnection rate in the limit $D \to 1$ and that the fractal reconnection rate is always higher than the one predicted by the Sweet-Parker model. Furthermore, although in the limit $R_m \to \infty$ the reconnection rate predicted by the Petschek-like model results the more efficient, there exists always a certain range of the Lundquist number R_m, depending on the fractal dimension D, for which the fractal reconnection model is more efficient than the Petschek-like model. The crucial point of a correct estimation and applicability of the above expression stands in the correct evaluation of D_{in} and D_{out}, which depends on the topology of the current sheet.

In passing we note that when the above scenario is applied using typical length scales estimated by *in-situ* observations of magnetic reconnection in space plasmas, one gets the reconnection rates typically observed and in agreement with the estimated Hall reconnection rate $M_A \approx 0.09$ (Huba & Rudakov, 2004) assuming a diffusion region shaped as a filamentary structure mainly aligned to the inflow region (direction *i* in Figure 2).

The fractal reconnection model described here is not based on first principles, because the non-space filling, self-similar nature of the reconnection region is simply assumed.

The important work necessary for further development will be to give a dynamical sense to the quantities D_{in} and D_{out}, that here might appear just as convenient fitting parameters. Studies have been made to regard irregular filamentary structures in plasmas as descending from calculable fluid-model processes (Zheng & Furukawa, 2010).

The feeling is however that it would be very interesting to deduce the fractal nature of the reconnection region from kinetic or microscopic-statistical theories, rather than extracting it from extreme behaviours of the plasma as a fluid.

5. Conclusion

Dissipation consists of the irreversible transfer of energy from the proper MHD variables to the particle degrees of freedom of the plasma, considered as "microscopic" (and usually treated via Thermodynamics). Depending on the spatial and temporal scales on which dissipation takes place, it may activate some "sub-fluid level" of the theory, which interpolates between the continuous system, representing the traditional MHD, and the discrete one, describing the plasma through the motion of its particles. This "sub-fluid" level should probably consist of mesoscopic coherent structures existing because of dissipative process, and evolving through a stochastic (strongly noisy) dynamics. Consequently, the self-consistent theory describing this intermediate level of plasma description is expected to be a theory of SCSs.

In this Chapter, three models to approach this "SCS Theory" have been exposed: metriplectic algebrization of MHD, stochastic field theory and fractal magnetic reconnection.

Each of the three models tries to mimic one aspect of the complete theory of SCSs. The metriplectic MHD presents the non-Hamiltonian algebrization; the SFT for the resistive MHD is characterized by the presence of noise yielding a path integral approach; the fractal model of reconnection admits the irregular nature of MHD fields, involving the fractional calculus.

A large amount of work must still be done to imagine how those three approaches could be combined in a unique framework, the invoked "SCS Theory", reducing to the three models in different limits: this further research is for sure out of the subject of the work here, in which a flavour had to be given about some characteristics that this "SCS Theory" should have.

As a final remark, we underline that the self-consistent "SCS Theory" should present a sort of scale-covariance, because all the phenomena concerning plasma ISCs do involve multi-scale dynamics. The technique of Renormalization Group will then be naturally applied to such a thory (see e.g. Chang et al., 1992 and references therein). A first direct application of such technique, using the exact full dynamic differential renormalization group for critical dynamics can be found in Chang et al. (1978). The use of Renormalization Group techniques to predict physical quantities to be compared with real spacecraft data is already well established (see e.g. Chang, 1999; Chang et al., 2004), and the results are very encouraging, confirming our idea that any "SCS Theory" has to be based on scale-covariance.

6. Acknowledgements

The authors are grateful to Philip J. Morrison of the Institute for Fusion Studies of the University of Texas in Austin, for useful discussions and criticism. The work of Massimo Materassi has been partially supported by EURATOM through the "Contratto di Associazione Euratom-Enea-CNR". The work of Emanuele Tassi was partially supported by the Agence Nationale de la Recherche (ANR GYPSI n. 2010 BLAN 941 03). This work was also supported by the European Community under the contract of Association between EURATOM, CEA, and the French Research Federation for fusion study. The views and opinions expressed herein do not necessarily reflect those of the European Commission. Emanuele Tassi acknowledges also fruitful discussions with the "Equipe de Dynamique Nonlinéaire" of the Centre de Physique Théorique of Marseille.

Giuseppe Consolini and Massimo Materassi did this work as a part of ISSI Team n. 185 "Dispersive cascade and dissipation in collisionless space plasma turbulence - observations and simulations", lead by E. Yordanova.

7. References

Bălescu, R.; (1997). Statistical Dynamics: matter out of equilibrium, Imperial College Press, ISBN: 1860940455, London.

Bihlo, A. (2008). Rayleigh-Bénard Convection as a Nambu-metriplectic problem, *Journal of Physics A*, 292001, Vol. 41.

Birn, J.; Priest, E.; (2007). Reconnection of magnetic fields. Magnetohydrodynamics and collisionless theory and observations, Cambridge University Press, ISBN-10 0-521-85420-2, Cambridge, United Kingdom

Biskamp, D.; (1993). Nonlinear Magnetohydrodynamics, Cambridge University Press, ISBN-10 0-521-59918-0, Cambridge, United Kingdom.

Biskamp, D.; (2000). Magnetic reconnection in plasmas - Cambridge Monographs in Plasma Physics, Cambridge University Press, ISBN-13 978-0-521-02036-7.

Borovsky, J.E., and Funsten, H.O.; (2003). MHD turbulence in the Earth's plasma sheet: Dynamics, dissipation and driving, *Journal of Geophysical Research*, 108, 1284, doi: 1.1029/2002JA009625.

Bruno, R., Carbone, V., Veltri, P., et al.; (2001). Identifying intermittency events in the solar wind, *Planetary and Space Science*, 49, 1201-1210.

Bruno, R., Carbone, V.; (2005). The solar wind as a turbulence laboratory, *Living Reviews in Solar Physics*, 2, 4.

Chang, T.S.; Nicoll, J.F.; Young J.E.; (1978). A closed-form differential renormalization-group generator for critical dynamics, *Physics Letters*, 67A, 287-290.

Chang, T.; (1992). Low-dimensional behavior and symmetry breaking of stochastic systems near criticality -- Can these effects be observed in space and in the laboratory?, *IEEE Transactions on Plasma Science*, 20, 691-694.

Chang, T.; (1999). Self-organized criticality, multi-fractal spectra, sporadic localized reconnections and intermittent turbulence in the magnetotail, *Physics of Plasmas*, 6, 4137-4145.

Chang, T., Tam, S.W.Y., Wu, C.C. and Consolini, G.; (2003). Complexity forced and/or self-organized criticality, and topological phase transitions in space plasmas, *Space Science Review*, 107, 425-444.

Consolini, G.; (2002). Self-organized criticality: a new paradigm for the magnetotail dynamics, *Fractals*, 10, 275-283.

Consolini, G., Kretschmar, M., Lui, A.T.Y., et al.; (2005). On the magnetic field fluctuations during magnetospheric tail current disruption: a statistical approach, *Journal of Geophysical Research*, 110, A07202.

Courbage, M. & Prigogine, I.; (1983). *Intrinsic randomness and intrinsic irreversibility in classical dynamical systems*, Proceedings of the National Academy of Science of USA, Vol. 80, pp. 2412-2416, Physics.

De Michelis, P., G. Consolini, M. Materassi, R. Tozzi; (2011). An information theory approach to storm-substorm relationship, *Journal of Geophysical Research*, vol. 116, A08225, doi:10.1029/2011JA016535.

Dungey, J.W.; (1953). Conditions for the occussrence of electrical discarges in astrophysical plasmas. *Philosphical Magazine*, 44, 725-728

Feynman, R. P. and Hibbs, A. R.; (1965). Quantum Mechanics and Path Integrals, McGraw-Hill Book Company, ISBN: 0-07-020650-3, New York.

Flierl, G.R. and Morrison, P.J.; (2011). Hamiltonian-Dirac Simulated Annealing: Application to the Calculation of Vortex States, *Physica D*, pp. 212-232, Vol. 240.

Goodall, D. H. J; (1982). High speed cine film studies of plasma behaviour and plasma surface interactions in tokamaks, *Journal of Nuclear Materials*, Volumes 111-112, 11-22.

Goldstein, H., Poole, C. P. Jr., Safko, J.L.; (1980). *Classical Mechanics*, Addison-Wesley series in physics, ISBN: 978-0201657029, Newark (CA) USA.

Haken, H.; (1983). Synergetics, an introduction, Springer-Verlag, ISBN: 3-8017-1686-4, Berlin, Heidelberg, New York.

Hennaux, M. & Teitelboim, C.; (1992). Quantization of gauge systems, Princeton University Press, ISBN: 0-691-08775-X, Princeton (New Jersey, USA).

Holm, D.D.; Marsden, J.E.; Ratiu, T. & Weinstein, A. (1985). Nonlinear Stability of Fluid and Plasma Equilibria, *Physics Reports*, pp. 1-116, Vol. 123, No. 1 & 2.

Huba, J.D.; Rudakov, L.I. (2004), Hall magnetic reconnection rate, *Physical review Letters*, 93, 175003

Jaynes, E.T.; (1957). Information Theory and Statistical Mechanics, *Physical Review*, vol. 106, no. 4, 620-630.

Kelley, M. C.; (1989). The Earth's ionosphere, Academic Press Inc., ISBN: 978-0-12-088425-4, Burlington (MA) USA.

Klimas, A.J., Vassiliadis, D.V., Baker, D.N., et al.; (1996). The organized nonlinear dynamics of the magnetosphere, Journal of Geophysical Research, 101, 13089- 13114.

Klimontovich, Yu. L.; (1967). Statistical theory of nonequilibrium processes, Izdat. MGU Moskva, 1964; English translation Pergamon, Oxford, 1967.

Kretzschmar, M., and Consolini, G.; (2006). Complexity in the Earth's magnetotail plasma sheet, *Advances in Space Research*, 37, 552-558.

Lazarian, A., Vishniac, E.T. (1999). Reconnection in a weakly stochastic field, *Astrophysical Journal*, 517, 700-718

Lazarian, A., Vishniac, E.T. and Cho, J.; (2004). Magnetic field structure and stochastic reconnection in a partially ionized gas, *Astrophysical Journal*, 603, 180-197.

Lui, A.T.Y.; (1996). Current disruption in the Earth's magnetosphere: observations and models, *Journal of Geophysical Research*, 101 (A6), 13067-13088.

Marsden, J.E., Ratiu, T.S.; (1999). Introduction to Mechanics and Symmetry, Springer-Verlag, ISBN 0-387-98643-X, Berlin, Heidelberg, New York.

Materassi, M., Consolini, G.; (2007). Magnetic reconnection in real plasma: A fractal approach, *Physical Review Letters*, 99, 175002

Materassi, M., Consolini, G.; (2008). Turning the resistive MHD into a stochastic field theory, *Nonlinear Processes in Geophysics*, 15 (4), pp. 701-709.

Materassi, M.; (2010). Stochastic Lagrangian fot the 2D Visco-Resistive Magneto-Hydrodynamics, *Plasma Physics and Controlled Fusion*, 52 (2010) 075004.

Materassi, M., L. Ciraolo, G. Consolini, N. Smith; (2011). Predictive Space Weather: an information theory approach, *Advances in Space Research*, 47, pp. 877-885, doi:10.1016/j.asr.2010.10.026.

Materassi, M. & Tassi, E. (2011), Metriplectic Framework for Dissipative Magneto-Hydrodynamics, *Physica D*, http://www.sciencedirect.com/science/article/pii/S0167278911003708, also available as arXiv:1110.4404v1 at xxx.lanl.gov.

Milovanov, A.V., Zelenyi, L.M., Zimbardo, G., et al.; (2001). Self-organized branching of magnetotail current systems near the percolation threshold, *Journal of Geophysical Research*, 106, 6291.

Morrison, P. J. & Greene, J. M.; (1980). Noncanonical Hamiltonian Density Formulation of Hydrodynamics and Ideal Magnetohydrodynamics, *Physical Review Letters*, pp. 790-793, Vol. 45, No. 10, September 8.

Morrison, P.J.; (1982). Poisson Brackets for Fluids and Plasmas, *Mathematical Methods in Hydrodynamics and Integrability in Dynamical Systems*, AIP Conference Proceedings, pp. 13-46, ISBN 0-88318-187-8, La Jolla.

Morrison, P.J.; (1984). Some Observations Regarding Brackets and Dissipation, Center for Pure and Applied Mathematics Report PAM-228, University of California, Berkeley.

Morrison, P.J.; (1986). A Paradigm for Joined Hamiltonian and Dissipative Systems, Physica D, pp. 410-419, Vol. 18.

Morrison, P.J. & Hazeltine, R. D.; (1984). Hamiltonian Formulation of Reduced Magnetohydrodynamics, *Physics of Fluids*, Vol. 27, No. 4, April.

Morrison, P. J.; (1998). Hamiltonian Description of the Ideal Fluid, *Review of Modern Physics*, pp. 467-521, Vol. 70, No. 2, April.

Morrison, P. J.; (2009a). On Hamiltonian and Action Principle Formulations of Plasma Dynamics, in *New Developments in Nonlinear Plasma Physics: Proceedings for the 2009 ICTP College on Plasma Physics*, eds. B. Eliasson and P. Shukla, American Institute of Physics Conference Proceedings No.1188 (American Institute of Physics, New York, 2009), 329-344.

Morrison, P.J.; (2009b). Thoughts on Brackets and Dissipation: Old and New, *Journal of Physics: Conference Series*, pp. 1-12, Vol. 169, 012006.

Nguyen, S.H.Q. & Turski, Ł. A. (2009). Recursive Properties of Dirac and Metriplectic Dirac Brackets with Applications, *Physica A*, pp. 91-103, Vol. 388.

Nguyen, S.H.Q. & Turski, Ł. A. (2001). On the Dirac approach to constrained dissipative dynamics, J. Phys. A: Math. Gen., pp. 9281-9302, Vol. 34.

Parker, E. N.; (1957). Sweet's mechanism form erging magnetic fields in conducting fluids, *Journal of Geophysical Research*, 62, 509. (1963) Astro-Physical Journal Supplement 8,177.

Petscheck, H.E. (1964). Magnetic field annihilation, in *AAS/NASA Symposium on the Physics of Solar Flares*, W.N. Hess, ed. (Washington, DC: NASA SP-50), 425-439

Phythian, R.; (1977). The functional formalism of classical statistical dynamics, *Journal of Physics A*, 10.

Priest, E. R., Forbes, T.; (2000). Magnetic reconnection: MHD theory and applications, Cambridge University Press, ISBN: 9780511525087, New York.

Rechester, A. B. & Rosenbluth, M. N.; (1978). Electron heat transport in a tokamak with destroyed magnetic surfaces, *Physical Review Letters*, volume 40, number 1, 38-41.

Retinò, A., Sundkvist, D., Vaivads, A., et al.; (2007). In situ evidence of magnetic reconnection in turbulent plasma, *Nature - Physics*, 3, 235-238.

Sundkvist, D., Retinò, A., Vaivads, A., et al.; (2007). Dissipation in turbulent plasma due to reconnection in thin current sheets, *Physical Review Letters*, 99, 025004.

Sweet, P.A.; (1958). In *Electromagnetic Phenomena in Cosmical Physics*, edited by B. Lehnert (Cambridge University Press, Cambridge, England.

Tarasov, V.E. (2005). Continuos medium model for fractal media, *Physics Letters A*, 336, 167-174

Tarasov, V.E. (2006). Magnetohydrodynamics of fractal media, *Physics of Plasmas*, 13, 052107

Tetrault, D.; (1992a). Turbulent relaxation of magnetic fields, 1. Coarse-grained dissipation and reconnection, *Journal of Geophysical Research*, 97, 8531-8540.

Tetrault, D.; (1992b). Turbulent relaxation of magnetic fields, 2. Self-organization and intermittency, *Journal of Geophysical Research*, 97, 8541-8547.

Thiffeault, J-L. & Morrison, P. J.; (2000). Classification of Casimir Invariants of Lie-Poisson Brackets, *Physica D*, 136, 205–244.

Uritsky, V.M., Klimas, A.J., Vassiliadis, D., et al.; (2002). Scale-free statistics of spatiotemporal auroral emissions as depicted by Polar UVI images: The dynamic magnetosphere is an avalanching system, *Journal of Geophysical Research*, 107, 1426, doi: 10.1029/2001JA000281.

Vörös, Z., Baumjohann, W., Nakamura, R., et al.; (2004). Magnetic turbulence in the plasma sheet, *Journal of Geophysical Research*, 109, A11215, doi: 10.1029/2004JA010404.

Yankov, V.V. (1997). Magnetic field dissipation and fractal model of current sheet, *Physics of Plasmas*, 4, 571-574

Yordanova, E., Bergman, J., Consolini, G., et al.; (2005). Anisotropic scaling features and complexity in magnetospheric-cusp: a case study, *Nonlinear Processes in Geophysics*, 12, 817-825.

Zaslavsky, G. M.; (2002). Chaos, fractional kinetics, and anomalous transport, *Physics Reports*, 371, 461-580.

Zelenyi, L.M. & Milovanov, A.V.; (2004). Fractal topology and strange kinetics: from percolation theory to problems in cosmic electrodynamics, *Fiziki Uspekhi*, 47, 749-788.

Zheng, L. J. & Furukawa, M.; (2010). Current-interchange tearing modes: Conversion of interchange-type modes to tearing modes, *Physics of Plasmas*, 17, 052508.

3

Implicit Numerical Methods for Magnetohydrodynamics

Ravi Samtaney
*King Abdullah University of Science and Technology, Thuwal
Kingdom of Saudi Arabia*

1. Introduction

A fluid description of the plasma is obtained by taking velocity moments of the kinetic equations (Vlasov or Fokker-Planck equations) for electrons and ions and employing certain closure assumptions. A hierarchy of MHD models can be derived. Generally, if the time scales of interest are larger than the electron-ion collision time scales, then one may model the plasma as a single fluid. Furthermore, the fluid description of a plasma is valid when the length scales under investigation are larger than the Debye length; and the frequencies are smaller than the cyclotron frequency. The Debye length argument can also be cast in terms of a frequency: namely the plasma frequency. In addition, it is a standard assumption that the speeds involved are much smaller than the speed of light. The oft-used term "resistive MHD" is a single-fluid model of a plasma in which a single velocity and pressure describe both the electrons and ions. The resistive MHD model of a magnetized plasma does not include finite Larmor radius (FLR) effects, and is based on the simplifying limit in which the particle collision length is small compared with the macroscopic length scales.

1.1 Scope of this chapter

The scientific literature has numerous instances of methods and techniques to solve the MHD system of equations. To limit the scope of this chapter, we focus our discussion to single fluid resistive and ideal MHD. Although single fluid resistive (or ideal) MHD is in a sense the simplest fluid model for a plasma, these equations constitute a system of nonlinear partial differential equations, and hence pose many interesting challenges for numerical methods and simulations. In particular, there is a vast amount of literature devoted to numerical methods and simulations of resistive MHD wherein the time stepping method is explicit or semi-implicit. For example, in simulating MHD flows with shocks, shock-capturing methods from hydrodynamics have been tailored to MHD and have been very successfully used (see for example Reference Samtaney et al. (2005)). Such aforementioned shock-capturing methods almost exclusively employ explicit time stepping. This is entirely sensible given that the flow speeds are of the same order as, or exceed the fast wave speeds. In several physical situations, the diffusive time scales are much larger than the advective time scale. In these cases, the Lundquist number is large ($S >> 1$) and the diffusion terms are usually much smaller than the hyperbolic or wave-dominated terms in the equations. Usually the diffusion terms become important in thin boundary layers or thin current sheets within the physical domain. We are interested in computing such flows but with the additional constraint that the wave speeds are also much larger than the fluid advective speeds. In such cases, implicit time stepping methods are preferred to overcome the stiffness induced by the fast waves and march the flow simulations forward in time at time steps dictated by accuracy rather than stability constraints.

A brief outline of this chapter is as follows. In Section 2 we will provide a brief survey of implicit numerical methods for MHD. Following this, we will focus on two broad classes of nonlinearly implicit methods: Newton-Krylov (Section 3) and FAS or nonlinear multigrid (Section 4). Instead of writing a survey of a large number of implicit methods, we will present details of implicit methods with explanations of two different Newton-Krylov approaches (differing in the preconditioning approach) and on one nonlinear multigrid implementation for MHD. The chapter will close with a section on simulation test cases, and finally a conclusion section.

1.1.1 Rationale for implicit treatment

In compressible MHD we encounter the fast magnetosonic, Alfvén, and the slow magnetosonic waves. Typically, plasma confinement devices, such as tokamaks, stellarators, reversed field pinches etc. are characterized by a long scale length in the direction of the magnetic field, and shorter length scale phenomena in the direction perpendicular to the field. For example, in a tokamak, the magnetic field is dominantly along the toroidal direction and consequently the long length scale is mostly along the toroidal direction whereas the short scales are in the radial-poloidal plane. It is known that the Alfvén wave is a transverse wave with fastest propagation along the magnetic field. The fast magnetosonic, i.e., the fast compressional wave, is also anisotropic with the fastest propagation perpendicular to the magnetic field. In explicit methods, the time step is restricted by the familiar CFL condition. Several MHD phenomena are studied for long-time behavior where long-time is of the order of resistive or a combination of resistive-Alfvén time scales. For such investigations, the CFL condition implies an overly restrictive time step which translates to an enormous number of time steps. It is advantageous and desirable to design numerical schemes which allow us to have time steps larger than that imposed by the CFL condition, and yet the computational cost of each time step is only slightly larger than the explicit case. Finally, we remark that as we progress towards petascale computing and beyond to exascale, it is well recognized that breakthroughs and discoveries in science will be well-enabled by massive computations. However, hardware capability alone will not be sufficient and must be complemented by a large increase in efficiency by the development of clever algorithms. Implicit methods may prove beneficial as simulations increase in scale, since explicit methods can succumb to poor parallel weak scaling (Keyes et al. (2006)).

1.2 Resistive MHD equations

The single-fluid resistive MHD equations couple the equations of hydrodynamics and resistive low-frequency Maxwell's equations, and may be written in conservation form as

$$\frac{\partial U}{\partial t} + \underbrace{\nabla \cdot F(U)}_{hyperbolic\ terms} = \underbrace{\nabla \cdot F_d(U)}_{diffusive\ terms},$$

$$\frac{\partial U}{\partial t} + R(U) = 0 \tag{1}$$

where $R(U) \equiv \nabla \cdot F(U) - \nabla \cdot F_d(U)$, and the solution vector $U \equiv U(\mathbf{x}, t)$ is

$$U = \{\rho, \rho \boldsymbol{u}, \boldsymbol{B}, e\}^T$$

and the hyperbolic flux vector $F(U)$ and the diffusive fluxes $F_d(U)$ are given by

$$\begin{aligned} F(U) &= \Big\{\rho \boldsymbol{u}\,,\ \rho \boldsymbol{u}\boldsymbol{u} + \left(p + \frac{1}{2}\boldsymbol{B}\cdot\boldsymbol{B}\right)\bar{\bar{\mathbf{I}}} - \boldsymbol{B}\boldsymbol{B}\,, \\ &\qquad \boldsymbol{u}\boldsymbol{B} - \boldsymbol{B}\boldsymbol{u}\,,\ \left(e + p + \frac{1}{2}\boldsymbol{B}\cdot\boldsymbol{B}\right)\boldsymbol{u} - \boldsymbol{B}(\boldsymbol{B}\cdot\boldsymbol{u})\Big\}^T, \\ F_d(U) &= \Big\{0\,,\ Re^{-1}\bar{\bar{\boldsymbol{\sigma}}}\,,\ S^{-1}\left(\eta\nabla\boldsymbol{B} - \eta(\nabla\boldsymbol{B})^T\right), \\ &\qquad Re^{-1}\bar{\bar{\boldsymbol{\sigma}}}\cdot\boldsymbol{u} + \frac{\gamma}{\gamma-1}\frac{\kappa}{Re\,Pr}\nabla T + \frac{\eta}{S}\left(\frac{1}{2}\nabla(\boldsymbol{B}\cdot\boldsymbol{B}) - \boldsymbol{B}(\nabla\boldsymbol{B})^T\right)\Big\}^T. \end{aligned} \tag{2}$$

In the above equations ρ is the density, $\boldsymbol{u}$ is the velocity, $\boldsymbol{B}$ is the magnetic field, p and T are the pressure and temperature respectively, and e is the total energy per unit volume of the plasma. The plasma properties are the resistivity η, the thermal conductivity κ, and the viscosity μ, which have been normalized, respectively, by a reference resistivity η_R, a reference conductivity κ_R, and a reference viscosity μ_R. The ratio of specific heats is denoted by γ and generally fixed at 5/3 in most MHD simulations. The non-dimensional parameters in the above equations are the Reynolds number, defined as $Re \equiv \rho_0 U_0 L/\mu_R$, the Lundquist number, defined as $S \equiv \mu_0 U_0 L/\eta_R$, and the Prandtl number, denoted by Pr, which is the ratio of momentum to thermal diffusivity. The non-dimensionalization is carried out using a characteristic length scale, L, and the Alfvén speed $U_0 = B_0/\sqrt{\mu_0 \rho_0}$, where B_0, ρ_0, and μ_0 are the characteristic strength of the magnetic field, a reference density, and the permeability of free space, respectively. The equations are closed by the following equation of state

$$e = \frac{p}{\gamma - 1} + \frac{\rho}{2}\boldsymbol{u}\cdot\boldsymbol{u} + \frac{1}{2}\boldsymbol{B}\cdot\boldsymbol{B},$$

and the stress-strain tensor relation

$$\bar{\bar{\boldsymbol{\sigma}}} = \mu\left(\nabla\boldsymbol{u} + (\nabla\boldsymbol{u})^T\right) - \frac{2}{3}\mu\nabla\cdot\boldsymbol{u}\bar{\bar{\mathbf{I}}}.$$

Finally, a consequence of Faraday's law is that an initially divergence-free magnetic field must lead to a divergence-free magnetic field for all times, which corresponds to the lack of observations of magnetic monopoles in nature. This solenoidal property is expressed as $\nabla \cdot \boldsymbol{B} = 0$.

In the limit of zero resistivity, conductivity and viscosity, the equations of resistive MHD reduced to those of ideal MHD. These equations are similar to those written above with $\kappa = \mu = \eta = 0$. Ideal MHD equations are hyperbolic PDEs (although not strictly hyperbolic).

2. Brief survey of implicit methods for MHD

2.1 Early approaches

An implicit treatment of the fast magnetosonic wave coupled with arguments of large length scales dominantly in a certain direction allows one to investigate long-time scale phenomena in MHD in a computationally efficient manner. The approach discussed here was developed by Harned & Kerner (1985). We begin our discussion with a model problem which exposes the philosophy behind the implicit treatment of the fastest waves in MHD. Consider the following hyperbolic system of equations.

$$\begin{aligned}\frac{\partial u}{\partial t} &= a\frac{\partial v}{\partial x}, \\ \frac{\partial v}{\partial t} &= a\frac{\partial u}{\partial x}.\end{aligned} \tag{3}$$

This can be rewritten as

$$\frac{\partial^2 u}{\partial t^2} = a^2\frac{\partial^2 u}{\partial x^2}. \tag{4}$$

We then subtract a term from either side of the above equation as

$$\frac{\partial^2 u}{\partial t^2} - a_0^2\frac{\partial^2 u}{\partial x^2} = a^2\frac{\partial^2 u}{\partial x^2} - a_0^2\frac{\partial^2 u}{\partial x^2}, \tag{5}$$

where a_0 is a constant coefficient chosen mainly from stability considerations. Furthermore, a_0 is something which mimics the behavior of a, perhaps in some limit. The underlying idea of the semi-implicit methods discussed here is this: the term containing a_0 on the left hand side of eqn. (4) is treated implicitly, while the same term on the right hand side of eqn. (4) is treated explicitly. Moreover, the cost of solving the linear system stemming from the implicit treatment of the term containing a_0 should be small relative to the total cost of evolving the entire system. Harned and Kerner generalized the above approach to the MHD system in a slab geometry, with implicit treatment of the fast compressional wave. Furthermore, their method was applicable to a case where the scale lengths in the z-direction are much longer than those in the x-y plane. The fastest time scale is then due to the fast compressional wave in the x-y plane. The method for the implicit treatment of the shear Alfvén wave was proposed by Harned & Schnack (1986). The procedure is somewhat similar to the one adopted for the fast compressional wave except the linear term which is subtracted on the velocity evolution equation has a different form. The implicit treatment of the shear Alfvén wave is, in general, more problematic and required certain ad-hoc heuristics to be employed for stability (See Reference Harned & Schnack (1986) for details).

The above approaches may be classified as linearly implicit. An example of a nonlinearly implicit method is the work of Jones, Shumlak & Eberhardt (1997) on an upwind implicit method for resistive MHD. Their method applied to ideal MHD equations may be written as:

$$\frac{\partial U}{\partial t} = -R(U) = -\left(\frac{\partial F}{\partial x} + \frac{\partial G}{\partial y}\right) \tag{6}$$

where $R(U)$ is the divergence of the hyperbolic fluxes (here in this 2D discussion, F and G denote the fluxes in the $x-$ and $y-$ directions, respectively). The above equation (or rather system of equations) is discretized in time as

$$\frac{1}{2\Delta t}(3U_{i,j}^{n+1} - 4U_{i,j}^{n} + U_{i,j}^{n-1}) = -R_{i,j}^{n+1} \tag{7}$$

The above equation is implicit and is solved iteratively. Let $U^{n+1,k}$ denote the k-th iteration of the solution at the $n+1$-th time level. Rewrite the above equation as

$$\left(\frac{\partial U}{\partial t}\right)_{i,j}^{k+1} = -R_{i,j}^{n+1,k+1}, \tag{8}$$

where

$$\left(\frac{\partial U}{\partial t}\right)_{i,j}^{k+1} \equiv \frac{1}{2\Delta t}(3U_{i,j}^{n+1,k+1} - 4U_{i,j}^{n} + U_{i,j}^{n-1}). \tag{9}$$

A truncated Taylor series expansion yields:

$$\left(\frac{\partial U}{\partial t}\right)_{i,j}^{k+1} = \left(\frac{\partial U}{\partial t}\right)_{i,j}^{k} + \frac{\partial\left(\frac{\partial U}{\partial t}\right)_{i,j}^{k}}{\partial U}\Delta U_{i,j}^{k} \tag{10}$$

$$R_{i,j}^{k+1} = R_{i,j}^{k} + \frac{\partial R_{i,j}^{k}}{\partial U}\Delta U_{i,j}^{k}, \tag{11}$$

where $\Delta U_{i,j}^{k} = U_{i,j}^{n+1,k+1} - U_{i,j}^{n+1,k}$. The partial derivative of the divergence of the hyperbolic fluxes with respect to the solution vector is difficult to evaluate for second order upwind schemes. Hence, at this stage an approximation is made, i.e., such terms are evaluated with a first order scheme. The first order hyperbolic flux divergence, denoted as $\bar{R}$ is at point (i,j) is coupled to the neighboring four points in 2D. $\hat{R}_{i,j} \equiv \hat{R}(U_{i,j}, U_{i+\frac{1}{2},j}, U_{i-\frac{1}{2},j}, U_{i,j+\frac{1}{2}}, U_{i,j-\frac{1}{2}})$. Substituting all back into equation (8) gives

$$\left[\frac{\partial \hat{R}_{i,j}^{k}}{\partial U} + \frac{3I}{2\Delta t}\right]\Delta U_{i,j}^{k} = -\left[R_{i,j}^{k} + \left(\frac{\partial U}{\partial t}\right)_{i,j}^{k}\right] \tag{12}$$

The above equation is linear and iterated until ΔU_{ij}^{k} is driven to zero. The matrix in the linear system above is a large banded matrix and will be generally expensive to invert. Instead Jones et al. recommend the use of further approximations and using a lower-upper Gauss-Seidel (LU-SGS) technique. The hyperbolic fluxes R are evaluated with the Harten's approximate Riemann solver (Harten (1983)), applied with the framework of the eight-wave scheme developed by Powell et al. (1999). One philosophical concern about implicit upwinding methods is as follows. Generally, upwind methods are based on the solution of a Riemann problem at cell faces; such a solution is self-similar in time, i.e. depends only on x/t for times until the waves from neighboring cell faces Riemann problems start interacting. In traditional explicit upwind methods, this problem is avoided because we are operating within the CFL limit. In an implicit method, the CFL limit is violated and waves from neighboring Riemann solvers will interact. One may adopt the viewpoint that upwind methods are, in a sense, providing dissipation proportional to each wave and decrease the dispersion error which are the bane of central difference schemes. Adopting this viewpoint, one may ignore the interactions between neighboring Riemann problems.

2.2 Modern approaches

We, somewhat arbitrarily, classify *fully implicit* numerical methods (in distinction from semi-implicit or linearly implicit) as "modern". The main feature which distinguishes these approaches from semi-implicit or linear implicit is the ability to allow for very large time steps. The modern era of fully nonlinearly implicitly solvers was ushered in the since the early-mid nineties. Broadly the fully implicit methods can be classified as: (a) Newton-Krylov and (b) nonlinear multigrid (also known as FAS, i.e., full approximation scheme). An early example of an implicit Newton-Krylov-Schwarz method applied to aerodynamics was by Keyes (1995). Several papers subsequently appeared in the mid-late nineties and in early part of this century in fluid dynamics. Newton-Krylov methods found applicability in MHD in the early 2000s. In the subsequent sections, we will elaborate on both the Newton-Krylov and nonlinear multigrid as applied to MHD.

3. Newton-Krylov (NK) methods for MHD

3.1 General approach

The entire ideal MHD (or resistive MHD and beyond) can be written as a nonlinear function as follows:

$$\mathcal{F}(U^{n+1}) = 0, \tag{13}$$

where U^{n+1} is the vector of unknowns at time step $n+1$. For example, if we use a θ-scheme, one can write the nonlinear function as:

$$\mathcal{F}(U^{n+1}) = U^{n+1} - U^n + \theta \Delta t R(U^{n+1}) + (1-\theta)\Delta t R(U^n) = 0, \tag{14}$$

where $R(U)$ is the divergence of the fluxes (see equation 1). For compressible MHD, on a two dimensional $N \times M$ mesh, the total number of unknowns would then be $8MN$. The above nonlinear systems can be solved using an inexact Newton–Krylov solver. Apply the standard Newton's method to the above nonlinear system gives

$$\delta U^k = -\left[\left(\frac{\partial \mathcal{F}}{\partial U}\right)^{n+1,k}\right]^{-1} \mathcal{F}, \tag{15}$$

where $J(U^{n+1,k}) \equiv \left(\frac{\partial \mathcal{F}}{\partial U}\right)^{n+1,k}$ is the Jacobian; and $\delta U^k \equiv U^{n+1,k+1} - U^{n+1,k}$, and k is the iteration index in the Newton method. For the two dimensional system the Jacobian matrix is $8MN \times 8MN$ which, although sparse, is still impractical to invert directly.

In NK methods, the linear system at each Newton step is solved by a Krylov method. In Krylov methods, an approximation to the solution of the linear system $J\delta U = -\mathcal{F}$ is obtained by iteratively building a Krylov subspace of dimension m defined by

$$\mathcal{K}(r_0, J) = span\{r_0, Jr_0, J^2 r_0, \cdots, J^{m-1} r_0\}, \tag{16}$$

where r_0 is the initial residual of the linear system. The Krylov method can be either: one in which the solution in the subspace minimizes the linear system residual, or two in which the residual is orthogonal to the Krylov subspace. Within Newton-Krylov methods the two most commonly used Krylov methods are GMRES (Generalized Minimum Residual) and BiCGStab (Bi conjugate gradient stabilized) which can both handle non-symmetric linear systems. GMRES is very robust but generally is heavy on memory usage, while BiCGStab has a lower memory requirement, it is less robust given that the residual is not guaranteed to decrease monotonically.

Steps in a typical NK solver are the following:

1. Begin by guessing the solution $U^{n+1,0}$. Typically the initial guess is $U^{n+1,0} = U^n$.
2. For each Newton iteration $k = 1, 2, \cdots$
   (a) Using a Krylov method, approximately solve for δU^k,
   $J(U^k)\delta U^k = -\mathcal{F}(U^{n+1,k})$ so that $||J(U^k)\delta U^k + \mathcal{F}(U^{n+1,k})|| < Itol$.
   Each Krylov iteration requires:
   a. One matrix-vector multiply with J
   b. One preconditioner solve
3. Update the Newton iterate, $U^{n+1,k+1} = U^{n+1,k} + \lambda\delta U^k$
4. Test for convergence $||\mathcal{F}(U^{n+1,k+1})|| < ftol$.

It the approximate solution $U^{n+1,k}$ is "close" to the true solution U^* of the nonlinear system, the convergence is quadratic, i.e.,

$$||U^{n+1,k+1} - U^*|| \leq C||U^{n+1,k} - U^*||^2, \tag{17}$$

where C is a constant independent of $U^{n+1,k}$ and U^*. This result assumes that the linear system is solved exactly. If the linear systems are solved inexactly as in the Newton-Krylov method, then $Itol$, the linear system tolerance, has to be carefully chosen. In inexact NK, $Itol = \eta^k||\mathcal{F}^k||$. Quadratic convergence is retained if $\eta^k = C||\mathcal{F}^k||$. If we impose the condition that $lim_{k\to\infty}\eta^k = 0$ then convergence is super-linear, and if η^k is constant for all k then convergence is linear. Since Newton's method may be viewed as a linear model of the original nonlinear system, the model is a better approximation as the solution is approached. When "far" from the solution, it is not essential to solve the linear system to machine-zero convergence. The following choices for η^k are recommended which take into account how well the nonlinear system is converging.

$$\eta^k = \frac{\left|||\mathcal{F}^k - ||J^{k-1}\delta U^{k-1} + \mathcal{F}^{k-1}||\right|}{||\mathcal{F}^{k-1}||} \tag{18}$$

$$\eta^k = \gamma_1 \left(\frac{||\mathcal{F}^k||}{||\mathcal{F}^{k-1}||}\right)_2^{\gamma}, \tag{19}$$

where $\gamma_1 = 0.9$ and $\gamma_2 = 2$ as recommended by Eisenstat & Walker (1996). The first of these choices is how well the linear model agreed with the nonlinear system at the prior step, while the second uses a measure of the rate of convergence of the linear system.

In examining the Krylov methods, we notice that these require only matrix-vector products. Thus it is never necessary to store the entire Jacobian matrix. Hence the term "Jacobian-Free Newton-Krylov" (abbreviated JFNK) is frequently encountered in the literature. Furthermore, for complicated nonlinear systems such as those arising in MHD, the Jacobian entries are not even known analytically. Instead one can conveniently evaluate the Jacobian vector product using first order finite differences as follows:

$$J(U^k)\delta U^k \approx \frac{\mathcal{F}(U^{n+1,k}+\sigma\delta U^k)-\mathcal{F}(U^{n+1,k})}{\sigma}, \tag{20}$$

where σ is typically used as the square root of machine zero. The above expression assumes that $\mathcal{F}$ is sufficiently differentiable, a property which is easily violated in upwind methods with its myriad switches, and limited-reconstruction methods. The beauty of the Newton-Krylov method as outlined above is that it only relies on the evaluation of the nonlinear function $\mathcal{F}$. For a detailed review of the field of JFNK see the review paper by Knoll & Keyes (2004).

3.2 Preconditioners

Since all operations in the Newton-Krylov context require only linear complexity operations, the key component required for scalability of fully implicit simulations using this technology is an optimal preconditioning strategy for the inner Krylov linear solver (Kelley (1995); Knoll & Keyes (2004)). In Newton-Krylov algorithms, at each Newton iteration a Krylov iterative method is used to solve Jacobian systems of the form

$$J(U)V = -\mathcal{F}(U), \qquad J(U) \equiv I + \gamma\frac{\partial}{\partial U}(R(U)), \qquad \gamma = \theta\Delta t. \tag{21}$$

The number of iterations required for convergence of a Krylov method depends on the eigenstructure of J, where systems with clustered eigenvalues typically result in faster convergence than those with evenly distributed eigenvalues (Greenbaum (1997); Greenbaum et al. (1996); Trefethen & Bau (1997)). Unfortunately, for a fixed Δt, as the spatial resolution is refined the distribution of these eigenvalues spreads, resulting in increased numbers of Krylov iterations and hence non-scalability of the overall solution algorithm. The role of a preconditioning operator P is to transform the original Jacobian system (21) to either

$$JP^{-1}PV = -f \text{ (right prec.)}, \qquad \text{or} \qquad P^{-1}JV = -P^{-1}f \text{ (left prec.)}.$$

The Krylov iteration is then used to solve one of

$$(JP^{-1})W = -f, \qquad \text{or} \qquad (P^{-1}J)V = X,$$

where $X = -P^{-1}f$ is computed prior to the Krylov solve or $V = P^{-1}W$ is computed after the Krylov solve. Scalable convergence of the method then depends on the spectrum of the preconditioned operator (JP^{-1} or $P^{-1}J$), as opposed to the original Jacobian operator J. Hence, an optimal preconditioning strategy will satisfy the two competing criteria:

1. $P \approx J$, to help cluster the spectrum of the preconditioned operator.
2. Application of P^{-1} should be much more efficient than solution to the original system, optimally with linear complexity as the problem is refined and with no dependence on an increasing number of processors in a parallel simulation.

We note that the approximations used in the preconditioner should have no effect on the overall accuracy of the nonlinear system. It can be shown that JFNK method applied with right preconditioning preserves the conservation properties of the equations written in conservation form (Chacón (2004)) regardless of the nonlinear convergence tolerances. However, one cannot prove this for left preconditioning unless the solution is converged to machine precision (Chacón (2004)).

Preconditioners can be divided into two broad classes:

- Algebraic preconditioners: The nature of such preconditioners is of the "black-box" type. These represent a close representation of the Jacobian and are obtained using relatively inexpensive algebraic techniques such as stationary iterative techniques, incomplete LU decomposition, multigrid techniques etc. These preconditioners typically require forming and storing the Jacobian matrix.
- "Physics-based" preconditioners: These preconditioners are derived from other techniques such Picard iteration, or by semi-implicit techniques. They do not require forming and storing the entire Jacobian matrix and can be harnessed for Jacobian-free implementations. The form of the preconditioners here generally tend to exploit the structure of the PDEs themselves and in this sense this type of preconditioning is "physics-based".

3.3 JFNK method for resistive MHD I

In this section, we essentially reproduce the work by Chacón (2008a), wherein a JFNK approach for resistive MHD with physics based preconditioners has been developed. The approach, given below, essentially relies on the trick of "parabolization" and using a Schur complement approach. Parabolization refers to the technique by which a hyperbolic system is converted to a parabolic one which is then amenable to multigrid techniques.

3.3.1 A model illustration

Consider the following hyperbolic system

$$\begin{aligned} \frac{\partial u}{\partial t} &= a\frac{\partial v}{\partial x}, \\ \frac{\partial v}{\partial t} &= a\frac{\partial u}{\partial x}. \end{aligned} \tag{22}$$

Differencing with backward Euler we get

$$\begin{aligned} u^{n+1} &= u^n + a\left(\frac{\partial v}{\partial x}\right)^{n+1}, \\ v^{n+1} &= v^n + a\left(\frac{\partial u}{\partial x}\right)^{n+1}. \end{aligned} \tag{23}$$

Substitute the second equation into the first to obtain:

$$\left(I - a^2\Delta t^2\frac{\partial^2}{\partial x^2}\right)u^{n+1} = u^n + a\Delta t\left(\frac{\partial v}{\partial x}\right)^n, \tag{24}$$

which is is much better conditioned because the parabolic operator is diagonally dominant. Multigrid techniques usually perform well on elliptic and parabolic operators do poorly on hyperbolic operators which are diagonal submissive.

We now turn to parabolization by the Schur complement approach.

$$\begin{bmatrix} D_1 & U \\ L & D_2 \end{bmatrix} = \begin{bmatrix} I & UD_2^{-1} \\ 0 & I \end{bmatrix} \begin{bmatrix} D_1 - UD_2^{-1}L & 0 \\ 0 & D_2 \end{bmatrix} \begin{bmatrix} I & 0 \\ D_2^{-1}L & I \end{bmatrix}. \tag{25}$$

Stiff off-diagonal blocks L and U are now shifted to the diagonal via the Schur complement $D_1 - UD_2^{-1}L$. Applied to the model system above, $D_1 - UD_2^{-1}L = \left(I - a^2\Delta t^2 \frac{\partial^2}{\partial x^2}\right)$.

3.3.2 Application to resistive MHD

We begin by examining the linearized resistive MHD equations. These are written as

$$\delta\rho = L_\rho(\delta\rho, \delta\boldsymbol{v}) \tag{26}$$

$$\delta p = L_p(\delta p, \delta\boldsymbol{v}) \tag{27}$$

$$\delta\mathbf{B} = L_B(\delta\mathbf{B}, \delta\boldsymbol{v}) \tag{28}$$

$$\delta\boldsymbol{v} = L_v(\delta\boldsymbol{v}, \delta\mathbf{B}, \delta\rho, \delta p), \tag{29}$$

which illustrates the couplings between the various unknowns. In NK the Jacobian has the following coupling

$$J\delta U = \begin{bmatrix} D_\rho & 0 & 0 & U_{\rho v} \\ 0 & D_p & 0 & U_{pv} \\ 0 & 0 & D_B & U_{Bv} \\ L_{\rho v} & L_{pv} & L_{Bv} & D_v \end{bmatrix} \begin{bmatrix} \delta\rho \\ \delta p \\ \delta\mathbf{B} \\ \delta\boldsymbol{v} \end{bmatrix}, \tag{30}$$

which shows that the momentum equations are intimately coupled with other equations but that the density is only coupled with the velocity nonlinearly and so on. The diagonal blocks are of the "advection-diffusion" type and clearly amenable to multigrid and easily inverted. The off-diagonal terms denoted by L and U contain the hyperbolic couplings. The above Jacobian is rewritten as

$$J\delta U = \begin{bmatrix} M & U \\ L & D_v \end{bmatrix} \begin{pmatrix} \delta\boldsymbol{u} \\ \delta\boldsymbol{v} \end{pmatrix}, \tag{31}$$

where

$$\delta\boldsymbol{u} = \begin{pmatrix} \delta\rho \\ \delta p \\ \delta\mathbf{B} \end{pmatrix}, \quad M = \begin{pmatrix} D_\rho & 0 & 0 \\ 0 & D_p & 0 \\ 0 & 0 & D_B \end{pmatrix}. \tag{32}$$

The matrix M above is relatively easy to invert and is amenable to multigrid. The Schur complement analysis of the above 2×2 system is given below:

$$\begin{bmatrix} M & U \\ L & D_v \end{bmatrix}^{-1} = \begin{bmatrix} I & 0 \\ -LM^{-1} & I \end{bmatrix} \begin{bmatrix} M^{-1}L & 0 \\ 0 & P_S^{-1} \end{bmatrix} \begin{bmatrix} I & -M^{-1}U \\ 0 & I \end{bmatrix}. \tag{33}$$

where $P_S = D_v - LM^{-1}U$ is the Schur complement. The exact Jacobian inverse require M^{-1} and P_S^{-1}. The following predictor-corrector algorithm is proposed.

$$\delta u^* = -M^{-1}\mathcal{F}_u \quad (Predictor) \tag{34}$$

$$\delta v* = -P_S^{-1}[\mathcal{F}_v - L\delta v^*] \quad (Velocity\,update) \tag{35}$$

$$\delta u = \delta u^* - M^{-1}U\delta v \quad (Corrector). \tag{36}$$

Multigrid is impractical for P_S because of the M^{-1} factor and hence some simplifications are desirable. For the velocity update and the corrector part in the above equations, we can treat $M^{-1} \approx \Delta t$. This gives

$$\delta u^* = -M^{-1}\mathcal{F}_u \tag{37}$$

$$\delta v* = -P_{SI}^{-1}[\mathcal{F}_v - L\delta v^*] \tag{38}$$

$$\delta u = \delta u^* - \Delta t U\delta v, \tag{39}$$

where $P_{SI} = D_v - \Delta t LU$ and is block-diagonally dominant. Multigrid is employed to compute the inverse of P_{SI} and M.

3.4 NK method for resistive MHD II

In this section, we discuss yet another NK approach to resistive MHD. This section is essentially based on the work by Reynolds et al. (2006) in which they have developed a fully implicit Jacobian-Free NK method for compressible MHD. The main difference between this section and the previous one is in the preconditioning strategy employed during the Krylov step.

The resistive MHD equations are rewritten in a form which allows a method-of-lines approach. Reynolds et al. use a BDF method (up to fifth order accurate):

$$g(\mathbf{U}^n) \equiv \mathbf{U}^n - \Delta t_n \beta_{n,0} R(\mathbf{U}^n) - \sum_{i=1}^{q_n} \left[\alpha_{n,i}\mathbf{U}^{n-i} + \Delta t_n \beta_{n,i} R(\mathbf{U}^{n-i})\right], \tag{40}$$

where $R(\mathbf{U})$ is defined using the divergence of the fluxes (both hyperbolic and diffusion terms) as in equation (1). The time-evolved state $\mathbf{U}^n$ solves the nonlinear residual equation $g(\mathbf{U}) = 0$. q_n determines the method's order of accuracy and at $q_n = \{1,2\}$ the method is stable for any Δt_n, with stability decreases as q_n increases. $\alpha_{n,i}$ and $\beta_{n,i}$ are fixed parameters for a given method order q_n. In this approach $\Delta t_n, q_n$ are adaptively chosen at each time step to balance solution accuracy, solver convergence, and temporal stability (Hindmarsh (2000)).

Alternatively one may also use a θ-scheme

$$g(U^n) = U^n - U^{n-1} + \Delta t \left(\theta R(U^n) + (1-\theta)R(U^{n-1})\right), \tag{41}$$

where $\theta = 0.5$ corresponds to a Crank-Nicholson approach. The inexact Jacobian-Free NK approach is adopted to solve the nonlinear function $g(\mathbf{U}^n)$. The divergence of the fluxes in (1) is discretized using the following finite difference form

$$\left(\frac{\partial f}{\partial x}\right)_{i,j,k} = \frac{\tilde{f}_{i+\frac{1}{2},j,k} - \tilde{f}_{i-\frac{1}{2},j,k}}{\Delta x}, \tag{42}$$

where f may represent either the hyperbolic or the parabolic fluxes, and Δx is the mesh spacing in the x-direction (assumed uniform). The quantity $\tilde{f}_{i+\frac{1}{2},j,k}$ is referred to as the numerical flux through the face $\{i+\frac{1}{2}, j, k\}$ and is computed as a linear combination of the fluxes at cell centers as

$$\tilde{f}_{i+\frac{1}{2},j,k} = \sum_{\nu=-m}^{n} a_\nu f_{i+\nu,j,k}. \tag{43}$$

Reynolds et al. give the options for several spatial difference schemes. For a second-order central difference implementation, $m = 0$, $n = 1$ and $a_0 = a_1 = \frac{1}{2}$; for a fourth-order central difference approximation, $m = 1$, $n = 2$, and $a_{-1} = a_2 = \frac{-1}{12}$, $a_0 = a_1 = \frac{7}{12}$; and for tuned second-order central differences, $a_{-1} = a_2 = -0.197$, $a_0 = a_1 = 0.697$ (Hill & Pullin (2004)). These central difference approximations are free of dissipation errors, except perhaps near domain boundaries. They do, however, suffer from dispersion errors. Consequently, physical phenomena that are not well resolved can suffer from ringing. The dispersion errors can be minimized by using schemes such as the tuned-second order scheme, mentioned above, which has lower dispersion error than the central difference schemes. The numerical approximation to the divergence $\nabla \cdot \boldsymbol{B}$ is written as

$$\begin{aligned}\nabla \cdot \boldsymbol{B} = {} & \frac{\tilde{B}^x_{i+\frac{1}{2},j,k} - \tilde{B}^x_{i-\frac{1}{2},j,k}}{\Delta x} + \frac{\tilde{B}^y_{i,j+\frac{1}{2},k} - \tilde{B}^y_{i,j-\frac{1}{2},k}}{\Delta y} + \frac{\tilde{B}^z_{i,j,k+\frac{1}{2}} - \tilde{B}^z_{i,j,k-\frac{1}{2}}}{\Delta z} \\ & + \mathcal{O}(\Delta x^p) + \mathcal{O}(\Delta y^p) + \mathcal{O}(\Delta z^p)\end{aligned} \tag{44}$$

where B^α is the α-component of the magnetic field, and the terms $\tilde{B}^\alpha$ are evaluated as shown in equation (43), and p is the order of the spatial derivatives. If the numerical approximation of $\nabla \cdot \boldsymbol{B}$ is ensured to be zero at $t = 0$ then it can be easily shown that the numerical fluxes, as computed above, ensure the solenoidal property of the magnetic field in the discrete sense is automatically satisfied. This conservation property of preserving the solenoidal nature of the magnetic field in an implicit method is generally very desirable.

3.4.1 Preconditioner formulation

The preconditioner strategy, overall, uses an operator split approach to separate the wave-dominated portion from the diffusion portion. Instead of solving $J\,\delta\mathbf{U} = -g$, we solve the related system $(JP^{-1})(P\,\delta\mathbf{U}) = -g$, i.e., the right preconditioning approach is adopted.

Since MHD stiffness results from fast hyperbolic and diffusive effects, we set

$$P^{-1} = P_h^{-1} P_d^{-1} = J(\mathbf{U})^{-1} + \mathcal{O}(\Delta t^2).$$

This operator-splitting approach, widely used as a stand-alone solver, is used to accelerate convergence of the more stable and accurate implicit NK approach.

P_h: **Ideal MHD Preconditioner:** The ideal MHD preconditioner discussed here essentially exploits the local wave structure of the underlying hyperbolic portion of the PDEs. Hence this approach may be dubbed a "wave-structure"-based preconditioner. For linear multistep time integration approaches, it is convenient to first rewrite the nonlinear problem (40) in the form

$$f(U) = U + \gamma\left[\partial_x F(U) + \partial_y G(U) + \partial_z H(U)\right] + g = 0, \tag{45}$$

where the terms $F(U)$, $G(U)$ and $H(U)$ denote the x, y and z directional hyperbolic fluxes, and the term g incorporates previous time-level information into the discretized problem. This nonlinear problem has Jacobian

$$J(U) = I + \gamma\left[\partial_x(J_F(U)(\cdot)) + \partial_y(J_G(U)(\cdot)) + \partial_z(J_H(U)(\cdot))\right], \tag{46}$$

with, e.g., $J_F(U) = \frac{\partial}{\partial U}F(U)$. Using the notation $(\cdot)$ to denote the location at which the action of the linear operator takes place, e.g.

$$\left[I + \gamma\partial_x(J_F(U)(\cdot))\right]V = V + \gamma\partial_x(J_F(U)V).$$

Omitting the explicit dependence on U from the notation, and introducing nonsingular matrices L_F, L_G and L_H, re-write the Jacobian system (46) as

$$\begin{aligned}
J &= I + \gamma\left[\partial_x\left(J_F L_F^{-1} L_F(\cdot)\right) + \partial_y\left(J_G L_G^{-1} L_G(\cdot)\right) + \partial_z\left(J_H L_H^{-1} L_H(\cdot)\right)\right] \\
&= I + \gamma\Big[J_F L_F^{-1}\partial_x\left(L_F(\cdot)\right) + \partial_x\left(J_F L_F^{-1}\right)L_F(\cdot) \\
&\quad + J_G L_G^{-1}\partial_y\left(L_G(\cdot)\right) + \partial_y\left(J_G L_G^{-1}\right)L_G(\cdot) \\
&\quad + J_H L_H^{-1}\partial_z\left(L_H(\cdot)\right) + \partial_z\left(J_H L_H^{-1}\right)L_H(\cdot)\Big] \\
&= I + \gamma\Big[J_F L_F^{-1}\partial_x\left(L_F(\cdot)\right) + J_G L_G^{-1}\partial_y\left(L_G(\cdot)\right) + J_H L_H^{-1}\partial_z\left(L_H(\cdot)\right) \\
&\quad + \partial_x\left(J_F L_F^{-1}\right)L_F(\cdot) + \partial_y\left(J_G L_G^{-1}\right)L_G(\cdot) + \partial_z\left(J_H L_H^{-1}\right)L_H(\cdot)\Big].
\end{aligned}$$

The preconditioning scheme in this approach is based on the assumption that the majority of the stiffness found in the Jacobian is a result of a small number of very fast hyperbolic waves. To develop an approach for separately treating only these fast waves, consider the preconditioning matrix, P, constructed using a directional and operator-based splitting of J,

$$\begin{aligned}
P &= \left[I + \gamma J_F L_F^{-1}\partial_x\left(L_F(\cdot)\right)\right]\left[I + \gamma J_G L_G^{-1}\partial_y\left(L_G(\cdot)\right)\right]\left[I + \gamma J_H L_H^{-1}\partial_z\left(L_H(\cdot)\right)\right] \\
&\quad \left[I + \gamma\partial_x\left(J_F L_F^{-1}\right)L_F + \gamma\partial_y\left(J_G L_G^{-1}\right)L_G + \gamma\partial_z\left(J_H L_H^{-1}\right)L_H\right] \\
&= J + O(\gamma^2).
\end{aligned} \tag{47}$$

Denote these components as $P = P_F P_G P_H P_{local}$. Through constructing the operator P as a product in this manner, the preconditioner solve consists of 3 simpler, 1-dimensional implicit advection problems, along with one additional correction for spatial variations in the directional Jacobians J_F, J_G and J_H. Hence, $Pu = b$ may be solved via the steps (i) $P_F\,\chi = b$, (ii) $P_G\,w = \chi$, (iii) $P_H\,v = w$, and (iv) $P_{local}\,u = v$. Note that the splitting (47) is not unique, and that in fact these operations can be applied in any order. The technique for efficient solution of each of the above systems is presented in the ensuing paragraphs.

Directional Preconditioner Solves:
First consider solution of the three preconditioning systems P_F, P_G and P_H from (47) of the form, e.g. (x-direction)

$$P_F\chi = b \quad \Leftrightarrow \quad \chi + \gamma\, J_F L_F^{-1} \partial_x \left(L_F \chi\right) = b. \tag{48}$$

To this point L_F, L_G, and L_H are still unspecified. These are $n \times n$ matrices ($n = 7$ or 8 for compressible MHD depending upon whether the seven- or eight-wave formulation is used) whose rows are the left eigenvectors of the respective Jacobians, giving the identities,

$$L_F J_F = \Lambda_F L_F, \qquad \Lambda_F = \mathrm{diag}(\lambda^1, \ldots, \lambda^n), \qquad J_F R_F = R_F \Lambda_F,$$

where $R_F \equiv L_F^{-1}$ are the right eigenvectors ($n \times n$ column matrix), and λ^k are the eigenvalues of J_F. Through pre-multiplication of (48) by L_F, gives

$$L_F\chi + \gamma\, L_F J_F R_F \partial_x \left(L_F\chi\right) = L_F b \qquad \Leftrightarrow \qquad L_F\chi + \gamma\, \Lambda_F \partial_x \left(L_F\chi\right) = L_F b.$$

Defining the vector of characteristic variables $w = L_F\chi$, decouple the equations as ,

$$w + \gamma \Lambda_F \partial_x w = L_F b \qquad \Leftrightarrow \qquad w^k + \gamma \lambda^k \partial_x w^k = \beta^k,\ k = 1, \ldots, n,$$

where w^k denotes the k-th element of the characteristic vector w, and $\beta = L_F b$.

Spatial discretization of each of the characteristic variables w^k in the same manner as the original PDE (1), results in a tightly-banded linear system of equations (tridiagonal, pentadiagonal, etc., depending on the method), to solve for the values w_j^k. For example the tridiagonal version due to a $O(\Delta x^2)$ finite-difference discretization is

$$w_j^k + \frac{\gamma \lambda_j^k}{2\Delta x}\left(w_{j+1}^k - w_{j-1}^k\right) = \beta_j^k. \tag{49}$$

Reynolds et al. use a second order centered finite-volume approximation, with resulting systems for each w^k that are tridiagonal. Moreover, the above approach results not only in tridiagonal systems for each characteristic variable w^k, but the systems are in fact *block tridiagonal*, where each block corresponds to only one spatial $\{x, y, z\}$ row that is decoupled from all other rows through the domain in the same direction. Thus solution of these linear systems can be very efficient, as the computations on each row may be performed independently of one another.

Furthermore, since the initial assumption was that the stiffness of the overall system resulted from a few very fast waves, one may not construct and solve the above systems for each characteristic variable w^k. In cases where the wave speeds can be estimated, a pre-defined cutoff to the number of waves included in the preconditioner can be set. This reduction allows for significant savings in preconditioner computation. For those waves that are not preconditioned, approximate them as having wave speed equal to zero, i.e. solving with the approximation $\hat{\Lambda}_F = diag(\lambda^1, \ldots, \lambda^q, 0, \ldots, 0)$. Omission of the $(n-q)$ slowest waves in this fashion amounts to a further approximation of the preconditioner to the original discretized PDE system. Writing $\hat{P}_F$ as the x-directional preconditioner based on q waves, consider $\|\chi - \hat{\chi}\|_p$, where χ solves $P_F\chi = b$ and $\hat{\chi}$ solves $\hat{P}_F\hat{\chi} = b$, i.e.

$$\chi + \gamma J_F R_F \partial_x (L_F \chi) = b, \qquad \hat{\chi} + \gamma \hat{J}_F R_F \partial_x (L_F \hat{\chi}) = b,$$

where $\hat{J}_F = R_F \hat{\Lambda}_F L_F$. Left-multiplying by L_F and proceeding as before, to obtain

$$w + \gamma \Lambda_F \partial_x w = L_F b, \qquad \hat{w} + \gamma \hat{\Lambda}_F \partial_x \hat{w} = L_F b,$$

$$\Leftrightarrow$$

$$w^k + \gamma \lambda^k \partial_x w^k = (L_F b)^k, \quad k = 1, \ldots, n$$

$$\hat{w}^k + \gamma \lambda^k \partial_x \hat{w}^k = (L_F b)^k, \quad k = 1, \ldots, q$$

$$\hat{w}^k = (L_F b)^k, \quad k = q+1, \ldots, n.$$

Since the eigenvector matrices L_F and R_F may be renormalized as desired, and the eigenvalues are ordered so that $\lambda_i \geq \lambda_j$, for $i < j$, the dominant error from preconditioning only the q fastest waves is approximately

$$\frac{|\gamma \lambda^{q+1} / \Delta x|}{1 - |\gamma \lambda^{q+1} / \Delta x|}.$$

Hence omission of waves with small eigenvalues compared to the dynamical time scale (i.e. $\gamma\lambda \ll 1$) will not significantly affect preconditioner accuracy.

Local Non-Constant Coefficient Correction Solve:

The remaining component of the split preconditioner (47) comprises the local system $P_{local} u = v$,

$$\begin{aligned} & \left[I + \gamma \partial_x (J_F R_F) L_F + \gamma \partial_y (J_G R_G) L_G + \gamma \partial_z (J_H R_H) L_H\right] u = v \\ \Leftrightarrow \quad & \left[I + \gamma \partial_x (R_F \Lambda_F) L_F + \gamma \partial_y (R_G \Lambda_G) L_G + \gamma \partial_z (R_H \Lambda_H) L_H\right] u = v. \end{aligned}$$

Note that for spatially homogeneous Jacobians, $\partial_x (R_F \Lambda_F) = 0$ (similarly for y and z), so this system reduces to $u = v$. Hence this component may optionally be included to correct for spatial inhomogeneity in J_F, J_G and J_H. In keeping with the previous discretization approaches, approximate this system as, e.g.

$$\gamma \partial_x (R_F \Lambda_F) L_F \approx \tfrac{\gamma}{2\Delta x} \left(R_{F,i+1} \Lambda_{F,i+1} - R_{F,i-1} \Lambda_{F,i-1}\right) L_{F,i}.$$

These solves are spatially decoupled (with respect to u), resulting in a block-diagonal matrix whose solution requires only $n \times n$ dense linear solves at each spatial location.

P_d: **Diffusive MHD Preconditioner:** P_d solves the remaining diffusive effects within the implicit system,

$$\partial_t \mathbf{U} - \nabla \cdot \mathbf{F}_v = 0.$$

Setting P_d to be the Jacobian of this operator,

$$\begin{aligned} P_d = J_v(\mathbf{U}) &= I - \overline{\gamma}\frac{\partial}{\partial \mathbf{U}}\left(\nabla \cdot \mathbf{F}_v\right) \\ &= \begin{bmatrix} I & 0 & 0 & 0 \\ 0 & I - \overline{\gamma} D_{\rho \mathbf{v}} & 0 & 0 \\ 0 & 0 & I - \overline{\gamma} D_{\mathbf{B}} & 0 \\ -\overline{\gamma} L_\rho & -\overline{\gamma} L_{\rho \mathbf{v}} & -\overline{\gamma} L_{\mathbf{B}} & I - \overline{\gamma} D_e \end{bmatrix} \end{aligned}$$

and then its structure may be exploited for efficient and accurate solution. To solve $P_d\, y = b$ for $y = [y_\rho, y_{\rho \mathbf{v}}, y_{\mathbf{B}}, y_e]^T$:

1. Update $y_\rho = b_\rho$
2. Solve $(I - \overline{\gamma} D_{\rho \mathbf{v}})\, y_{\rho \mathbf{v}} = b_{\rho \mathbf{v}}$ for $y_{\rho \mathbf{v}}$
3. Solve $(I - \overline{\gamma} D_{\mathbf{B}})\, y_{\mathbf{B}} = b_{\mathbf{B}}$ for $y_{\mathbf{B}}$
4. Update $\tilde{b}_e = b_e + \overline{\gamma}\left(L_\rho\, y_\rho + L_{\rho \mathbf{v}}\, y_{\rho \mathbf{v}} + L_{\mathbf{B}}\, y_{\mathbf{B}}\right)$
5. Solve $(I - \overline{\gamma} D_e)\, y_e = \tilde{b}_e$ for y_e.

Due to their diffusive nature, steps 2, 3 and 5 are solved using a system-based geometric multigrid solver. Step 4 may be approximated through one finite-difference, instead of constructing and multiplying by the individual sub-matrices:

$$L_\rho\, y_\rho + L_{\rho \mathbf{v}}\, y_{\rho \mathbf{v}} + L_{\mathbf{B}}\, y_{\mathbf{B}} = \frac{1}{\sigma}\left[\nabla \cdot \mathbf{F}_v(U + \sigma W) - \nabla \cdot \mathbf{F}_v(\mathbf{U})\right]_e + O(\sigma),$$

where $W = \left[y_\rho, y_{\rho \mathbf{v}}, y_{\mathbf{B}}, 0\right]^T$.

As far as implementation details are concerned, Reynolds et al. employ the SUNDIALS software library (Hindmarsh et al. (2005)). Within SUNDIALS, extensive use is made of the CVODE ordinary differential equations integration package, as well as KINSOL for nonlinear solution of algebraic systems.

3.5 Next steps

Once the preconditioner is in place, several heuristic ideas may be applied to further decrease computational time. Some of these ideas discussed in Reynolds et al. (2010) are: freezing the Jacobian for a few time steps, freezing the computations of the eigen-values and eigen-vector, preconditioning only the fastest waves, eliminating the local solve etc. Depending on the physical problem under investigation, these heuristic ideas can reduce the computational time significantly. Reynolds et al. (2011) have generalized their approach to tokamak geometry wherein the poloidal plane is discretized using a curvilinear mesh. The form of the equations solved are similar to the ones discussed by Samtaney et al. (2007). The complexity of generating the Jacobian for the Newton-Krylov method makes it an attractive candidate for automatic differentiation tools. This is, in fact, employed by Reynolds & Samtaney (2012) and Reynolds et al. (2011) for implicit solution of the resistive MHD in the tokamak geometry.

They report that auto-differentiation tools can lead to an improvement in the accuracy of the computed Jacobian compared with a finite difference approach.

4. Nonlinear multigrid method for MHD

The literature on using nonlinear multigrid for MHD is quite sparse. Here we focus on the recent work by Adams et al. (2010). Multigrid methods are motivated by the observation that a low resolution discretization of an operator can capture modes or components of the error that are expensive to compute directly on a highly resolved discretization. More generally, any poorly locally-determined solution component has the potential to be resolved with coarser representation. This process can be applied recursively with a series of coarse grids, thereby requiring that each grid resolve only the components of the error that it can solve efficiently. These coarse grids have fewer grid points, typically about a factor of two in each dimension, such that the total amount of work in multigrid iterations can be expressed as a geometric sum that converges to a small factor of the work on the finest mesh. These concepts can be applied to problems with particles/atoms or pixels as well as the traditional grid or cell variables considered here. Multigrid provides a basic framework within which particular multigrid methods can be developed for particular problems.

Geometric multigrid is useful because it has the potential to be very efficient especially if the geometric domains of interest are simple enough that explicit coarse grids can be practically constructed even if, for instance, unstructured grids are used. Geometric multigrid not only provides a powerful basis on which to build a specific solution algorithm, but also allows for the straightforward use of nonlinear multigrid, or full approximation scheme (FAS) multigrid (Brandt (1977)) and matrix-free implementations. Given that the MHD problems are nonlinear, FAS multigrid is very efficient in that it solves the nonlinear set of equation directly and obviates the need of an outer (Newton) iteration. This is a critical component in attaining textbook efficiency on nonlinear problems. Figure 1 shows the standard multigrid FAS V-cycle and uses the smoother $u \leftarrow S(A, u, b)$, the restriction operator R_k^{k+1}, which maps residuals and current solutions from the fine grid space k to the coarse grid space $k+1$ (the rows of R_k^{k+1} are the discrete representation, on the fine grid, of the coarse grid functions), and the prolongation operator P_{k+1}^k, which maps the current solution from the coarse grid to the fine grid. Common notation for this multigrid V-cycle is V(μ_1,μ_2), where μ_1 and μ_2 are the number of pre- and post-smoothing steps, respectively. The canonical model problem is the Laplacian, for which point-wise Gauss-Seidel smoothers combined with linear interpolation for the restriction and prolongation operators generate method that reduce error by about an order of magnitude per V(1,1) cycle. This is theoretically optimal in that this rate of residual reduction is independent of mesh size and the amount of work in a V-cycle is given by a geometric sum that converges to about five work units. This so-called textbook efficiency has been observed, if not proven, for multigrid methods in a wide variety of applications (see Trottenberg et al. (2000) and references therein for details).

A concept used to determine if a point-wise smoothing method exists is *h-ellipticity*. Brandt & Dinar (1979) first introduced h-ellipticity and it is described in Trottenberg et al. (2000). H-ellipticity is the minimum Fourier symbol of the high half (in at least one dimension) of the spectrum of a discrete operator divided by the maximum Fourier symbol

function $u \leftarrow MGV(A_k, u_k, f_k)$
    if coarse grid $k+1$ **exists**
        $u_k \leftarrow S(A_k, u_k, f_k)$ /* pre-smoothing */
        $r_k \leftarrow f_k - A_k u_k$
        $r_{k+1} \leftarrow R_k^{k+1}(r_k)$ /* restriction of residual to coarse grid */
        $w_{k+1} \leftarrow R_k^{k+1}(u_k)$ /* restriction of current solution to coarse grid */
        $u_{k+1} \leftarrow MGV(A_{k+1}, w_{k+1}, r_{k+1} + A_{k+1}w_{k+1})$ /* recursive multigrid application */
        $u_{k+1} \leftarrow u_{k+1} - w_{k+1}$ /* convert solution to an increment */
        $u_k \leftarrow u_k + P_{k+1}^k(u_{k+1})$ /* prolongation of coarse grid correction */
        $u_k \leftarrow S(A_k, u_k, f_k)$ /* restriction of residual to coarse grid */
    else
        $u_k \leftarrow A_k^{-1} f_k$ /* post-smoothing */
    return u_k

Fig. 1. Nonlinear FAS multigrid *V-cycle* algorithm

of the operator. An h-ellipicity bounded well above zero is a necessary and sufficient condition for the existence of a point-wise smoother for an operator with a symmetric stencil (Trottenberg et al. (2000)). An important result of h-ellipticity is that effective point-wise smoothers (eg, Gauss-Seidel and distributive Gauss-Seidel) can be constructed for upwind discretizations of hyperbolic systems with no restriction on the time step, whereas point-wise Gauss-Seidel is unstable for a central difference scheme for a large time step. Adams et al. observed textbook multigrid efficiency with standard multigrid methods (e.g., point-wise Gauss-Seidel smoothers) using a first-order upwinding method for ideal and resistive MHD. First-order accuracy is, however, generally not sufficient for many applications, and second-order schemes are required for efficiency. These stable low-order smoothers have been used extensively with a higher-order operator via a defect correction scheme, which is identical to preconditioning, but is more amenable to a nonlinear solve (Atlas & Burrage (1994); Böhmer et al. (1984); Dick (1991); Hemker (1986); Koren (1991)).

An additional requirement of an optimal solver is to be able to reduce the algebraic error to the order of the discretization (or truncation) error for steady-state problems. For transient problems the solver needs to reduce the algebraic error to below the *incremental* error – that is, the product of the truncation error of the time integration scheme and the spacial truncation error. Reducing algebraic error far below that of the incremental error is computationally wasteful, though potentially useful for debugging. There is generally no need to spend resources to reduce the algebraic error far below the incremental error. This observation leads to our definition of an optimal solver as one that can reduce the error to less than the incremental error with a few work units per time step. This is an ambitious goal in that it requires both scalability and small constants in the actual computational costs. In fact, this results in a solver in which the rate of reduction in the residual actually increases as the mesh is refined, because the truncation error decreases. This goal can be achieved by using a multigrid V-cycle within what is called an F-cycle iteration (Trottenberg et al. (2000)). Figure 2 shows the standard nonlinear multigrid F-cycle with defect correction to accommodate the nonlinear V-cycle with a lower-order operator ($\tilde{A}$ is the first-order upwinding operator) for which our point-wise Gauss-Seidel smoother is stable. The complexity of an F-cycle is asymptotically similar to a V-cycle, and it can be proven to result in a solution with algebraic error that is

```
function u ← MGF(A_k, u_k, f_k)
    if coarse grid k+1 exists
        r_k ← f_k − A_k u_k
        r_{k+1} ← R_k^{k+1}(r_k)                /* restriction of residual to coarse grid */
        w_{k+1} ← R_k^{k+1}(u_k)                /* restriction of residual to coarse grid */
        u_{k+1} ← MGF(A_{k+1}, w_{k+1}, r_{k+1} + A_{k+1} w_{k+1}) /* recursive multigrid application */
        u_{k+1} ← u_{k+1} − w_{k+1}             /* convert solution to an increment */
        u_k ← u_k + P_{k+1}^k(u_{k+1})          /* prolongation of coarse grid correction */
        u_k ← MGV(Ã_k, u_k, f_k − A_k u_k + Ã_k u_k) /* low-order V-cycle, defect correction */
    else
        u_k ← A_k^{-1} f_k                      /* accurate solve of coarsest grid */
    return u_k
```

Fig. 2. Nonlinear FAS multigrid *F-cycle* algorithm with defect correction

less than the incremental error on the model problem (Trottenberg et al. (2000)). Multigrid can thus achieve discretization error with a work complexity of a few residual calculations. An additional advantage of the FAS multigrid algorithm is that it is an effective global nonlinear solver in that it does not suffer from the problem of limited radius of convergence of a standard Newton method.

5. Numerical test cases

5.1 Linear wave propagation

Linear wave propagation refers to the initialization of low amplitude magnetosonic or Alfvén waves and computing their evolution using the nonlinear equations. If the amplitude is small ($\mathcal{O}(\epsilon)$), these waves will propagate linearly with nonlinear effects essentially being $\mathcal{O}(\epsilon^2)$. Linear waves may be initialized in 2D using the following procedure. First choose a background quiescent equilibrium state as $\tilde{U} = (\rho, \rho\boldsymbol{u}, \boldsymbol{b}, e)^T$, where $\rho = 1$, $\boldsymbol{u} = \boldsymbol{0}$, $\boldsymbol{b} = (\cos\alpha\cos\theta, \sin\alpha\sin\theta, 0)^T$. Here, $\theta = \tan^{-1}\frac{k_y}{k_x}$, in which the ratio $\frac{k_y}{k_x}$ gives the direction of wave propagation and α is the orientation of the constant magnetic field. We project these equilibrium conserved quantities to characteristic variables via $W = L\tilde{U}$, where L is the left eigenvector matrix of the linearized MHD system. The $k-$th linear wave is setup by perturbing the $k-$th characteristic, $w^k = w^k + \epsilon\cos(\pi k_x x + \pi k_y y)$. The initial condition is then set as $U(x, y, 0) = R(\tilde{U})W$, where R is the right eigenvector matrix. Periodic boundary conditions should be implemented in both the x- and y-directions. This procedure can be easily extended to three dimensions.

Chacón (2004) tested the evolution of a magnetosonic wave to verify that the method had low dissipation using a Newton-Krylov approach but without any preconditioning. Reynolds et al. (2006) also tested the evolution of a slow magnetosonic wave propagating 45 deg to the mesh, with a Newton-Krylov solver without preconditioning. Numerical tests at 256^2 mesh resolution in 2D indicated that even without preconditioning, the implicit NK method yielded over a factor of ten decrease in CPU time. Reynolds et al. (2010) reported a further benefit of over a factor of five decrease in CPU time when the wave-structure based preconditioner was employed for the linear wave propagation test. Furthermore, for linear waves aligned with the mesh, the preconditioned solves converged in one Krylov iteration

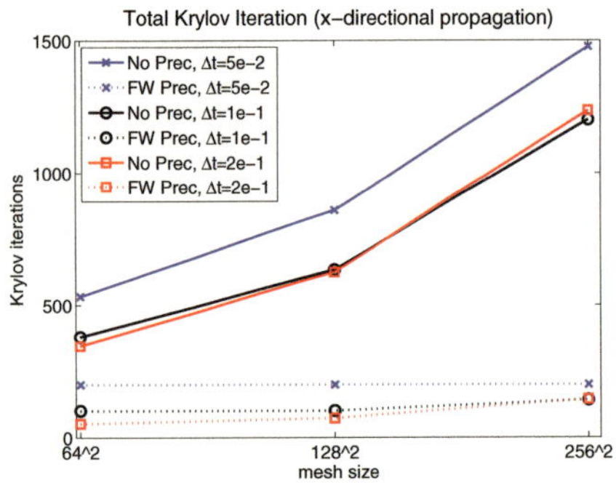

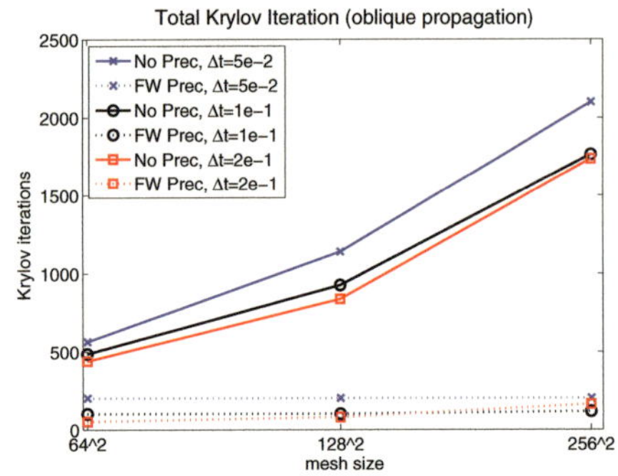

Fig. 3. Krylov iterations for the linear wave tests: x-directional (left) and oblique (right). Figure obtained from authors of Reference (Reynolds et al. (2010)).

for several tests, indicating that the wave-structure based preconditioner is optimal for such cases.

Here we reproduce results from Reynolds et al. (2010) of a slow magnetosonic wave of amplitude $\epsilon = 10^{-5}$ in a periodic domain chosen as $[0,2] \times [0,2]$. The wave is propagated until a final time of 10 units. The equilibrium state chosen in $\tilde{U} = (1, 0, \cos\alpha\cos\theta, \sin\alpha\sin\theta, 0, 0.1)^T$, $\alpha = -44.5^o$. Two different propagation directions are: $\theta = 0, 45^o$, i.e., the wave propagates aligned with the $x-$ axis, and along the diagonal. Results for the wave propagation are shown in Figure 3. The total number of Krylov iterations is plotted for different time step sizes and spatial discretizations (horizontal axis). For the linear wave propagating aligned with the mesh, the preconditioner is nearly exact, and hence the Krylov iterations remain nearly constant as the mesh is refined, as compared with the non-preconditioned tests that increase rapidly. For the oblique propagation case, the directional splitting does not appear to significantly affect the preconditioner accuracy, again resulting in nearly constant Krylov iterations with mesh refinement.

5.2 Magnetic reconnection in 2D

Magnetic reconnection (MR) refers to the breaking and reconnecting of oppositely-directed magnetic field lines in a plasma. In this process, magnetic field energy is converted to plasma kinetic and thermal energy. A test which has gained a lot of popularity in testing MHD codes is the so-called GEM reconnection challenge problem (Birn & et al. (2001a)). The initial conditions consist of a perturbed Harris sheet configuration as described in Birn & et al. (2001a). Reynolds et al. (2006) computed the GEM reconnection challenge problem with a Newton-Krylov method without preconditioning and reproduced the expected Sweet-Parker scaling for the reconnection rate for Lundquist numbers ranging from $S = 200 - 10^4$. Furthermore, for a mesh resolution of 512×256 their implicit method (without preconditioning) achieved a speedup of about 5.6 compared with an explicit method.

The GEM reconnection problem was also chosen for extensive testing by the nonlinear multigrid method developed by Adams et al. (2010). In fact, this work also extended the GEM problem by including a guide field in the third direction, thereby increasing the stiffness induced by more than a factor of five. Adams et al. also reported on the scalability of their approach by demonstrating good weak scaling up to 32K processors on a CRAY XT-5 supercomputer.

Figure 4 (reproduced from Adams et al. (2010)) from shows a time sequence of current density J_z field during reconnection. The goal is to develop solvers with a complexity equivalent to

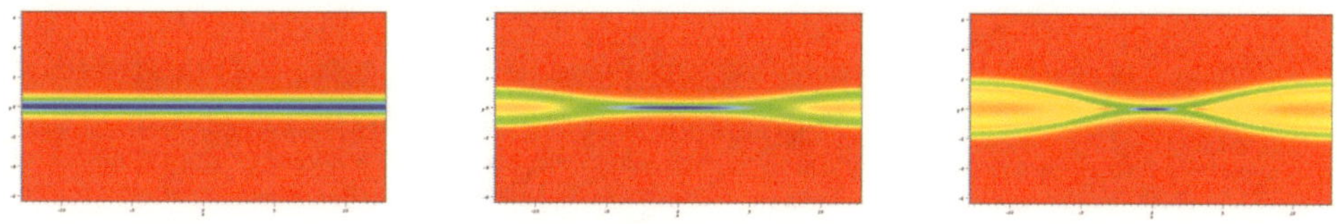

(a) (b) (c)

Fig. 4. Time sequence of current density, J_z during reconnection at time (a)t=0, (b)t=15 and (c)t=60. Parameters for this test are in Adams et al. (2010). Figure obtained from authors of Reference (Adams et al. (2010)).

a few residual calculations (work units) per time step, with the largest time step that can accurately resolve the dynamics of the problem. In this study, the solver is fixed at one iteration of FAS F-cycle with two defect corrected V(1,1) cycles at each level, as described in Section §4, and with a work complexity of about 18 work units per time step. There are three applications of the fine grid operator in residual calculations and defect correction in FAS multigrid, and three fine grid work units in the smoothers and residual calculations in each of the two V(1,1) cycle, plus lower-order work in restriction/prolongation and FAS terms. This results in about ten work units on the fine grid. Each successive grid is four times smaller (in 2D), and F-cycles process the second grid twice, the third grid three times, and so on, resulting in the equivalent of about eight additional work units for a total of 18 work units (there are actually fewer total work units in 3D because the coarse grids are relatively smaller). The smoother is nonlinear Gauss-Seidel with one iteration per grid point and red-black (or checkerboard) ordering. Even though Adams et al. (2010) use defect correction, they demonstrate a second-order rate of convergence on several important diagnostic quantities: these are the kinetic energy, reconnection rate and reconnected magnetic flux as shown in Figure 5,

5.3 Ideal Kelvin-Helmholtz instability

This test is generally a hard test for implicit solvers because the growth rate of the instability is high and the dynamics becomes nonlinear very quickly stressing all aspects of an implicit solver. On a 256^2 mesh, Chacón (2008b) reports a speedup in excess of three for the implicit solver compared with an explicit one, and nearly ten Krylov iterations per time step (and nearly five Newton steps per time step) for a time step which was 156 times larger than that for an explicit method. Reynolds et al. (2010) also performed the ideal Kelvin-Helmholtz instability (KHI) test with their wave preconditioner NK solver, and reported a speed up over a factor of three compared with simulations without using a preconditioner for a 256^2 mesh. They did not report comparisons with an explicit time stepping solver. In their 2D simulations, the number of Krylov iterations per time step ranged from 6-13 and the number of Newton steps ranged from 1-3 per time step. We hasten to add that this is not meant to be a comparison

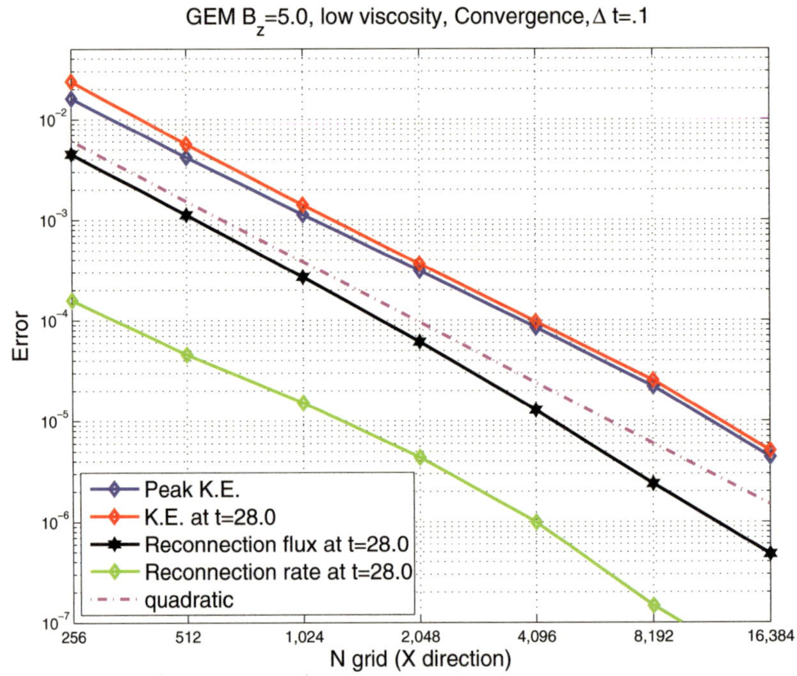

Fig. 5. Order of spatial accuracy for simulating magnetic reconnection usinga nonlinear multigrid approach. Error in peak kinetic energy and kinetic energy, reconnection flux rate and reconnection rate, high viscosity cases, $B_z = 0$ (left) and low viscosity $B_z = 5$ (right). Figure obtained from authors of Reference (Adams et al. (2010)).

between KHI simulations by Chacón and Reynolds et al. because their respective setup and solver tolerances were not necessarily identical. However, these test results are an indication of the type of speed up and code performance one may expect with strongly nonlinear MHD cases with a Newton-Krylov approach. Reynolds et al. (2010) also performed simulation tests on the 3D version of the ideal KHI.

Here we reproduce results from Reynolds et al. (2010) for the 2D KHI test. We set the computational domain to $\left[-\frac{5}{4}, \frac{5}{4}\right] \times \left[-\frac{1}{2}, \frac{1}{2}\right] \times \left[-\frac{5}{4}, \frac{5}{4}\right]$, with periodic boundary conditions in the x- and z-directions, and homogeneous Neumann boundary conditions in the y-direction. We initialize the constant fields $\rho = 1$, $\boldsymbol{b} = (0.1, 0, 10)^T$, $p = 0.25$, and $u_y = u_z = B_y = 0$. We then set $u_x = \frac{1}{2}\tanh(100y) + \frac{1}{10}\cos(0.8\pi x) + \frac{1}{10}\sin(3\pi y) + \frac{1}{10}\cos(0.8\pi z)$. This problem employs the resistive MHD equations, with resistivity, viscosity, and heat conduction coefficients set to 10^{-4}, and all runs are taken to a final time of $T_f = 2$. As previous results on this problem suggest that the instability growth rate is independent of the size of the resistivity, such small parameters are natural since the instability is predominantly driven by nonlinear (hyperbolic) effects (Jones, Gaalaas, Ryu & Frank (1997); Knoll & Brackbill (2002)). Moreover, for these parameters $T_f = 2$ is well within the nonlinear evolution regime for this problem. Snapshots of the x and z components of the (initially homogeneous) magnetic field at $t = 2$ are shown in Figure 6 for a 256^2 mesh simulation computed with a time step of $\Delta t = 0.0025$. Throughout this simulation, the number of nonlinear iterations ranged from 1

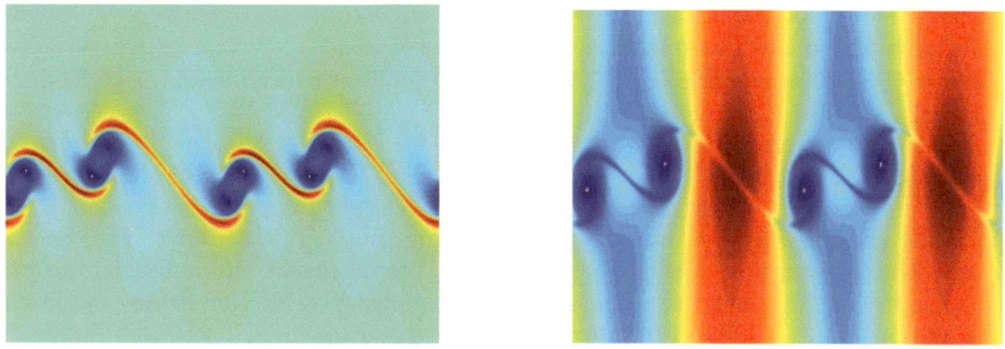

Fig. 6. Snapshots of B_x (left) and B_z (right) in the 2D Kelvin–Helmholtz test at $t = 2$. Figure obtained from authors of Reference (Reynolds et al. (2010)).

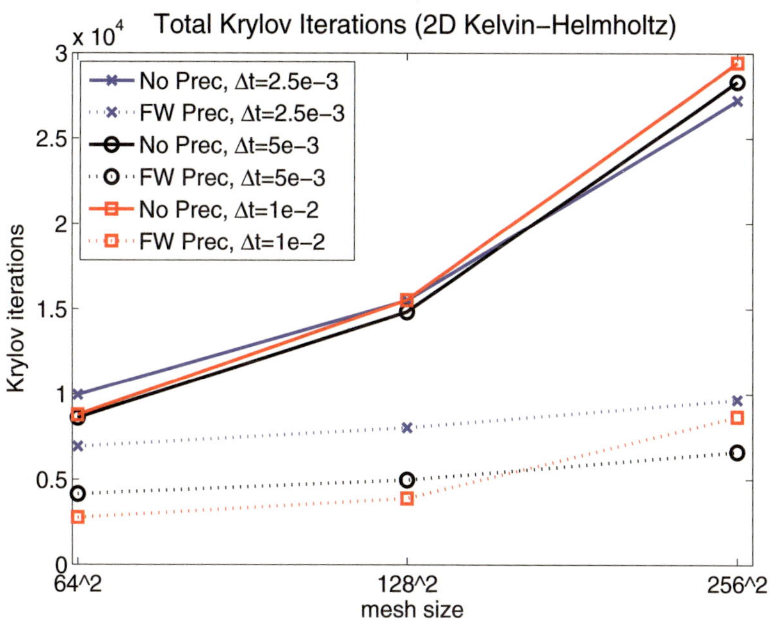

Fig. 7. Krylov iterations for the 2D Kelvin–Helmholtz tests. Figure obtained from authors of Reference (Reynolds et al. (2010)).

to 3, with the associated preconditioned Krylov iteration counts in the range of 6–13 per time step. Solver results for these tests are shown in Figure 7. For all time step sizes and all spatial discretizations used, the preconditioner results in significantly fewer linear iterations, with the disparity growing as the mesh is refined.

5.4 Other examples

There are a variety of other test cases reported in the literature ranging from ideal to resistive MHD. Chacón (Chacón (2004; 2008a;b)) reports on 2D tearing instability test cases and demonstrates a speedup ranging from $8-15$ for a 128^2 mesh Chacón (2008a). Another example of a good verification test case in 3D is that of 3D island coalescence (Chacón (2008a)). Reynolds et al. (2006) reported on a 3D ideal MHD problem which models pellet fueling in tokamaks.

6. Conclusion

In this chapter, we discussed the need for implicit algorithms for resistive magnetohydrodynamics. We highlighted two broad classes of nonlinear methods: Newton-Krylov and nonlinear multigrid. We illustrated two Newton-Krylov approaches for MHD which are essentially very similar in the overall approach, but differed in the preconditioning strategies for expediting the iterative solution steps in the Krylov linear solver stage of the overall method. One preconditioning strategy is based on a "parabolization" approach while the other utilizes the local wave structure of the underlying hyperbolic waves in the MHD PDEs. The literature on the use of nonlinear multigrid for MHD is essentially sparse and therein we focused on a defect-correction approach coupled with a point-wise Gauss-Seidel smoother utilizing a first order upwind approach. Both approaches are valid and have their place, but it is clear that the nonlinear multigrid approach for MHD is still relatively new and could be further developed.

6.1 Future challenges

In this chapter, we have focused exclusively on methods for single fluid resistive MHD. Future challenges will lie in the area of implicit methods for more complicated extended MHD models with FLR effects, several of which exhibit dispersive wave phenomena such as Whistler, Kinetic Alfvén waves, and gyroviscous waves. These dispersive high frequency waves essentially make the stable explicit time step proportional to the square of the mesh spacing, i.e., $\Delta t \propto \Delta x^2$; and hence the benefit from implicit methods is much more than those for single fluid MHD. Some progress in using Newton Krylov approaches for Hall-MHD has been reported by Chacón & Knoll (2003). However more work is required for general geometry, and inclusion of all dispersive wave families. Research in the area of nonlinear multigrid is essentially unexplored for extended MHD. Another interesting challenge in developing implicit methods for MHD is the combination of JFNK or FAS methods with adaptive mesh refinement (AMR). Some progress towards JFNK with AMR has been reported by Philip et al. (2008) on reduced incompressible MHD in 2D. Combining implicit methods with AMR will help mitigate not only the temporal stiffness issues but also help effectively resolve the range of spatial scales in MHD.

7. References

Adams, M. F., Samtaney, R. & Brandt, A. (2010). Toward textbook multigrid efficiency for fully implicit resistive magnetohydrodynamics, *J. of Comput. Phys.* 229: 6208–6219.

Atlas, I. & Burrage, K. (1994). A high accuracy defect correction multigrid method for the steady incompressible Navier Stokes equations, *J. Comput. Phys.* 114: 227–233.

Birn, J. & et al. (2001a). Geospace Environmental Modeling (GEM) magnetic reconnection challenge, *J. Geophys. Res.* 106: 3715–3719.

Böhmer, K., Gross, W., Schmitt, B. & Schwarz, R. (1984). Defect corrections and Hartree–Fock method, *in* K. Böhmer & H. J. Stetter (eds), *Defect Correction Methods: Theory and Applications*, Computing Suppl. 5, Springer–Verlag, Vienna, pp. 193–209.

Brandt, A. (1977). Multi-level adaptive solutions to boundary value problems, *Math. Comput.* 31: 333–390.

Brandt, A. & Dinar, N. (1979). Multigrid solutions to elliptic flow problems, *in* S. Parter (ed.), *Numerical Methods for Partial Differential Equations*, Academic Press, New York, pp. 53–147.

Chacón, L. (2004). A non-staggered, conservative, divB=0, finite volume scheme for 3D implicit extended magnetohydrodynamics in curvilinear geometries, *Computer Physics Comm.* 163: 143–171.

Chacón, L. (2008a). An optimal, parallel, fully implicit Newton-Krylov solver for three-dimensional visco-resistive magnetohydrodynamics, *Phys. Plasmas* 15: 056103–056103–12.

Chacón, L. (2008b). Scalable parallel implicit solvers for 3D magnetohydrodynamics, *Journal of Physics: Conference Series* 125: 012041.

Chacón, L. & Knoll, D. A. (2003). A 2D high-β Hall MHD implicit nonlinear solver, *J. Comput. Phys.* 188: 573–592.

Dick, E. (1991). Second order formulation of a multigrid method for steady Euler equations through defect correction, *J. Comput. Appl. Math.* 35: 159–168.

Eisenstat, S. C. & Walker, H. F. (1996). Choosing the forcing terms in an inexact Newton method, *SIAM J. Sci. Comput.* 17(1): 16–32.

Greenbaum, A. (1997). *Iterative Methods for Solving Linear Systems*, SIAM, Philadelphia.

Greenbaum, A., Pták, V. & Strakous, Z. (1996). Any nonincreasing convergence curve is possible for gmres, *SIAM J. Matrix Anal. Appl.* 17(3): 465–469.

Harned, D. S. & Kerner, W. (1985). Semi-implicit method for three-dimensional compressible magnetohydrodynamic simulation, *J. Comput. Phys.* 60: 62–75.

Harned, D. S. & Schnack, D. D. (1986). Semi-implicit method for long time scale magnetohydrodynamic computations in three dimensions, *J. Comput. Phys.* 65: 57–70.

Harten, A. (1983). High-resolution schemes for hyperbolic conservation-laws, *J. Comput. Phys.* 49(3): 357–393.

Hemker, P. W. (1986). Defect correction and higher order schemes for the multigrid solution of the steady Euler equations, *in* W. Hackbusch & U. Trottenberg (eds), *Multigrid Methods II*, Springer–Verlag, Berlin, pp. 149–165.

Hill, D. J. & Pullin, D. I. (2004). Hybrid tuned center-difference-WENO method for large eddy simulations in the presence of strong shocks, *J. Comput. Phys.* 194: 435–450.

Hindmarsh, A. (2000). The PVODE and IDA algorithms, *Technical Report UCRL-ID-141558*, LLNL.

Hindmarsh, A., Brown, P., Grant, K., Lee, S., Serban, R., Shumaker, D. & Woodward, C. (2005). SUNDIALS, SUite of Nonlinear and DIfferential/ALgebraic equation Solvers, *ACM Trans. Math. Software* 31: 363–396.

Jones, O. S., Shumlak, U. & Eberhardt, D. S. (1997). An implicit scheme for nonideal magnetohydrodynamics, *J. Comput. Phys.* 130(2): 231–242.

Jones, T. W., Gaalaas, J. B., Ryu, D. & Frank, A. (1997). The MHD Kelvin Helmholtz instability II. the roles of weak and oblique fields in planar flows, *Ap.J.* 482: 230–244.

Kelley, C. T. (1995). *Iterative Methods for Linear and Nonlinear Equations*, Frontiers in Applied Mathematics, SIAM, Philadelphia.

Keyes, D. E. (1995). Aerodynamic applications of Newton-Krylov-Schwarz solvers, *in* M. Deshpande, S. Desai & R. Narasimha (eds), *Proceedings of the 14th International Conference on Numerical Methods in Fluid Dynamics*, Springer,New York, pp. 1–20.

Keyes, D. E., Reynolds, D. R. & Woodward, C. S. (2006). Implicit solvers for large-scale nonlinear problems, *J. Phys.: Conf. Ser.* 46: 433–442.

Knoll, D. A. & Brackbill, J. U. (2002). The Kelvin-Helmholtz instability, differential rotation, and three-dimensional, localized, magnetic reconnection, *Phys. Plasmas* 9(9): 3775–3782.

Knoll, D. A. & Keyes, D. E. (2004). Jacobian-free Newton-Krylov methods: a survey of approaches and applications, *J. Comp. Phys.* 193: 357–397.

Koren, B. (1991). Low-diffusion rotated upwind schemes, multigrid and defect correction for steady, multi-dimensional Euler flows, *in* W. Hackbusch & U. Trottenberg (eds), *Multigrid Methods III*, Vol. 98 of *International Series of Numerical Mathematics*, Birkhäuser, Basel, pp. 265–276.

Philip, B., Chacón, L. & Pernice, M. (2008). Implicit adaptive mesh refinement for 2D reduced resistive magnetohydrodynamics, *J. Comput. Phys.* 227: 8855–8874.

Powell, K. G, Roe, P. L., Linde, T. J., Gombosi, T. I. & DeZeeuw, D. L. (1999). A solution-adaptive upwind scheme for ideal magnetohydrodynamics, *J. Comp. Phys.* 154: 284–300.

Reynolds, D. R. & Samtaney, R. (2012). Sparse jacobian construction for mapped grid visco-resistive magnetohydrodynamics, *6th International Conference on Automatic Differentiation,Fort Collins, CO, USA, July 23 - 27* . To be submitted.

Reynolds, D. R., Samtaney, R. & Tiedeman, H. C. (2011). A fully implicit Newton-Krylov-Schwarz method for tokamak MHD: Jacobian construction and preconditioner formulation, *22nd International Conference on Numerical Simulation of Plasmas, Long Branch, NJ, USA, September 7-9* .

Reynolds, D. R., Samtaney, R. & Woodward, C. S. (2010). Operator-based preconditioning of stiff hyperbolic systems, *SIAM J. Sci. Comput.* 32: 150–170.

Reynolds, D. R., Samtaney, R. & Woodward, C. S. (2006). A fully implicit numerical method for single-fluid resistive magnetohydrodynamics, *J. Comp. Phys.* 219: 144–162.

Samtaney, R., Colella, P., Ligocki, T. J., Martin, D. F. & Jardin, S. C. (2005). An adaptive mesh semi-implicit conservative unsplit method for resistive MHD, Journal of Physics: Conference Series. SciDAC 2005, pp. 40–48.

Samtaney, R., Straalen, B. V., Colella, P. & Jardin, S. C. (2007). Adaptive mesh simulations of multi-physics processes during pellet injection in tokamaks, *Journal of physics: Conference series. SciDAC 2007*, Vol. 78, p. 012062.

Trefethen, L. N. & Bau, D. III (1997). *Numerical Linear Algebra*, SIAM.

Trottenberg, U., Oosterlee, C. W. & Schüller, A. (2000). *Multigrid*, Academic Press, London.

4

Dynamics of Magnetic Relaxation in Spheromaks

Pablo L. Garcia-Martinez
[1]*Laboratoire de Physique des Plasmas, Ecole Polytechnique, Palaiseau cedex*
[2]*CONICET and Centro Atómico Bariloche (CNEA), San Carlos de Bariloche*
[1]*France*
[2]*Argentina*

1. Introduction

The first attempts to get energy from the controlled fusion of two light atoms nuclei date back to the beginning of the fifties of the last century. The crucial difficulty to achieve this goal is that particles need to have a large amount of thermal energy in order to have a significant chance of overcoming the Coulomb repulsion. At such high temperatures the atoms are fully ionized conforming a plasma. Such a hot plasma can not be in contact with solid walls because it will be rapidly cooled down. Two main methods have been developed to confine plasmas: the magnetic confinement and the inertial confinement. Here we are concerned with the magnetic confinement approach.

Under certain conditions some magnetic configurations studied in the context of plasma confinement become unstable and undergo a process called magnetic (or plasma) relaxation. This process generally causes the system to evolve toward a self-organized state with lower magnetic energy and almost the same magnetic helicity. A key physical mechanism that operates during plasma relaxation is the localized reconnection of magnetic field lines. It was demonstrated that magnetic relaxation can be employed to form and sustain configurations relevant to magnetic confinement research.

The theoretical description of magnetic relaxation is given in terms of a variational principle (Taylor, 1974). Despite the remarkable success of this theory to describe the final self-organized state toward which the plasma evolves, it does not provide any information on the dynamics of the plasma during relaxation. Since the process of relaxation always involves fluctuations that degrade plasma confinement it is very important to understand their dynamics.

The dynamics of the fluctuations induced during the relaxation process can be studied in the context the magnetohydrodynamic (MHD) model. In this Chapter, we will study the dynamics of the relaxation in kink unstable spheromak configurations. To that end we will solve the time-dependent non-linear MHD equations in three spatial dimensions.

The rest of the Chapter is organized as follows. In Section 2 we give a general introduction to magnetic confinement of high temperature plasma which is the main motivation of this study. The physical background of this work is the MHD model which is presented in Section 3. In Section 4 we describe the magnetic relaxation theory and its relationship with plasma self-organization. The role of the magnetic helicity and magnetic reconnection is also discussed. In Section 5 we present a study of the dynamics of magnetic relaxation in kink unstable spheromak configurations. These configurations are of special interest because they approximate quite well the measurements in spheromaks during sustainment (Knox et al., 1986);(Willet et al., 1999). Previous works have shown the existence of a partial relaxation behavior in marginally unstable configurations (Garcia-Martinez & Farengo, 2009a); (Garcia-Martinez & Farengo, 2009b). In this work we analyze this process in detail and we show, in particular, that this behaviour is connected to the presence of a rational surface near the magnetic axis. The main conclusions are summarized in Section 6.

2. Magnetic confinement of high temperature plasmas

The charged particles which constitute a high temperature plasma are subjected to the Lorentz force. The objective of magnetic confinement is to create magnetic field configurations to constrain the motion of the particles trying to keep them trapped far from the container's wall. In order to accomplish this goal the following four conditions must be fulfilled:

1. The configuration must be in magnetohydrodynamic (MHD) equilibrium.
2. The configuration must be stable (or it should be possible to mitigate or control potential instabilities).
3. Methods to produce, heat and sustain the configuration must be available.
4. The losses due to transport of heat and particles must be low enough to allow the system to have an adequate confinement time.

Here we will discuss some general aspects of the first three points. A more detailed discussion on these topics may be found, for instance, in the book of Wesson (2004).

2.1 MHD equilibrium

It is said that a magnetic configuration is in static MHD equilibrium if the Lorentz force cancels out exactly the pressure force

$$\mathbf{J} \times \mathbf{B} = \nabla p. \tag{1}$$

This force balance is part of the momentum equation of the MHD model that will be presented in Sec. 3. The magnetic configurations employed for plasma confinement almost always have toroidal topology. In this situation, each magnetic field line describes a toroidal magnetic surface. These toroidal magnetic surfaces are nested around a circle called magnetic axis (see Fig. 1). The separatrix is the outermost closed surface that does not touch the vessel. Three axisymmetric toroidal configuration schemes are shown in Fig. 1. It is a common practice to decompose the magnetic field into its toroidal and poloidal components. If we place a cylindrical coordinate system at the center of the torus, aligning the z-axis with the axis of symmetry (of revolution) of the torus, the toroidal direction coincides with the azimuthal direction and the poloidal plane lies in the r-z plane. In the right column of Fig. 1 we show the profiles of the toroidal and poloidal magnetic fields as a function of the distance between the magnetic axis and the separatrix for each configuration. Let's review the main features of these configurations.

- Tokamak. The toroidal magnetic field is much larger than the poloidal one. This intense toroidal field is imposed by a set of large external coils while the poloidal field comes from

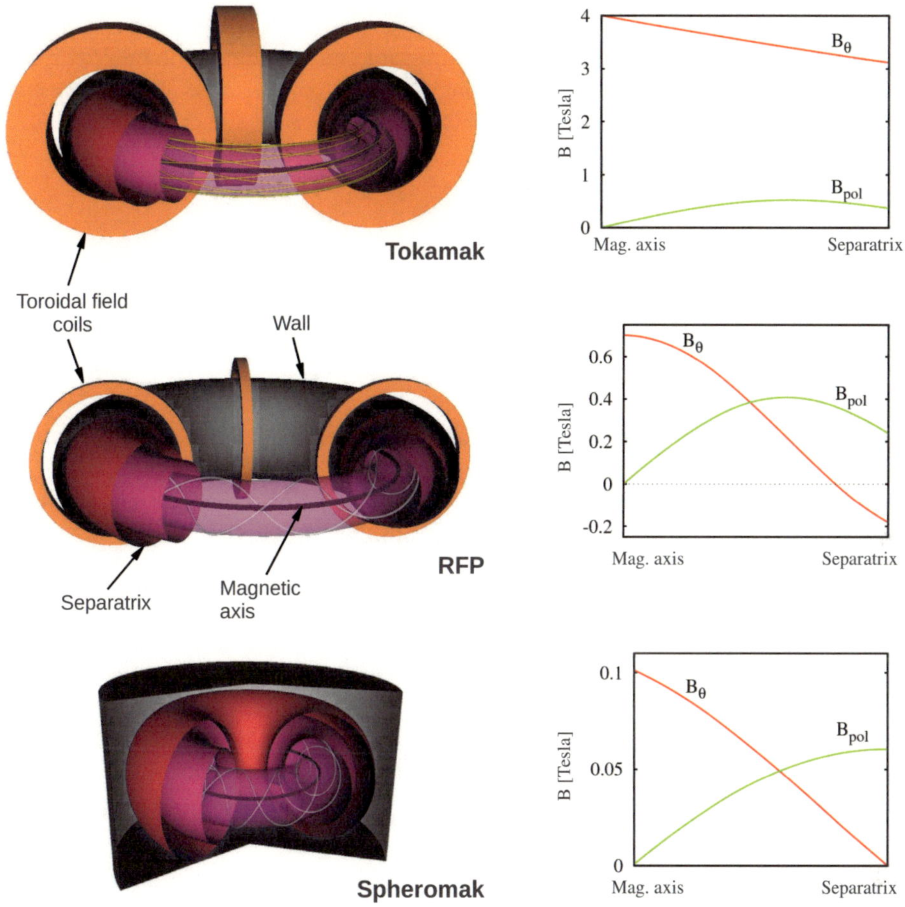

Fig. 1. Three examples of toroidal axisymmetric configurations used in magnetic confinement research: the tokamak, the reversed field pinch (RFP) and the spheromak.

the toroidal current that flows through the plasma. Typically, this current is produced by the electric field induced by the temporal variation of the magnetic flux linked by the torus.

- RFP (reversed field pinch). It is also an axisymmetric toroidal device whose aspect ratio (ratio of the major radius and the minor radius of the torus) is generally larger than that of the tokamak. The toroidal and poloidal fields have similar strengths. This makes the system much more prone to develop MHD instabilities. The magnetic field generation is analogous to that of the tokamak but using smaller coils for the toroidal field. The toroidal field reverses (changes its sign) near the separatrix opposing the externally applied field as a result of a magnetic relaxation process.
- Spheromak. It belongs to the family of compact tori. These are toroidal magnetic configurations formed inside a simply connected volume. The lack of elements being linked by the plasma represents a great advantage from a constructive and economical point of view. The two components of the magnetic field have similar strength. This

configuration is formed as a result of a relaxation process that self-organizes the magnetic field, closely related to that occurring in the RFP.

In all these three systems as well as in other important configurations the magnetic surfaces spanned by the magnetic field lines play a central role in confinement. We examine this in more detail. Let $\psi(r,z)$ be the poloidal flux function defined as

$$\psi(r,z) = \int_{S(r,z)} \mathbf{B} \cdot d\mathbf{s} \tag{2}$$

were $S(r,z)$ is the circle of radius r centered at the position z of the vertical axis. If the configuration is axisymmetric we can express the poloidal flux function as

$$\psi(r,z) = 2\pi \int_0^r B_z(\chi,z)\, \chi d\chi. \tag{3}$$

With this definition ψ reaches its maximum value at the magnetic axis ($\psi(r_{\mathrm{ma}}, z_{\mathrm{ma}}) = \psi_{\mathrm{ma}}$). The magnetic surfaces, or flux surfaces, can be determined by the equation $\psi(r,z) = C$ where C is a constant. Note that this useful labeling system for the flux surfaces breaks down when the axisymmetry is lost (due to an instability for example).

The poloidal flux function acts as a stream function for the poloidal field since

$$\mathbf{B}_p = \nabla \times \left(\frac{\psi(r,z)}{2\pi r} \hat{\theta} \right) \tag{4}$$

where $\hat{\theta}$ is the unit vector pointing in the toroidal direction. Note that the poloidal flux function is closely related to the toroidal component of the magnetic vector potential $\mathbf{A}$ since Eq. (4) implies that

$$\psi = 2\pi r A_\theta. \tag{5}$$

This relationship has important consequences for the confinement of the plasma particles. Due to the axisymmetry, the canonical angular momentum $P_\theta = mrv_\theta + qrA_\theta$ turns out to be a constant of the motion of each particle (m and Ze being the mass and the charge of the particle, respectively). In terms of the poloidal flux we can see that

$$P_\theta = mrv_\theta + \frac{Ze}{2\pi}\psi \tag{6}$$

is a constant of motion. If the magnetic field is strong enough the term mrv_θ may become very small compared with $Ze\psi/(2\pi)$. In that case the particles are constrained to move along surfaces of constant ψ, i.e. along magnetic surfaces. For this reason, an effective way of confining charged particles can be obtained by creating a set of nested toroidal magnetic surfaces. The rupture of flux surfaces caused by asymmetries in the field generation or instabilities certainly has a detrimental effect on confinement.

2.2 Stability

An equilibrium is unstable if it is possible to find a small perturbation that growths when is applied. Otherwise, the equilibrium is stable. The instabilities observed in magnetically confined plasmas can be classified into two groups: the microinstabilities and the macroinstabilities. The first group is responsible for the turbulence at small scales and it is generally related to finite Larmor radius effects (the gyroradius of charged particles turning around the magnetic field) and asymmetries in the velocity distribution function of the different species that compound the plasma. On the other hand, the macroinstabilities involve fluctuations having a length scale comparable to that of the whole system and can, in their simplest version, be described by the MHD model presented in Section 3. Their appearance generally leads to the termination of the discharge and the destruction of the configuration. In this Chapter we will deal with this kind of instabilities.

The usual procedure to study the MHD stability of an equilibrium is based on the analysis of the energy increment δW introduced by a small perturbation to the equilibrium (Friedberg, 1987). Using the linearized equations of the MHD model it is possible to compute the growth rate of each perturbation. If all possibles modes decay then the equilibrium is MHD stable.

According to the source of energy that feeds the instability, the macroinstabilities can be divided in:

- External modes. In this case the energy of the instability comes from the interaction between the plasma and the boundary (the separatrix) or the external magnetic fields. Two typical examples appearing in spheromaks are the shift and the tilt instabilities. The first one consists in the displacement of the configuration as a whole while the second one involves the rigid rotation of the magnetic surfaces. The flux conserver (the chamber of conducting walls inside which the spheromak is formed) plays a crucial role in suppressing these instabilities. For instance, in a cylindrical flux conserver the tilt instability can be avoided if the elongation of the cylinder (ratio between height and radius) is lower than 1.6.
- Current driven modes. They are activated by non uniform current distributions. The most common example of this kind of instabilities is the kink mode, which may be either an internal (it does not affect the separatrix) or an external mode. In tokamaks this instability is closely related to a phenomenon called sawtooth oscillations that limits in practice the maximum value of toroidal current. In spheromaks and RFP's the kink mode triggers the relaxation process that forms and sustains the configuration.
- Pressure modes. Pressure gradients combined with an adverse magnetic field line curvature may act as a source of energy to develop instabilites (called ballooning or interchange modes).

In Section 5 we will consider internal kink modes in spheromak configurations. A comprehensive description of the MHD modes relevant to magnetic confinement configurations can be found elsewhere (Friedberg, 1987);(Wesson, 2004).

2.3 Formation and sustainment

Once an MHD equilibrium with good stability properties has been devised it is necessary to find appropriate methods to form and sustain the configuration. The formation methods depend on the configuration under consideration. In fact, a given configuration can be obtained using different formation schemes. In most cases, after the formation process the plasma has a temperature sensibly lower than that required for fusion. Moreover, the resistive dissipation which is ubiquitous on real plasmas causes the currents and the magnetic fields to decay, so the configuration would be lost in the resistive time scale. It is then imperative to apply adequate methods to drive currents and heat the plasma. Some common methods that have already been successfully implemented are:

- Current induction by a primary coil. A primary coil is located at the center of the torus and plays the role of the primary of an electric transformer while the plasma itself is the secondary (the primary coil was not sketched in Fig. 1). This is the usual approach to induce the toroidal current in tokamaks and RFP's. It does not allow to operate in truly steady state and it can not be used in compact tori.
- Radio frequency waves. Energy can be transferred to the plasma from an external source of electromagnetic waves (antenna). The electric field of the waves transfers momentum to the particles inducing currents and heating the plasma by collisions. Within a multi-species plasma there exists a number of resonant frequencies that enhance the coupling between the plasma and the antenna (ionic and electronic cyclotron resonances, hybrid resonances, etc.).
- Neutral beam injection. Neutral atoms injected are not deflected by the magnetic field and can penetrate the plasma until they become ionized through collisions. Once ionized these particles follow orbits determined by the magnetic field and their energy. This process heats the plasma and drives localized currents.
- Rotating magnetic fields. Plasma electrons may be dragged, and thus a current may be induced, by externally applied rotating magnetic fields.
- Helicity injection. When a current is established along the magnetic field some amount of magnetic helicity (see Sec. 4) is injected in the magnetic configuration. The driven current may destabilize the configuration triggering a relaxation process that redistributes the current. This is the main method used in spheromak sustainment and is the subject of study of this Chapter.

2.4 The spheromak configuration

Early experiments in toroidal pinch configurations exhibited, under certain conditions, the spontaneous reversal of the toroidal field near the wall of the chamber. This unexpected feature was succesfully explained in terms of the relaxation theory proposed by Taylor (1974). According to this theory, MHD fluctuations cause the plasma to minimize its magnetic energy while conserving the total magnetic helicity (see Sec. 4).

Some years later, it was realized that the minimum energy state, for a given amount of magnetic helicity, inside a sphere is a system of nested toroidal magnetic flux surfaces (Rosenbluth & Bussac, 1979). The idea of a configuration relevant for fusion research that would be self produced (or self-organized) inside a simply connected volume attracted the attention of the scientific community. Several experiments were designed in order to check this theoretical prediction. The success of these experiments was considered a proof of the remarkable robustness of the relaxation theory (Bellan, 2000).

Despite the initial enthusiasm, it was later realized that the relaxation process involves MHD fluctuations that strongly degrade the confinement. Because of these fluctuations the confinement peformance of the spheromak is much lower than that of the tokamak or the

RFP. Little is known about the dynamics of these fluctuations since the relaxation theory is only able to predict the final state of the plasma but it can not provide any detail on how this state is attained (Jarboe, 2005).

3. The MHD model

The MHD model describes the macroscopic behavior of a plasma in many situations of interest in a relatively simple manner. Its validity relies, however, in a number of assumptions that have to be borne in mind in order to understand what kind of phenomena can be explained by the model and what effects lie outside this description.

3.1 Basic assumptions of the MHD model

The MHD model regards the plasma as a quasi-neutral electrically conducting fluid. The first and most fundamental assumption of this description is to regard the ensemble of ions and electrons conforming the plasma as a single continuum medium. This is valid when the length scales associated with the magnetic field gradients is much larger than the internal length scales of the plasma (such as the ionic and electronic gyroradii). This condition holds in virtually every laboratory plasma dedicated to fusion research.

The second important assumption is to consider that the plasma is in thermodynamic equilibrium so the particles have a Maxwellian distribution of velocities. This is a good approximation as long as the shortest time scale of the process under consideration is much longer than the collision time and the shortest length scale of the system is larger than the mean free path of the particles. In other words, the plasma should be in a collisional regime (this condition is required to derive the fluid equations from the kinetic equations (Braginskii, 1965)). The collisionality hypothesis is usually not satisfied at the highest temperatures obtained in modern tokamak experiments. However, spheromak plasmas are much colder ($T \sim 10^2$ eV) so that this assumption is still reasonable. Moreover, there are several arguments supporting the validity of the MHD model even in collisionless systems (Friedberg, 1987); (Priest & Forbes, 2000).

Finally, in the context of the MHD model the plasma is assumed to be electrically neutral (or quasi-neutral since the charges are present but exactly balanced). This is approximately true when the length scales under consideration are larger than the Debye shielding of electrons.

3.2 MHD equations

Now we seek for the equations that describe the evolution of the two main quantities that govern the dynamics of such an MHD system: the velocity field and the magnetic field. The equation for the evolution of the plasma velocity $\mathbf{u}$, expresses the balance of linear momentum

$$\rho\left(\frac{\partial \mathbf{u}}{\partial t} + \mathbf{u}\cdot\nabla\mathbf{u}\right) = -\nabla p + \mathbf{J}\times\mathbf{B} + \mu\nabla\cdot\Pi \tag{7}$$

where ρ is the mass density and p is the thermodynamic pressure. The second term on the right hand side is the Lorentz force, where $\mathbf{J}$ is the current density and $\mathbf{B}$ is the magnetic field. We note that due to quasi-neutrality the current density is produced by the relative motion between ions and electrons. The last term in Eq. (7) is the viscous force where $\mu = \rho\nu$ is the dynamic viscosity, ν is the cinematic viscosity and the tensor Π is given by

$$\Pi = \nabla \mathbf{u} + \nabla \mathbf{u}^T - \frac{2}{3}(\nabla \cdot \mathbf{u})\mathbf{I}. \tag{8}$$

If the flow is incompressible ($\nabla \cdot \mathbf{u} = 0$) Π reduces to $\nabla \mathbf{u}$.

Let us mention some basic aspects of the Lorentz force term. Using the low-frequency Ampère law $\mathbf{J} = \nabla \times \mathbf{B}$ (we rescale $\mathbf{B}$ and $\mathbf{J}$ in such a way that $\mu_0 = 1$) and the vector identity $(\nabla \times \mathbf{B}) \times \mathbf{B} = (\mathbf{B} \cdot \nabla)\mathbf{B} - \nabla(B^2/2)$, we can decompose this term into two contributions

$$\mathbf{J} \times \mathbf{B} = (\mathbf{B} \cdot \nabla)\mathbf{B} - \nabla\left(\frac{B^2}{2}\right). \tag{9}$$

The first term on the right represents a magnetic tension force in the direction of $\mathbf{B}$ which has a restoring effect when the magnetic field lines are bent. The second term is regarded as a magnetic pressure that acts in all directions. Clearly, both forces must cancel out along the magnetic field lines since the term $\mathbf{J} \times \mathbf{B}$ can not accelerate the fluid in the direction of $\mathbf{B}$.

The equation for the magnetic field evolution comes from the Maxwell equations and a constitutive law that relates the electric field to the magnetic field and the current density (the Ohm's law). We begin with the Faraday's law in the low-frequency limit (i.e. neglecting the displacement current)

$$\nabla \times \mathbf{E} = -\frac{\partial \mathbf{B}}{\partial t}. \tag{10}$$

On the other hand, the Ohm's law relates the current density to the electric field in the frame of reference of the conducting medium $\mathbf{E}' = \eta \mathbf{J}'$, where η is the electric resistivity (the reciprocal of the conductivity) and the prime denotes that the quantities have to be measured in the plasma's reference frame. When this equation is expressed in the lab's frame (from which the plasma moves at velocity $\mathbf{u}$) it adopts the form

$$\mathbf{E} = -\mathbf{u} \times \mathbf{B} + \eta \mathbf{J} \tag{11}$$

where relativistic effects have been neglected ($u \ll c$, where c is the speed of light).

Combining Eqs. (10) and (11) together with the identity $\nabla \times \nabla \times \mathbf{B} = \nabla(\nabla \cdot \mathbf{B}) - \nabla^2 \mathbf{B}$ and the constraint $\nabla \cdot \mathbf{B} = 0$, we obtain the MHD induction equation

$$\frac{\partial \mathbf{B}}{\partial t} = \nabla \times (\mathbf{u} \times \mathbf{B}) + \eta \nabla^2 \mathbf{B} \tag{12}$$

where spatial uniformity of η was assumed. Although not considered in this work, we point out that, whenever present, resistivity gradients may give rise to the so-called current interchange effect which constitutes an effective mechanism of current exchange between flux tubes (Zheng & Furukawa, 2010). Note that the terms $\mathbf{J} \times \mathbf{B}$ and $\mathbf{u} \times \mathbf{B}$ introduce a strong non-linear coupling between Eqs. (7) and (12).

3.3 Diffusion of magnetic field lines and frozen-in-flux condition

The two terms in the right hand side of Eq. (12) account for two very different physical effects. The quotient between the magnitudes of these effects can be estimated as

$$\frac{|\nabla \times (\mathbf{u} \times \mathbf{B})|}{|\eta \nabla^2 \mathbf{B}|} \sim \frac{u_0/L}{\eta/L^2} = \frac{u_0 L}{\eta} \equiv R_m \tag{13}$$

where u_0 and L are typical velocity and length scales and R_m is the magnetic Reynolds number. Thus, when $R_m \sim 0$ the magnetic field simply diffuses and the configuration decays in the resistive time scale $\tau_r = L^2/\eta$.

The opposite limit ($R_m \gg 1$) is more representative of the actual situation in most laboratory (in the context of magnetic confinement) and space plasmas. In this limit (called ideal limit) the induction equation reduces to,

$$\frac{\partial B}{\partial t} = \nabla \times (\mathbf{u} \times \mathbf{B}) \qquad (R_m \gg 1). \tag{14}$$

This equation implies the conservation of the magnetic flux through any closed surface that moves with the local velocity of the fluid. If we regard the magnetic field lines as very thin flux tubes and we imagine closed curves surrounding them that move with the fluid, we realize that the plasma drags the field lines as it moves. It is said that the field lines are frozen in the plasma (*frozen-in-flux condition*). Since each field line is simply convected by the flow (assumed to be smooth and continuous) its connectivity is preserved. This means that in the ideal MHD approximation the changes in the topology of the magnetic field are not possible. This idea, which is intimately related to the Kelvin's circulation theorem for inviscid flows, was first introduced by Hannes Alfvén in 1943. More details on the frozen in flux condition may be found elsewhere (Biskamp, 2000);(Priest & Forbes, 2000).

3.4 Closing the system of equations

The system formed by Eqs. (7) and (12) and the constraint $\nabla \cdot \mathbf{B} = 0$, has too many unknowns and can not be solved. Even if the current density can easily be expressed in terms of the magnetic field using Ampère's law, we still need to introduce some information concerning the density and the pressure. We describe four common approaches to accomplish this.

Firstly, the zero-β approximation gives the simplest option. The nondimensional parameter $\beta = 2p_0/B_0^2$ measures the ratio between the thermodynamic pressure and the magnetic pressure. A very low β value (which is the case in most confinement experiments) means that the dynamics of the plasma is mainly dictated by the magnetic field (via the Lorentz force) while the pressure gradient has little influence. Thus, we may simply remove the term ∇p from Eq. (7) and assume that $\rho = \rho_0$ is a constant.

The second option is to consider an incompressible flow. In this case ρ is still a constant but the pressure gradient is no dropped. In this case the required information is completed by the incompressible condition $\nabla \cdot \mathbf{u} = 0$. The pressure can not be directly computed with this equation but it can be indirectly inferred. It plays the role of a Lagrange multiplier. Although this is a less crude and more consistent option, it is not generally a good approximation for low β plasmas.

Thirdly, we could allow compressible flows by using the continuity equation

$$\frac{\partial \rho}{\partial t} + \nabla \cdot (\rho \mathbf{u}) = 0 \tag{15}$$

and assuming some simple relationship between the pressure and the density. For instance, $p = c_s^2 \rho$ where c_s is the speed of sound (isothermal approximation) or $p = c_{ad}\rho^\gamma$ where c_{ad} is a constant and γ is the polytropic index (polytropic approximation).

Finally, a more elaborated option can be obtained if we incorporate, besides Eq. (15), an equation for the energy balance

$$\frac{\partial w}{\partial t} = -\nabla \cdot \mathbf{q} - Q_{c,r} \tag{16}$$

where w is the total energy density defined as

$$w = \rho\left(\frac{u^2}{2} + e\right) + \frac{B^2}{2} \tag{17}$$

e is the internal energy per unit of mass, the term $Q_{c,r}$ accounts for conductive and radiative losses and $\mathbf{q}$ is the energy flux given by

$$\mathbf{q} = \left[\rho\left(e + \frac{u^2}{2}\right) + p\right]\mathbf{u} + \mathbf{E} \times \mathbf{B} - \mu(\mathbf{u} \cdot \Pi). \tag{18}$$

The three terms on the right denote respectively the energy flux due to convection, the electromagnetic energy flux (Poynting's vector) and the viscous dissipation of energy. In this context we also need an equation of state of the form $p = p(\rho, e)$. The most common choice is the ideal gas law $p = (\gamma - 1)\rho e$ where γ is the ratio of specific heats.

3.5 Scales and dimensionless numbers

The results presented in Sec. 5 are obtained by numerically solving the MHD equations. It is a common practice to use a nondimensionalized version of the considered equations. The removal of the units is achieved by the choice of suitable scales that can be condensed in few nondimensional numbers. We will see which are the chosen scales and the relevant nondimensional quantities in this study.

Since spheromaks are very low-β plasmas, the zero-β approximation is used to close the system. The resulting equations can be nondimensionalized with a length scale a (the radius of the cylinder inside which we will solve the equations) and a velocity scale $c_A = B/\sqrt{\rho}$, which is known as the Alfvén velocity. Perpendicular perturbations travel along the magnetic field lines at this velocity. The time will be normalized by the Alfvén time $\tau_A = a/c_A$, which represents the typical time scale of the MHD fluctuations.

Using these scales we obtain two nondimensional numbers: the Lundquist number $S = ac_A/\eta$ and ν/ac_A which is usually expressed in terms of the magnetic Prandtl $P_m = \nu/\eta$. The Lundquist number can be rewritten as

$$S = \frac{a^2/\eta}{a/c_A} = \frac{\tau_r}{\tau_A} \tag{19}$$

where τ_r is the time scale of resistive dissipation. In the simulations presented in Sec. 5 we have used $P_m = 1$ and $S = 2 \times 10^4$.

4. Plasma relaxation

It is common to observe magnetized fluids and plasmas as well as other continuum media to exist naturally in states with some kind of large scale order. These states are to some extend independent of the initial conditions, that is to say, they are preferred configurations toward which the system evolves if the correct boundary conditions are imposed. Moreover, if the system is perturbed it tends to return to the same preferred state recovering the large scale order. The large scale order of some quantity always comes together with the disorder at small scales of another quantity. These preferred configurations are called *self-organized* sates and the dynamical process of achieving these states is called *self-organization* (Hasegawa, 1985). Plasma relaxation is an example of self-organization.

Self-organized (or relaxed) states can not be deduced from force balance or stability considerations alone. The theory of magnetic relaxation always relies on some variational principle, that is to say, the minimization of some quantity subjected to one or more constraints. Possibly the simplest and surely the most widespread option adopted to describe plasma relaxation was introduced by Taylor (1974). While a rather obvious choice was made for the quantity to minimize (the magnetic energy) a very clever option was made for the constraint. Among all the ideal MHD invariants Taylor chose the total magnetic helicity. The total magnetic helicity quantifies several topological properties of the system and even when magnetic reconnection can change the topology of the magnetic field lines, the total helicity of the system is still a robust invariant. These ideas are further developed below.

4.1 Magnetic helicity

The total magnetic helicity H of the magnetic field $\mathbf{B}$ within the volume V is

$$H = \int_V \mathbf{A} \cdot \mathbf{B} \, dV \tag{20}$$

where $\mathbf{A}$ is the potential vector ($\mathbf{B} = \nabla \times \mathbf{A}$). A relevant question may be posed at this time regarding how this quantity is modified by a change in the gauge of $\mathbf{A}$. It is clear that in order to have a meaningful definition, Eq. (20) should be gauge invariant. The helicity change ΔH introduced when $\mathbf{A}$ is replaced by $\mathbf{A} + \nabla\chi$ is

$$\Delta H = \int_V \nabla\chi \cdot \mathbf{B} \, dV = \int_V \nabla \cdot (\chi \mathbf{B}) \, dV \tag{21}$$

where we have used the fact that $\nabla \cdot \mathbf{B} = 0$. Applying the divergence theorem in a simply connected volume V this becomes

$$\Delta H = \int_S \chi \mathbf{B} \cdot \mathbf{ds} \tag{22}$$

where S is the surface that encloses V and $\mathbf{ds}$ is the outward-pointing normal surface element. Therefore, the definition (20) is gauge invariant only if the normal component of the magnetic field vanishes at the boundary of V, which was assumed to be simply connected. We will respect these two conditions throughout this work. When the normal component of the magnetic field does not vanish at the boundary or the volume V is not simply connected a generalized definition, the so-called *relative helicity*, must be employed (Finn & Antonsen, 1985).

To see how H can measure topological properties of the system we will consider the concept of flux tube. The magnetic flux through an open and orientable surface S is

$$\Phi = \int_S \mathbf{B} \cdot \mathbf{ds} = \oint_C \mathbf{A} \cdot \mathbf{dl} \tag{23}$$

where C is the path along the perimeter of S in the counterclockwise direction. Note that the Eq. (23) holds even if the gauge of $\mathbf{A}$ is changed.

We present an example given by Moffat (1978). Consider two linked flux tubes like those shown in Fig. 2 (a). We assume that there are no other contributions to the magnetic field. In

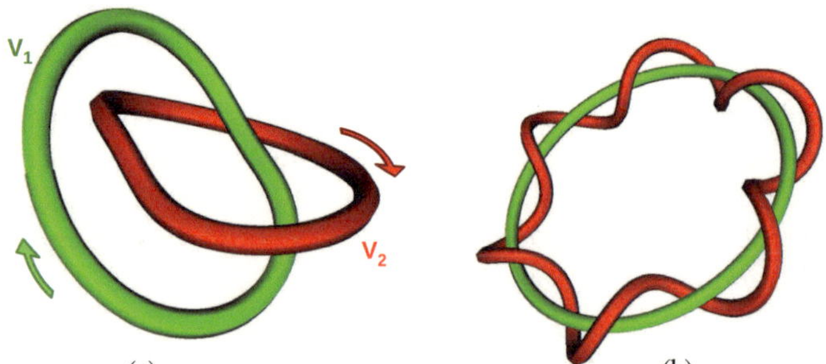

Fig. 2. Linked flux tubes.

this simplified case the total helicity can be computed as $H = H_1 + H_2$, with $H_i = \int_{V_i} \mathbf{A} \cdot \mathbf{B}\, dV$, for $i = 1, 2$. To compute H_i we note that $dV = \mathbf{dl} \cdot \mathbf{ds}$, where $\mathbf{dl}$ is the element of length along the tube and $\mathbf{ds}$ its cross section. By construction, $\mathbf{dl}$ and $\mathbf{ds}$ are parallel to $\mathbf{B}$, so we can rearrange the integrand as $\mathbf{A} \cdot \mathbf{B}\, dV = \mathbf{A} \cdot \mathbf{B}\, \mathbf{dl} \cdot \mathbf{ds} = (\mathbf{A} \cdot \mathbf{dl})(\mathbf{B} \cdot \mathbf{ds})$, and thus

$$H_i = \oint_{C_i} \int_{S_i} (\mathbf{A} \cdot \mathbf{dl})(\mathbf{B} \cdot \mathbf{ds}). \tag{24}$$

Since the magnetic flux is constant along each curve C_i the last equation can be written as

$$H_i = \Phi_i \oint_{C_i} \mathbf{A} \cdot \mathbf{dl}. \tag{25}$$

On the other hand, the contour C_1 encloses the magnetic flux Φ_2 and vice versa, so from Eq. (23) it is clear that

$$\oint_{C_1} \mathbf{A} \cdot \mathbf{dl} = \Phi_2 \quad \text{and} \quad \oint_{C_2} \mathbf{A} \cdot \mathbf{dl} = \Phi_1 \tag{26}$$

and thus $H_1 = H_2 = \Phi_1 \Phi_2$ which finally gives

$$H_{\text{link}} = 2\Phi_1 \Phi_2 \tag{27}$$

for the helicity due to the linking of the tubes. In general, if each tube winds around the other N times (in Fig. 2 (b) we have $N = 6$) one obtains $H_1 = H_2 = N\Phi_1\Phi_2$ (Moffat, 1978). N is the *linking number* of the tubes.

The helicity can also measure the self-twisting of a single tube. So far we did not pay attention to the tube's cross section shape. In order to compute the helicity of a twisted flux tube it is convenient to consider structures having elongated cross sections, like the ribbon shown in Fig. 3 (a). This ribbon is untwisted and has no helicity. If we cut this ribbon, we rotate one end

(a) (b) (c)

Fig. 3. (a) An untwisted ribbon-like flux tube, (b) a twisted flux tube and (c) the same twisted tube marked with different colors.

by 2π and we join both ends together again we obtain the twisted ribbon shown in Fig. 3 (b). The helicity of this structure may be computed using Eq. (27) obtained for two linked tubes applying the following reasoning. Regard the twisted tube as two adjacent tubes carrying one half of the total magnetic flux (see Fig. 3 (c)). The helicity of this system has a contribution coming from the linking of the two tubes H^1_{link} and also a contribution coming from the self twisting of the smaller tubes. If Φ is the total magnetic flux of the original twisted tube, the helicity due to the linking of each half is

$$H^1_{\mathrm{link}} = 2\left(\frac{\Phi}{2}\right)^2. \tag{28}$$

Note that this mental process to convert helicity due to twisting into helicity due to linking can be recursively applied to obtain

$$H^n_{\mathrm{link}} = 2^n\left(\frac{\Phi}{2^n}\right)^2 = \frac{\Phi^2}{2^n} \tag{29}$$

for each contribution due to linking. Finally, the helicity of the twisted tube is obtained by adding all these contributions

$$H = \lim_{N\to\infty}\sum_{n=1}^{N} H^n_{\mathrm{link}} = \Phi^2\sum_{n=1}^{\infty}\frac{1}{2^n} = \Phi^2. \tag{30}$$

4.2 Localized magnetic reconnection

Magnetic reconnection is ubiquitous in almost all space and laboratory plasmas. It consists in a rearrangement of the topology of the magnetic field due to a change in the connectivity of the magnetic field lines. This process plays an important role in several confinement devices (such us the tokamak, the RFP and the spheromak) as well as in several astrophysical phenomena (Earth magnetosphere, solar corona, solar wind, accretion disks, etc.). Since the majority of those plasmas have a very high magnetic Reynolds number (and a high Lundquist number as well) the ideal MHD model should provide an adequate level of description for the physical system. However, as already mentioned in Sec. 3.3, topological changes in the magnetic field are not allowed in the ideal limit. What actually happens is that the coupled non linear evolution of the magnetic field and the flow inevitably develops *current sheets*, i.e. localized regions where the magnetic field gradients become very large. Within these highly localized regions the ideal approximation breaks down and the last term of Eq. (12) becomes relevant causing the magnetic field to diffuse and change the connectivity of the field lines.

The fundamental *ansatz* of the plasma relaxation theory is that these localized reconnection events do not change the total helicity of the system. Even when dissipation is involved in this process, it is assumed that only magnetic energy is affected. How can magnetic helicity be conserved during a localized reconnection event is schematically shown in Fig. 4. Two

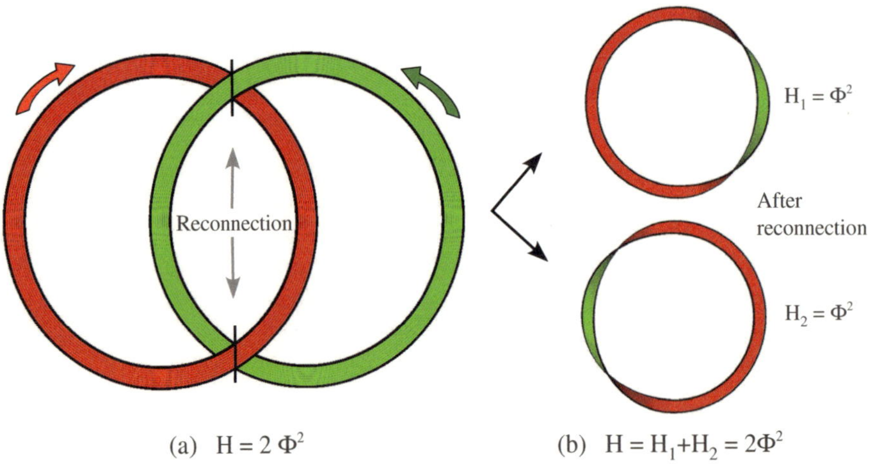

Fig. 4. (a) Two linked ribbon-like flux tubes undergo a localized reconnection process that give rise to two separate but twisted tubes (b). The global helicity of the system is conserved.

untwisted flux tubes that are initially linked can be locally reconnected giving rise to a pair of separate but twisted tubes, in such a way that the total helicity of the system is conserved.

Moreover, plasma relaxation is based on the fact that localized reconnection events allow topological changes and dissipate magnetic energy much faster than the total helicity. There exists a number of arguments to justify this behavior (see Sec. 9.1.1 of Priest & Forbes (2000) or Montgomery et al. (1978)). Let's consider, for instance, how these two quantities (magnetic energy and helicity) decay in the presence of a small uniform resistivity η. Magnetic energy,

$W = \int_V B^2/2\, dV$, decays at the rate

$$\frac{dW}{dt} = -\eta \int_V J^2\, dV. \tag{31}$$

This expression can be obtained by scalar multiplying Eq. (10) by $\mathbf{B}$ and integrating over a fixed volume at whose boundary the normal component of the Poynting vector vanishes. For the magnetic helicity we take the time derivative of (20) and obtain

$$\frac{dH}{dt} = \int_V \left(\frac{\partial \mathbf{A}}{\partial t} \cdot \mathbf{B} + \mathbf{A} \cdot \frac{\partial \mathbf{B}}{\partial t} \right) dV = \int_V \left(\frac{\partial \mathbf{A}}{\partial t} \cdot \mathbf{B} + \mathbf{A} \cdot \nabla \times \frac{\partial \mathbf{A}}{\partial t} \right) dV. \tag{32}$$

Using vector identities and the divergence theorem, the last expression can be rewritten as

$$\frac{dH}{dt} = 2 \int_V \frac{\partial \mathbf{A}}{\partial t} \cdot \mathbf{B}\, dV - \int_S \mathbf{A} \times \frac{\partial \mathbf{A}}{\partial t} \cdot \mathbf{ds} \tag{33}$$

where S is the surface that encloses V. If field lines do not penetrate the volume V the surface integral in Eq. (33) vanishes. In the absence of charge separation $\mathbf{E} = -\partial \mathbf{A}/\partial t$, and using Ohm's law $\mathbf{E} = \eta \mathbf{J}$, we finally obtain

$$\frac{dH}{dt} = -2\eta \int_V \mathbf{J} \cdot \mathbf{B}\, dV. \tag{34}$$

It is clear that in the absence of resistivity (the ideal limit) W and H are conserved and, for this reason, it is said that they are ideal invariants. By contrast, in a real plasma both energy and helicity will decay at a rate proportional to η. However, when turbulence is present, magnetic fluctuations produce many thin current sheets with thicknesses of order $\eta^{1/2}$ and current densities proportional to $B\eta^{-1/2}$. In such case, the energy decay rate becomes

$$\frac{dW}{dt} \propto \int_V B^2\, dV$$

which is independent of η. On the other hand, the total helicity decays as

$$\frac{dH}{dt} \propto 2\eta^{1/2} \int_V B^2 dV$$

so that as η tends to zero the helicity dissipation becomes negligible. Therefore, it is important to keep in mind that a plasma will relax (in the sense described here) only if there is a certain level of small scale fluctuations that gives rise to many localized current sheets. For this reason, relaxation theory does not apply to devices with a very low level of magnetic fluctuations, such as the tokamak.

Localized magnetic reconnection events may redistribute currents in the plasma by helicity transfer between flux tubes. Even when this idea must be applied with care because the helicity is by definition a global quantity, it is clear that the helicity of a single flux tube may change after reconnection with another flux tube. This helicity transfer process is certainly at work during toroidal current drive in spheromaks and other devices sustained by helicity injection.

4.3 Relaxed states

The magnetic relaxation theory is developed for systems in which the magnetic forces are dominant, i.e. whenever the parameter β is low. In such cases, the MHD equilibrium Eq. (1) reduces to the force-free condition

$$\nabla \times \mathbf{B} = \lambda(\mathbf{r})\, \mathbf{B} \tag{35}$$

where $\lambda(\mathbf{r})$ is some scalar function. As discussed in the previous Section, magnetic fluctuations induce localized reconnection events that relax the plasma toward the state of minimum magnetic energy maintaining the total helicity of the system (Taylor, 1974). Woltjer (1958) has shown that force-free fields with λ equal to a constant represent the state of lowest magnetic energy under the constraint of magnetic helicity conservation in a closed system (i.e. with no field lines intercepting the boundary). The proof uses the method of Lagrange multipliers. At a constrained minimum, the variation of magnetic energy is equal to a constant (the Lagrange multiplier) times the variation of helicity

$$\delta W = \frac{\lambda}{2} \delta H \tag{36}$$

where $\lambda/2$ is the Lagrange multiplier. Substituting $W = \int_V B^2/2\, dV$ for the magnetic energy and Eq. (20) for H yields

$$\int_V \left[2\, \mathbf{B} \cdot \delta\mathbf{B} - \lambda(\delta\mathbf{A} \cdot \mathbf{B} + \mathbf{A} \cdot \delta\mathbf{B})\right] dV = 0. \tag{37}$$

Using the identities

$$\mathbf{B} \cdot \delta\mathbf{B} = \delta\mathbf{A} \cdot \nabla \times \mathbf{B} - \nabla \cdot (\mathbf{B} \times \delta\mathbf{A})$$

and

$$\mathbf{A} \cdot \delta\mathbf{B} = \mathbf{B} \cdot \delta\mathbf{A} + \nabla \cdot (\delta\mathbf{A} \times \mathbf{A})$$

and the divergence theorem in Eq. (37) one obtains

$$\int_V 2\, (\nabla \times \mathbf{B} - \lambda\mathbf{B}) \cdot \delta\mathbf{A}\, d^3r = 0 \tag{38}$$

where we omitted the surface integrals because they vanish in the absence of field lines penetrating the volume under consideration. Since $\delta\mathbf{A}$ is arbitrary, the parenthesis of the integrand of Eq. (38) must be identically zero, which finally gives us the linear force-free condition

$$\nabla \times \mathbf{B} = \lambda\mathbf{B} \tag{39}$$

where λ is a constant. When we impose $\mathbf{B} \cdot \mathbf{ds} = 0$ at the boundary, we obtain an eigenvalue problem that has non trivial solution only for certain discrete values λ_n (which are real and positive).

Since $\nabla \times \mathbf{B} = \lambda_n\mathbf{B}$, we can write the magnetic field as $\mathbf{B} = \lambda_n\mathbf{A} + \nabla f$, where f is an arbitrary potential. Thus, we can compute the magnetic energy as

$$W = \frac{1}{2} \int_V \mathbf{B} \cdot (\lambda_n\mathbf{A} + \nabla f) dV = \frac{\lambda_n}{2} \int_V \mathbf{B} \cdot \mathbf{A}\, dV = \frac{\lambda_n}{2} H \tag{40}$$

since $\int_V \mathbf{B} \cdot \nabla f dV = \int_V \nabla \cdot (f\mathbf{B}) dV = \int_{\partial V} (f\mathbf{B}) \cdot \mathbf{ds} = 0$. Eq. (40) gives us an important meaning for the eigenvalue: λ_n is proportional to the quotient W/H. For this reason it is clear that for a given amount of helicity, the minimum energy state will be given by the lowest allowed value of λ_n (i.e. λ_1).

The most frequent model employed to describe a spheromak configuration is the relaxed state inside a cylindrical flux conserver. Using cylindrical coordinates, the condition $\mathbf{B} \cdot \mathbf{ds} = 0$ means $B_z = 0$ at $z = 0$ and $z = h$ and $B_r = 0$ at $r = a$, where h and a are the height and the radius of the cylinder. In this case the solution to Eq. (39) can be found analytically (Bellan, 2000). In terms of Bessel functions and trigonometric functions the solution is

$$B_r = B_0 \frac{\pi}{\gamma_1 h} J_1(\gamma_1 r) \cos(k_1(z-h)) \tag{41}$$

$$B_\theta = -B_0 \frac{\lambda_1}{\gamma_1} J_1(\gamma_1 r) \sin(k_1(z-h)) \tag{42}$$

$$B_z = -B_0 J_0(\gamma_1 r) \sin(k_1(z-h)) \tag{43}$$

where $\gamma_1 = x_{11}/a$, $k_1 = \pi/h$ and x_{11} is the first zero of J_1. Note that, since this is an eigenfunction (of the curl operator) it is defined up to a constant B_0. Note also that this solution has no toroidal dependence (i.e. it is axisymmetric). The corresponding eigenvalue which depends on the geometry of the flux conserver is

$$\lambda_1 = \sqrt{\frac{x_{11}^2}{a^2} + \frac{\pi^2}{h^2}}. \tag{44}$$

In Fig. 5 we show the magnetic field lines obtained after following the trajectories given by Eqs. (41) - (43) from four different positions.

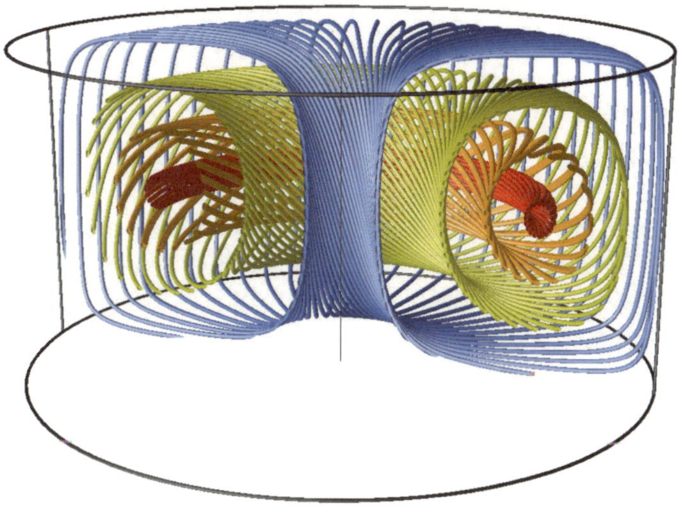

Fig. 5. Four magnetic field lines showing four nested magnetic flux surfaces. This fully relaxed state has the same value of λ (equal to λ_1) on each surface.

5. Dynamics of magnetic relaxation in spheromak configurations

The relaxation theory as formulated by Taylor (1974) is a variational principle that can not give details on the dynamical aspects of the process. All the considerations we have made regarding the important role of localized reconnection in helicity conservation are only heuristic arguments that try to explain the remarkable success of the theory at predicting the self-organized final state of the plasma.

There are a number of reasons that motivate the study of the dynamics of relaxation. For instance, in the context of spheromak research it is observed that during sustainment the system does not remain at the lowest energy state. Small deviations from the relaxed state as well as the ubiquitous presence of fluctuations are crucial issues that are out of the scope of relaxation theory (Knox et al., 1986);(Willet et al., 1999). In this work we study these aspects using numerical solutions of the non linear resistive MHD equations described in Sec. 3 as an initial and boundary value problem in three spatial dimensions. The nondimensional version described in Sec. 3.5 of these equations is used. The details of the numerical method are not presented here but can be found elsewhere (Garcia-Martinez & Farengo, 2009b).

In this Section we present a study of the dynamics of the kink mode in spheromak configurations. We will focus on the dynamics of systems that are only marginally unstable. Even when this may sound as a rather specific topic we will see that this is a simple setup in which we can study magnetic reconnection and helicity transfer between flux tubes. Firstly, we describe the kink unstable configurations used as initial condition and explain how they can be computed. Secondly, we study the dynamics of the kink instability in several cases and discuss in which cases it leads to a complete relaxation process (as described in the preceding Section) and in which cases the relaxation process is only partial. Thirdly, we introduce the concept of safety factor and resonant surfaces and explain their relevance to the partial relaxation behavior observed in marginally unstable configurations. Then, we analyze in detail the reconnection process that is driven by the dominant kink mode. Finally, we discuss simple models to describe this reconnection process.

5.1 Problem description

The minimum energy state for a given helicity inside a (not very elongated) cylindrical flux conserver, see Fig. 5, is stable against small MHD perturbations. There exists, however, a simple modification of this configuration which is MHD unstable, in particular *kink* unstable. Now we derive the equations that will allow us to compute as well as to better understand these modified configurations.

For simplicity we consider force-free configurations. In general this condition may be expressed as $\mathbf{J} = \lambda(\mathbf{r})\mathbf{B}$, where λ may be an arbitrary function. However, we will restrict our study to the case in which λ is a flux function, that is to say it takes the same value on each flux surface and can only change from one surface to another. This condition is expressed as

$$\nabla \times \mathbf{B} = \lambda(\psi)\mathbf{B}. \tag{45}$$

Since we consider axisymmetric configurations, we can express the poloidal magnetic field component ($\mathbf{B}_p$) in terms of ψ using Eq. (4), while for the toroidal component we have

$$J_z = (\nabla \times \mathbf{B})_z = \frac{1}{r}\frac{\partial}{\partial r}(rB_\theta) = \lambda B_z. \tag{46}$$

Using this, along with Eq. (3) we obtain

$$B_\theta = \frac{1}{2\pi r}\int_0^r \lambda B_z\, 2\pi\tilde{r}d\tilde{r} = \frac{1}{2\pi r}\int_0^\psi \lambda(\tilde{\psi})d\tilde{\psi}. \tag{47}$$

Thus, expressing the magnetic field in terms of ψ we can rewrite the toroidal component of Eq. (45) as

$$\frac{\partial^2\psi}{\partial r^2} - \frac{1}{r}\frac{\partial\psi}{\partial r} + \frac{\partial^2\psi}{\partial z^2} + \lambda(\psi)\int_0^\psi \lambda(\tilde{\psi})d\tilde{\psi} = 0 \tag{48}$$

which is the force-free version of the Grad-Shafranov equation. We are interested in solving Eq. (48) in the rectangle $\Omega : (r,z) = [0,a] \times [0,h]$, i.e. a cylinder of radius a and height h.

The most simple option for $\lambda(\psi)$ would be $\lambda = 0$, which corresponds to the vacuum solution (currentless magnetic field). The solution vanishes in this case if homogeneous boundary conditions ($\psi|_{\partial\Omega} = 0$) are applied.

A more interesting case is obtained by setting $\lambda = \lambda_n$ (constant) which gives

$$-\Delta^*\psi = \lambda_n^2\psi \tag{49}$$

where we have introduced the Grad-Shafranov operator defined as $\Delta^* = \partial^2/\partial r^2 - (1/r)\partial/\partial r + \partial^2/\partial z^2$. If we impose homogeneous boundary conditions we obtain an eigenvalue problem which has non trivial solutions only for a discrete set of real and positive values of λ_n. The lowest value (λ_1) is given by Eq. (44) and its associated eigenfunction is the minimum energy state described in detail in Sec. 4.3. Thus, if the appropriate boundary conditions are imposed, we can also regard the spheromak as the lowest eigenfunction of the Grad-Shafranov operator.

In this study we will consider initial equilibria having

$$\lambda(\psi) = \bar{\lambda}\left[1 + \alpha\left(2\frac{\psi}{\psi_{ma}} - 1\right)\right] \tag{50}$$

which is a linear $\lambda(\psi)$ profile with slope α and mean value $\bar{\lambda}$. When this linear profile is injected in Eq. (48) a generalized non-linear eigenvalue problem is obtained. Some mathematical considerations as well as a basic numerical scheme to solve this problem were given by Kitson & Browning (1990). Note that even if one is able to solve the non-linear Grad-Shafranov equation, the profile given by Eq. (50) includes ψ_{ma} which is not know *a priori*. The procedure adopted here is to set $\psi_{ma} = 1$, fix the desired value of α and iterate over $\bar{\lambda}$ until ψ is equal to one at the magnetic axis. With this procedure we obtain the values of $\bar{\lambda}$ listed in Table 1. Note that each α value uniquely defines a configuration.

In Fig. 6 (a) we show two linear $\lambda(\psi)$ profiles and (b) ψ contours and the λ colormap for the $\alpha = -0.4$ case. The reason why we have chosen negative values for α is the following. It is evident that for negative values of the slope the configuration will have larger λ values in the outer flux surfaces (at lower ψ values) and *vice versa*. Since λ is proportional to the current

α	0	-0.3	-0.4	-0.5	-0.6	-0.7	-0.8
$\bar{\lambda}$	4.95	5.08	5.18	5.32	5.51	5.78	6.23

Table 1. $\bar{\lambda}$ values for some prescribed α values.

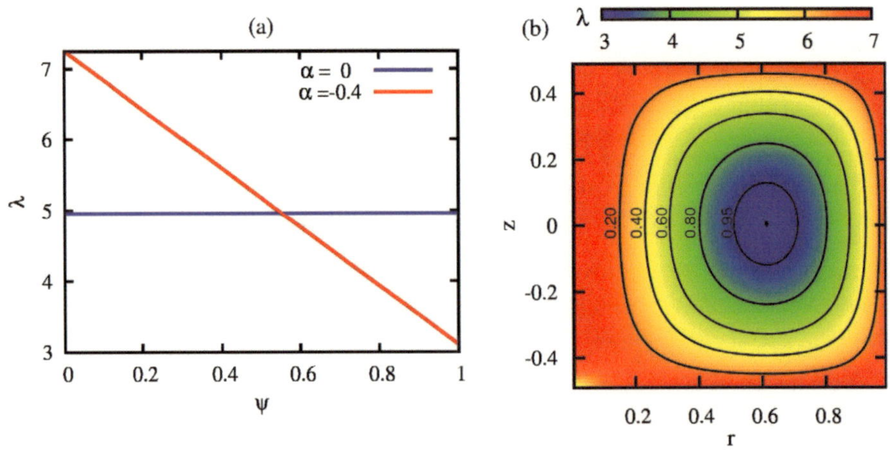

Fig. 6. (a) Two $\lambda(\psi)$ profiles. (b) ψ contours and λ colormap for the case with $\alpha = -0.4$. A hollow current profile is obtained for negative α values.

density it is said that the configuration has a *hollow* current profile. This is actually the case for spheromak configurations during sustainment.

Real spheromaks have some amount of open magnetic surfaces (i. e. there is some magnetic flux crossing the walls) along which current is driven. This injects magnetic helicity. Then the system relies on magnetic relaxation to drive the current in the inner flux surfaces. In order to sustain this current drive process in (quasi) steady state, some current (or λ) gradient is required. In fact, experiments show that sustained spheromaks are better approximated by a force-free state with $\alpha = -0.3$ rather than by the lowest energy state (having $\alpha = 0$) (Knox et al., 1986); (Willet et al., 1999).

5.2 Complete relaxation vs partial relaxation

Up to this point we know that the minimum energy state is MHD stable and that we can modify the configuration by giving the $\lambda(\psi)$ profile a non zero slope. Now we consider the stability of configurations having negative α values. A linear MHD stability analysis has determined that there exists a threshold value for the slope at which the system becomes unstable (Knox et al., 1986). Configurations with $\lambda(\psi)$ profiles that are steeper than the threshold (lower α values) are unstable while configurations with less steep profiles are stable. The value of this threshold (which lies between -0.3 and -0.4 for the geometry used here) was also verified using non-linear simulations of spheromak configurations (Garcia-Martinez & Farengo, 2009b).

The instability that arises has dominant toroidal number $n = 1$ (where n stands for the number of the coefficient of the Fourier decomposition in the toroidal direction). This current driven $n = 1$ mode is the kink mode. It is well known that the kink mode triggers the relaxation process in spheromaks during sustainment. It has been shown that when the initial unstable configuration has an α value close to the stability threshold, the relaxation process is not complete (Garcia-Martinez & Farengo, 2009a);(Garcia-Martinez & Farengo, 2009b). This means that the final state of the evolution is not a minimum energy state. In particular, the λ profile is not uniform. This partial relaxation behavior can be observed in Fig. 7. In the

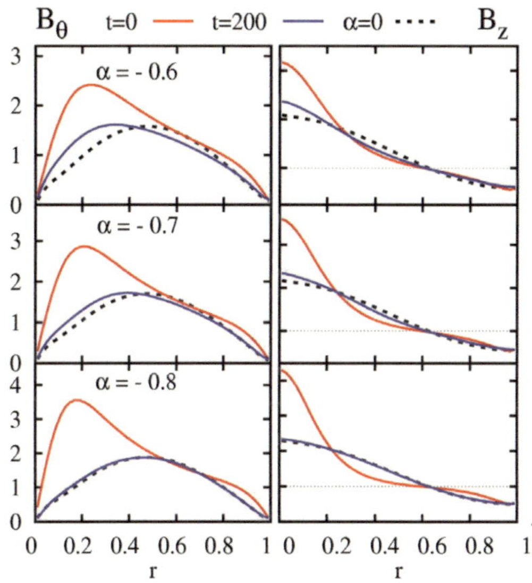

(a)

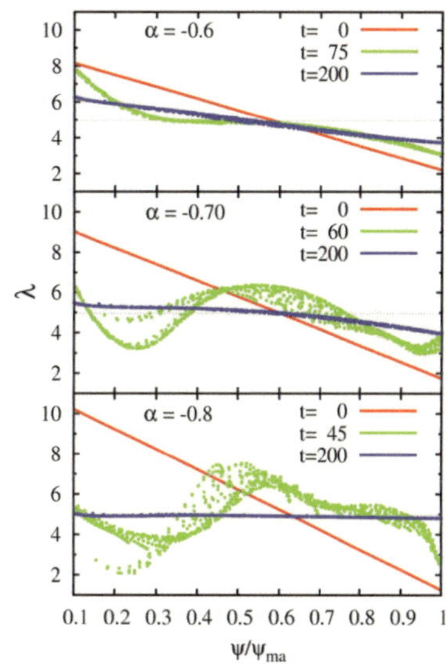

(b)

Fig. 7. (a) Toroidal and poloidal magnetic field profiles at $t = 0$ and $t = 200$ (final time). The dashed line shows the fully relaxed profiles. (b) $\lambda(\psi)$ profiles at three times for the same α values (Garcia-Martinez & Farengo, 2009b).

$\alpha = -0.6$ case it is clear that the final state does not have neither the same radial magnetic field profiles than the minimum energy state (shown in dashed lines) nor a uniform λ profile. On the other hand, the most unstable case, $\alpha = -0.8$, exhibits a fully relaxed final state.

Fig. 8 shows the evolution of the magnetic field lines during the kink instability. A magnetic island is formed due to the helical distortion of the magnetic axis. This island then moves toward the central position while the flux surfaces originally placed around the magnetic axis are gradually pushed outward. A localized magnetic reconnection layer can be observed in the region where the inner flux surfaces come into contact with the outer flux surfaces. This is indicated in the small box drawn in Fig. 8 (a). After a reconnection process, a system with axisymmetric nested flux surfaces is recovered (see Fig. 8 (e)).

The Poincaré maps showing the evolution of the $\alpha = -0.6$ case can be observed in Fig. 9. The overall behavior is analogous to the previously studied case. A magnetic island is formed at an outer position (relative to the magnetic axis) which then moves and occupies

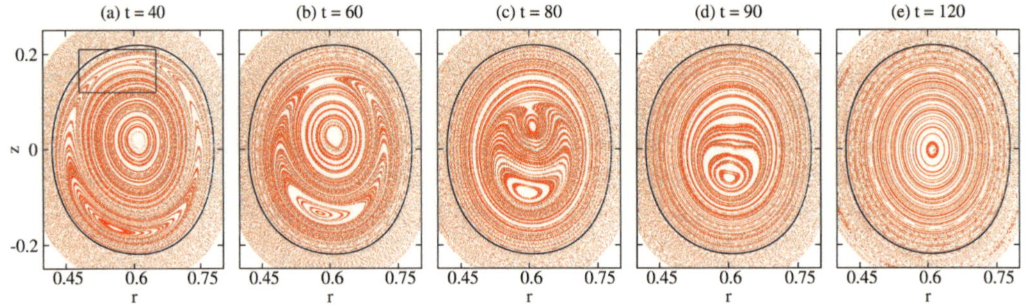

Fig. 8. Poincaré maps at several times showing the evolution of kink instability for the $\alpha = -0.4$ case. The black contour shows the initial position of the $q = 1$ surface.

the magnetic axis position. However, in this case a large region of stochastic magnetic field lines emerges between the two magnetic o-points and we are no longer able to identify a well defined localized reconnection layer.

The situation is even more drastic in the case with $\alpha = -0.7$ shown in Fig. 10. Most of the initially regular surfaces are quickly destroyed and large regions of stochastic field lines are observed. Though, a small coherent structure can still be devised even at times of strong activity (the saturation of the instability takes place at $t = 100$). After the instability saturation the toroidal modes decay and new regular nested flux surfaces are formed ($t = 200$).

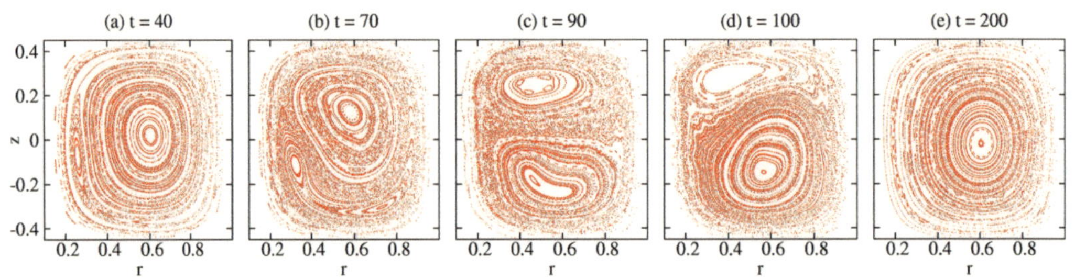

Fig. 9. Poincaré maps showing the evolution of the kink instability in the $\alpha = -0.6$ case.

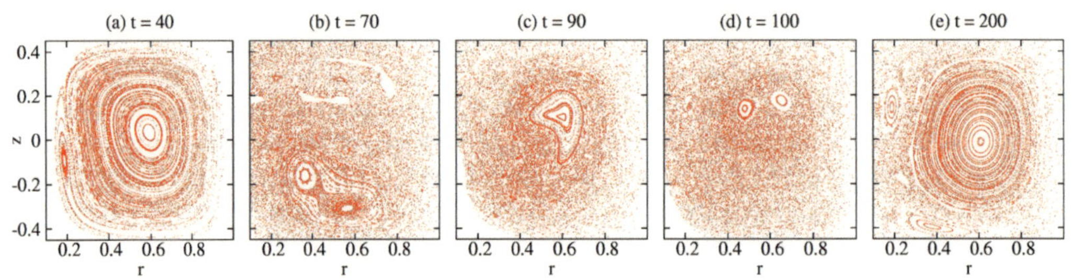

Fig. 10. Poincaré maps showing the evolution of the kink instability in the $\alpha = -0.7$ case.

In contrast with the marginally unstable case analyzed previously ($\alpha = -0.4$) where the activity was milder, here the larger level of fluctuations causes the field lines to wander through the whole domain. This facilitates the helicity transfer and enable a more effective flattening of the $\lambda(\psi)$ profile (as shown in Fig. 7).

It is important to keep in mind that these stochastic regions can be produced even by relatively low wave number magnetic fluctuations. In fact, few toroidal Fourier modes with a rather gentle dependence along the poloidal plane are enough to produce the disorder observed in Fig. 10 (d).

These observations are in agreement with the discussion presented in Sec. 4.2. As remarked there, a significant amount of small scale MHD activity (fluctuations) leading to the formation of numerous small current sheets is required to obtain the full relaxation behavior. In the marginal unstable case ($\alpha = -0.4$) the dominant kink mode produce a regular evolution in which a single localized current sheet is observed. This is not enough to produce a complete relaxation behavior with uniform λ in the final state, as it can be observed in Fig. 11.

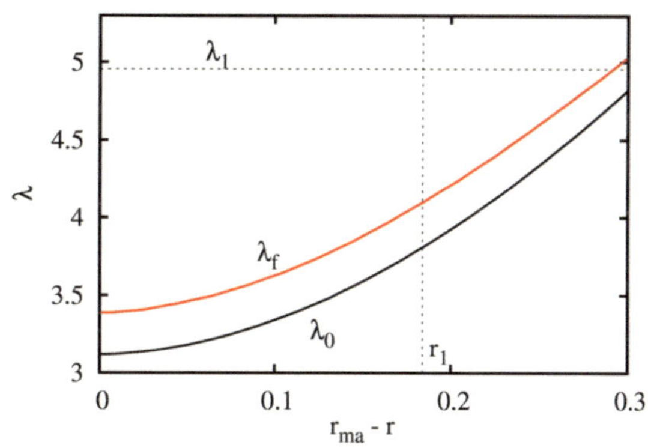

Fig. 11. λ profiles at $t = 0$ (λ_0) and at $t = 100$ (after reconnection, λ_f). In this plot the abscissa measures the distance to the magnetic axis. A partial relaxation behavior is evident, since λ_f is still far away from the eigenvalue λ_1.

As α is lowered (λ profile is steepened), the kink mode becomes stronger and activates higher order modes. Only when a significant level of activity is induced the Taylor's relaxation theory becomes applicable to obtain a good approximation of the final state of the system. Interestingly, the full relaxation behavior is recovered even for a modest separation of scales (Garcia-Martinez & Farengo, 2009a);(Garcia-Martinez & Farengo, 2009b).

5.3 Kink onset and resonant surfaces

Now we focus on the partial relaxation behavior of the marginally kink unstable configurations where relaxation theory is not applicable. A very useful concept developed in the context of the study of MHD modes (in particular the kink mode) is the safety factor q. The safety factor is the number of times a field line on a flux surface goes around toroidally for a single poloidal turn. Based on the equation for a field line

$$\frac{rd\theta}{ds} = \frac{B_\theta}{B_p} \tag{51}$$

where ds is the distance in the poloidal direction while moving a toroidal angle $d\theta$, the safety factor can be defined as

$$q = \frac{1}{2\pi} \oint \frac{1}{r} \frac{B_\theta}{B_p} ds \tag{52}$$

where the integral is taken over a single poloidal circuit. Note that q adopts the same value for every field line lying on the same flux surface and thus it is a flux function $q = q(\psi)$. In

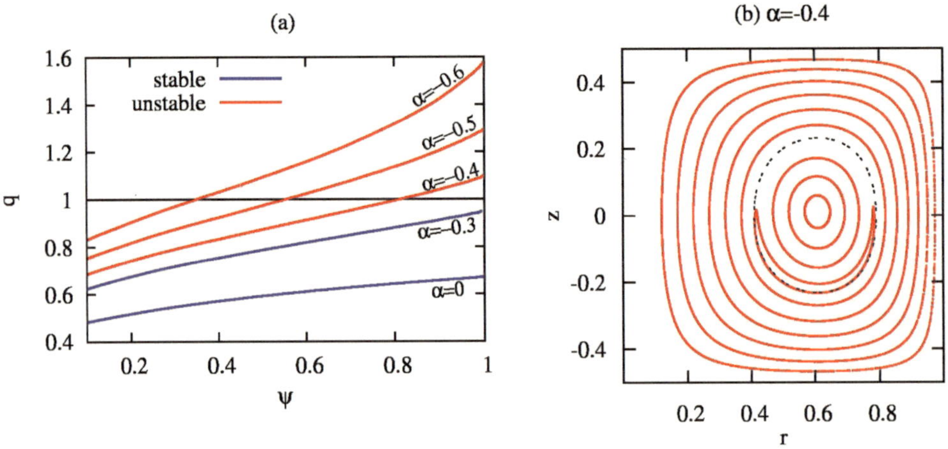

Fig. 12. (a) Safety factor profiles for several configurations. Note that configurations having a $q = 1$ surface are unstable. (b) Poincaré map showing ten field lines during the instability onset in the $\alpha = -0.4$ case. The dashed line shows the $q = 1$ surface, where the formation of a magnetic island is observed.

Fig. 12 (a) the q profiles for several configurations are shown. We already mentioned that the kink instability threshold lies between $\alpha = -0.3$ and $\alpha = -0.4$. In Fig. 12 (a) we can see that the kink instability is associated to the appearance of a rational surface with $q = 1$. Rational surfaces are those where $q = m/n$ being m and n integer numbers and thus q has a rational value. The field lines lying in such surfaces can not span a closed toroidal surface and are particularly prone to develop different MHD modes. That is why these surfaces are also called resonant surfaces. In Fig. 12 (b) we clearly see that it is at the $q = 1$ surface where the first modification to the flux surfaces occurs. This crescent shaped structure (which shows the onset of the island observed in Fig. 8) has a $n = 1$ toroidal dependence.

Note that, in the $\alpha = -0.4$ case, all the relevant MHD activity triggered by the kink takes place inside the $q = 1$ surface of the initial condition (Fig. 8). Thus, we can not expect this evolution to cause a complete relaxation process. However, some partial relaxation occurs due to the magnetic reconnection of flux surfaces having different λ values, as confirmed in Fig. 11. The magnetic reconnection process is further studied in the next Section.

5.4 Magnetic reconnection process

Here we describe the magnetic reconnection process that redistributes currents in the case with $\alpha = -0.4$. Consider the Poincaré map inside the box shown in Fig. 8 (a). This is

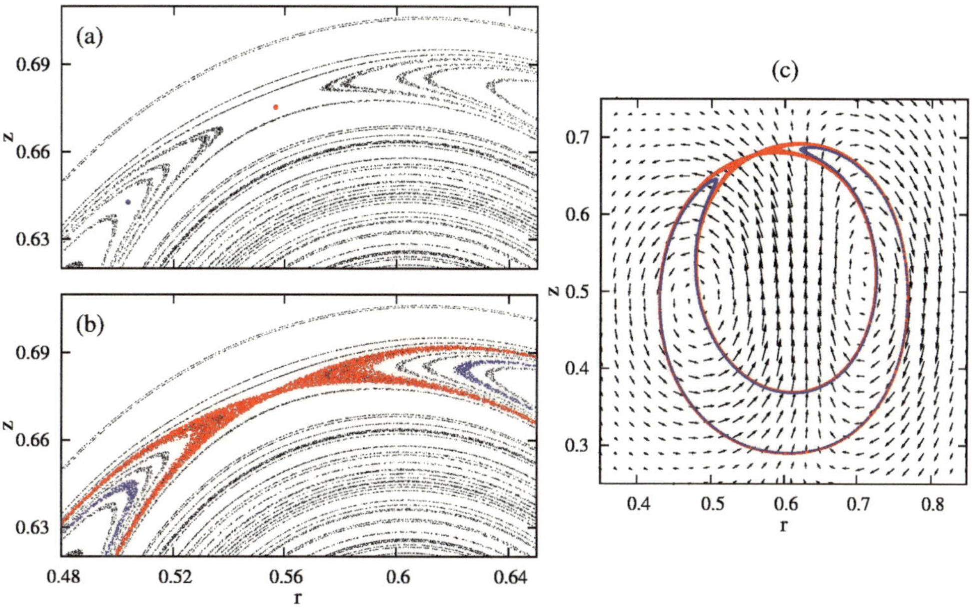

Fig. 13. (a) Poincaré map inside the box shown in Fig.8 (a). Two points, one red and one blue, are manually selected. (b) The same Poincaré plot including two additional magnetic lines followed from the selected points. (c) $n = 1$ component of the poloidal velocity (black vectors) and the two field lines also shown in (b).

shown in Fig. 13 (a). A reconnection layer is clearly identified, in the middle of which we have drawn a point (in red). We follow the magnetic field line that passes through this point for a long distance (ten thousand times the cylinder radius). The results for this single line are shown in Fig. 13 (b) and (c) (the red points) and in Fig. 14. The flow induced by the instability, shown with vectors in Fig. 13 (c), produces the helical distortion of the central flux surfaces. Eventually, one (or more) of these surfaces gets in contact with an outer surface. This is clearly observed in Fig. 14 where a single field line spans both surfaces. Note that the inner surface has a lower λ value than the outer one. At the helical reconnection layer λ adopts an intermediate value.

As a result of this reconnection a new magnetic structure is formed. This structure has a crescent shape cross section as shown by the blue dots in Fig. 13 (b) and (c). This is basically the closed surface that encloses the volume between the two reconnecting toroidal flux surfaces. Fig. 15 shows another visualization of this new magnetic entity. It has been constructed by following the magnetic field line that passes through the blue point indicated in Fig. 13 (a). It is interesting to note that this surface has a lower λ value in its inner face

Fig. 14. A single magnetic field line showing two reconnecting flux surfaces. Its color is proportional to the local λ value (the color scale is indicated on right). The outer surface has a higher λ than the inner surface. The helical reconnection layer adopts an intermediate value.

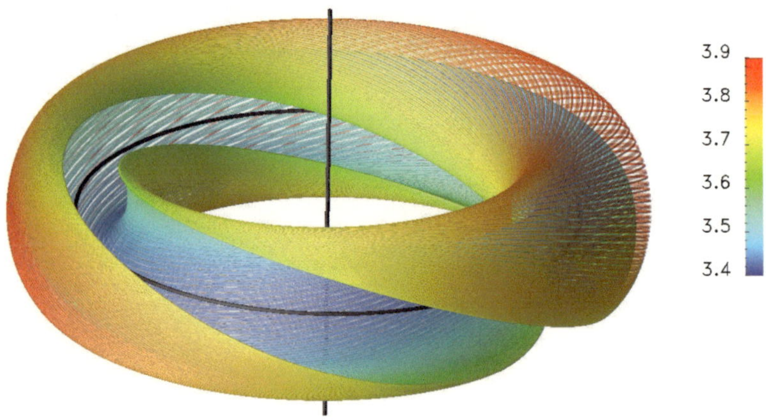

Fig. 15. Magnetic structure formed by the reconnection of the flux surfaces shown in Fig. 14. The color scale indicates local λ value.

(corresponding to the λ value of the original inner flux surface) and a higher λ value in its outer face. This clearly shows that the reconnection is a localized process. It is also evident that the mean λ value of this structure will lie between the λ values of the original surfaces.

With these considerations in mind we can reinterpret Fig. 8. The motion of the island toward the magnetic axis involves the reconnection of inner and outer surfaces having low and high λ values, respectively. The new surfaces formed adopt intermediate λ values. The result of this redistribution is shown in Fig. 11. Note that all this activity takes place in the region where $\psi \geq 0.8$ (the region inside the original location of the $q = 1$ surface). In Fig. 6 (a) we see that within this region $\lambda \lesssim 4$ and thus we can not expect a full relaxation process.

A final comment is made regarding the symmetry of this process. The kink mode has a $n = 1$ toroidal dependence and thus the reconnection layer shown in Fig. 13 has a dominant helical shape. However, we want to mention that there are also higher harmonics ($n > 1$) present in the reconnection process. This can be observed in Fig. 16 where the inner flux surface of Fig. 14 is shown. The high λ region (mainly yellow) shows the reconnection layer. It

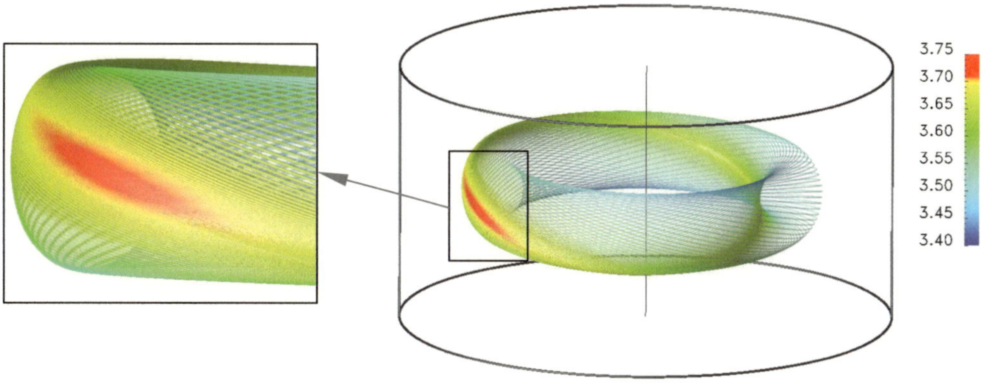

Fig. 16. Inner reconnecting magnetic flux surface. A zoom near the zone of higher λ value reveals the presence of higher toroidal harmonics ($n > 1$).

has a mainly helical structure, however, a zoom around the region with the highest λ values (shown in red) reveals the presence of higher toroidal components. It is not clear, at this point, if higher harmonics play an important role or this process could be recovered considering a two dimensional problem with helical symmetry.

5.5 Reconnection model for the resistive kink mode

The magnetic reconnection process described so far leads to a flux rearrangement in the region where $q > 1$. This process involves a rather regular evolution of the magnetic surfaces with only one helical current sheet. Without a significant level of MHD activity the magnetic relaxation theory becomes inapplicable. Now we seek for a simple but adequate model to describe the final state of the non-linear evolution of the resistive kink.

In the context of tokamak research, the evolution of the resistive kink has been intensively studied. In particular, it is believed that this mode is responsible for a phenomenon called *sawtooth oscillations* that limits in practice the maximum temperature reachable at the core. One of the first models to describe the final state of the non linear resistive kink mode was proposed by Kadomtsev (1975) (see also the explanation of Wesson (2004)). In this Section we describe the Kadomtsev's model and discuss its applicability to the results of our simulations. Then, a modification to the model that significantly improves the agreement with our results will be introduced.

The magnetic field lines on the $q = 1$ surface form a helix around the magnetic axis. The Kadomtsev's model describes the reconnection process in terms of the flux perpendicular to this helix, called helical flux ψ_h. This flux can be computed from the helical magnetic field

$$B_h = B_z(1-q) \tag{53}$$

as

$$\psi_h(r) = 2\pi \int_r^{r_{\mathrm{ma}}} B_z(x, z_{\mathrm{ma}})(1-q)x\,dx \tag{54}$$

where (r_{ma}, z_{ma}) is the position of the magnetic axis and this definition is to be used with $r \leq r_{ma}$. In what follows we will use the minor radius $\tilde{r} = r_{ma} - r$ as the abscissa. In order to

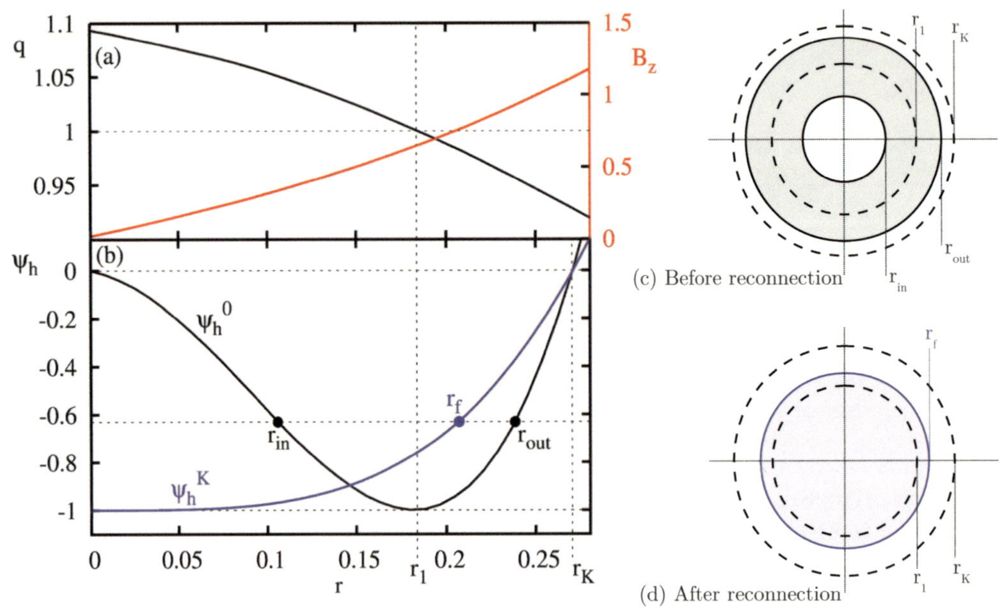

Fig. 17. (a) Initial q and poloidal field (B_z) profiles. (b) Initial (ψ_h^0) and final (ψ_h^K) helical flux predicted by Kadomtsev's model. ψ_h^K is obtained by assuming that the area enclosed by the two reconnecting surfaces before reconnection (c) is equal to the area inside the final reconnected surface (d).

not overload the notation we will drop the tilde. Fig. 17 (a) shows q and B_z as a function of the minor radius. Note that B_h changes its sign at r_1, where $q = 1$, producing a minimum in ψ_h^0 as shown in Fig. 17 (b). The Kadomtsev's model model provides a simple way to compute the helical flux after reconnection ψ_h^K (see Fig. 17 (b)) from which one can readily obtain the reconnected poloidal field profile.

The reconnection begins at the minimum value of ψ_h^0, i.e. at $\psi_h^0(r_1)$. It is assumed that this flux surface will form the new centre of the plasma and thus $\psi_h^K(0) = \psi_h^0(r_1)$. The reconnection then proceeds merging each pair of flux surfaces having the same ψ_h value. In the particular example of Fig. 17, the flux surfaces initially located at r_{in} and r_{out} will reconnect forming a new flux surface at r_f. The position of the final surface r_f is given by toroidal flux conservation. Assuming that the toroidal field does not change during the process, the area enclosed by the two initial surfaces should be equal to the area inside the final surface (see Fig. 17 (c) and (d)). This means that

$$r_f^2 = r_{out}^2 - r_{in}^2 \tag{55}$$

where we have simplified the problem by considering flux surfaces with circular cross section. The reconnection process ends at $\psi_h = 0$ so that the flux surfaces located outside r_K remain unaffected.

Several aspects of this model are in close agreement with the evolution of the marginally unstable case shown in Fig. 8. First of all, the fact that the reconnection process is restricted to the core, i.e. the region $q \lesssim 1$, and does not affect the whole configuration (as assumed by relaxation theory). Secondly, in Fig. 8 we effectively see that the small island formed at r_1 ($q = 1$) moves until it occupies the position of the magnetic axis. Thirdly, in our simulation we also observe what is called a complete reconnection process. Note that since ψ_h^{K} is monotonic this means that B_h after reconnection does not change its sign. This means in turn that q does not cross 1 (in fact q is equal to 1 at $r = 0$). The absence of $q = 1$ surfaces prevents the appearance of magnetic islands just after the reconnection and thus it is said that the Kadomtsev's model predicts complete reconnection. Accordingly, we do not observe any island (other that the magnetic axis) after the reconnection (see Fig. 8) and the resulting q profile does not cross 1, as observed in Fig. 18 (a). Despite this agreement in the overall

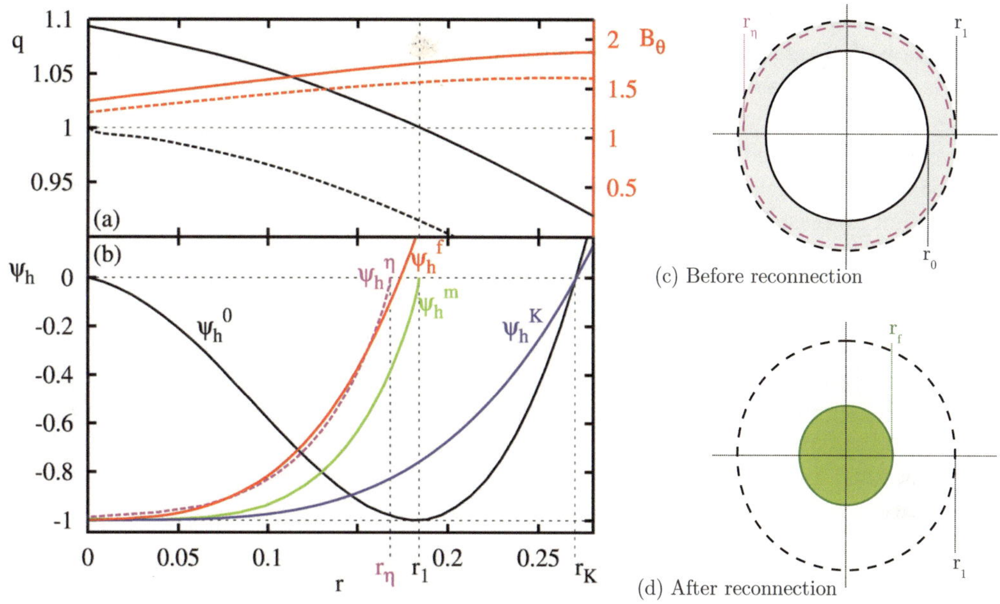

Fig. 18. (a) Initial (solid) and reconnected, i.e. at $t = 100$, (dashed) q and toroidal field profiles. (b) Several helical fluxes profiles: initial (ψ_h^0), final (ψ_h^{f}), predicted by Kadomtsev (ψ_h^{K}), predicted by the modified model (ψ_h^{m}) and predicted by the modified model with correction due to resistive decay (ψ_h^{η}). The modified model proposed involves the reconnection of the surface at r_0 (c) with the surface at r_{f} (d). The shaded regions have the same area.

behavior we will show that the results of the $\alpha = -0.4$ case are better described by introducing a modification to the Kadomtsev's model. In Fig. 18 (b) the initial helical flux ψ_h^0 and the final ψ_h^{K} predicted by Kadomtsev are compared with the actual final helical flux ψ_h^{f} (the red curve) obtained at $t = 100$ for the $\alpha = -0.4$ case. Note that the agreement is not good.

A better approximation can be obtained by looking at Fig. 8 more carefully and noting that the reconnection process takes place *inside* the $q = 1$ surface. Since little or no effect is observed outside r_1 we propose a modified procedure for the construction of the reconnected helical flux ψ_h^{m}. Again, the flux surface at r_1 is reconnected with the magnetic axis so $\psi_h^{\mathrm{m}}(0) = \psi_h^0(r_1)$.

Then, the flux surface initially placed at r_0, see Fig. 18 (c), reconnects and ends at r_f , Fig. 18 (d), in such a way that

$$r_f^2 = r_1^2 - r_0^2 \tag{56}$$

which expresses the conservation of the area of the shaded regions of Fig. 18 (c) and (d). With the initial helical flux ψ_h^0 and Eq. (56) it is possible to compute the reconnected helical flux predicted by the modified model ψ_h^m. This is shown by the green curve of Fig. 18 (b). While this prediction is much closer than the Kadomtsev's model to the actual final state there is still a significant difference. In what follows we will show that this difference is due to resistive dissipation.

Relations (55) and (56) express the toroidal flux conservation assuming that it does not decay, i.e. the toroidal fluxes inside r_K and r_1 do not change. However, as can be observed in Fig. 18 (a), the toroidal magnetic field is visibly reduced due to resistivity. One way to take into account this resistive decay is to change the reference radius with which we make the construction of ψ_h^m given by Eq. (56). In particular, we define r_η as the radius of the circle that contains at $t = 0$ the same amount of toroidal flux that is contained inside r_1 at $t = 100$. If we now compute the reconnected helical flux using Eq. (56) but changing r_1 by r_η we obtain ψ_h^η, shown by the dashed line of Fig. 18 (b). The agreement with the actual final helical flux is very good and this suggests that the modified model indeed captures the basic physics of the reconnection process.

6. Conclusions

In this Chapter we have presented a general picture of the magnetic confinement of high temperature plasmas. This has motivated the introduction of the MHD model which provides an adequate framework to study the macroscopic dynamics of fully ionized plasmas. We have focused our attention on the physical mechanism called plasma relaxation. In particular we have studied the magnetic relaxation process driven by the kink instability in spheromak configurations.

Experiments as well as previous theoretical works showed the existence of a partial relaxation behavior for marginally unstable configurations (they do not evolve toward the minimum energy state). This is in contrast to the well established relaxation theory that states that the plasma should relax to the minimum energy configuration. In this work we have explored these two regimes, namely complete relaxation and partial relaxation, by varying the slope of the initial $\lambda(\psi)$ profile. This controls the degree of instability of the initial configuration as well as the position of the rational surface having safety factor equal to one. The relevance of the position of this rational (or resonant) surface to the partial relaxation behavior was discussed. In particular, we showed that in marginally unstable cases this surface is not far from the magnetic axis and the MHD activity during relaxation remains inside this resonant surface (which is no longer resonant after relaxation). These results suggest that the $q = 1$ surface plays a major role in the evolution of spheromaks during sustainment because in that situation they operate around the kink instability threshold.

The analysis of more unstable cases showed that the full relaxation process predicted by the relaxation theory is only achieved when the magnetic fluctuations produce stochastic field line regions of size comparable of that of the whole system. This result clearly indicates that the relaxation theory as formulated by Taylor (1974) is applicable to highly unstable plasmas but it becomes useless to study the operation of configurations near an instability threshold.

The kink instability produces the helical deformation of the flux surfaces near the magnetic axis. This drives the reconnection of the inner flux surfaces with the outer ones. This process has been studied in detail. The reconnection layer has been identified as well as the new structure resulting from the reconnection of the two flux tubes. Taking the low (high) λ value of the inner (outer) tube on its inner (outer) side, these crescent shaped structures average the λ value inside the $q = 1$ surface. Even when the flux surfaces remain regular during this evolution, the process involves the full reconnection of all the magnetic tubes inside the $q = 1$ surface. This is of course undesired from the point of view of confinement and could partially explain the poor performance of spheromak operation (compared to tokamaks and RFP's). However further studies are required on this topic regarding the coupled dynamics between the kink and the external driving of the system. This could be done by applying appropriate boundary conditions to model the injection of helicity from a source (Garcia-Martinez & Farengo, 2010).

Finally, models for the reconnection process driven by the kink mode were discussed. The Kadomtsev's model was presented and showed to give a poor description of the actual simulation results. A modification to this model that greatly improves the agreement with simulations was proposed. A method to incorporate the correction due to the resistive decay of the configuration was described.

7. References

Bellan, P. (2000). *Spheromaks: a practical application of magnetohydrodynamic dynamos and plasma self-organization*, Imperial College Press, London.

Biskamp, D. (2000). *Magnetic reconnection in plasmas*, Cambridge University Press, Cambridge/New York.

Braginskii, S. (1965). Transport processes in a plasma, *Reviews of Plasma Physics*, 1: 205-311.

Finn, J. & Antonsen, T. (1985). Magnetic helicity: what it is, and what it is good for?, *Comments Plasma Phys. Controlled Fusion*, 33: 1139.

Friedberg, J. (1987). *Ideal MHD*, Plenum Press, New York.

Garcia-Martinez, P. & Farengo, R. (2009a). Selective decay in a helicity-injected spheromak, *J. Phys.: Conf. Ser.*, 166: 012010.

Garcia-Martinez, P. & Farengo, R. (2009b). Non-linear dynamics of kink-unstable spheromak equilibria, *Phys. Plasmas*, 16: 082507.

Garcia-Martinez, P. & Farengo, R. (2010). Spheromak formation and sustainment by tangential boundary flows, *Phys. Plasmas*, 17: 050701.

Hasegawa , A. (1985). Self-organization processes in continuous media, *Adv. Phys.*, 34: 1-42.

Heyvaerts J. & Priest, E. (1984). Coronal heating by reconnection in DC current systems: a theory based on Taylor's hypothesis. *Astron. Astrphys.*, 137: 63-78.

Jarboe, T. (2005). The spheromak confinement device, *Phys. Plasmas*, 12: 058103.

Kadomtsev, B. (1975). Disruptive instabilities in tokamaks, *Sov. J. Plasma Phys.*, 1: 389-391.

Kitson, D. & Browning, P. (1990). Partially relaxed magnetic field equilibria in a gun-injected spheromak, *Plasma Phys. Controlled Fusion*, 32 (14): 1265-1287.

Knox, S., Barnes, C., Marklin, G., Jarboe, T., Henins, I., Hoida, H. & Wright, B. (1986). Observations of spheromak equilibria which differ from the minimum-energy state and have internal kink distortions, *Phys. Rev. Lett.*, 56: 842-845.

Moffat, H. (1978). *Magnetic field generation in electrically conducting fluids*, Cambridge University Press, London/New York.

Montgomery, D., Turner, L. & Vahala, H. (1978). Three-dimensional magnetohydrodynamic turbulence in cylindrical geometry. *Phys. Fluids*, 21: 757-764.

Priest, E. & Forbes, T. (2000). *Magnetic reconnection: MHD theory and applications*, Cambridge University Press, New York.

Rosenbluth, M. & Bussac, M. (1979). MHD stability of spheromak, *Nuc. Fusion*, 19: 489-498.

Taylor, J. (1974). Relaxation of toroidal plasma and generation of reverse magnetic fields, *Phys. Rev. Lett.*, 33: 1139-1141.

Taylor, J. (1986). Relaxation and magnetic reconnection in plasmas, *Rev. Mod. Phys.*, 58: 741-763.

Wesson, J. (2004). *Tokamaks (Third Edition)*, Clarendon Press - Oxford, New York.

Willett, D., Browning, P., Woodruff, S. & Gibson, K. (1999). The internal magnetic structure and current drive in the SPHEX spheromak, *Plasma Phys. Controlled Fusion*, 41: 595-612.

Woltjer, L. (1958). A theorem on force-free magnetic fields, *Proceedings of the National Academy of Science*, 44: 489-491.

Zheng, L. & Furukawa, M. (2010). Current-interchange tearing modes: Conversion of interchange-type modes to tearing modes, *Phys. Plasmas*, 17: 052508.

5

MHD Activity in an Extremely High-Beta Compact Toroid

Tomohiko Asai and Tsutomu Takahashi
Nihon University
Japan

1. Introduction

1.1 Field-reversed configuration (FRC)

A field-reversed configuration (FRC) plasma is extremely high beta confinement system and the only magnetic confinement system with almost 100% of a beta value (Tuszewski, 1988; Steinhauer, 2011). The plasma is confined by the only poloidal magnetic field generated by a self-plasma current. The FRC has several potentials for a fusion energy system. As the one of the candidate for an advanced fusion reactor, for example, D-^{3}He fusion (Momota, 1992), FRC plasma is attractive. Recently, this plasma also has an attraction as target plasmas for an innovative fusion system, Magnetized Target Fusion (MTF) (Taccetti, 2003), Colliding and merging two high-β compact toroid (Guo, 2011; Binderbauer, 2010; Slough, 2007[a]) and Pulsed High Density FRC Experiments (PHD) (Slough, 2007[b]).

The plasma belongs to a compact toroid system. Here, 'compact' denotes a simply connected geometry, i.e., the absence of a central column. The system consists of a toroidal magnetic confinement system with little or no toroidal magnetic field. The typical magnetic structure of the FRC plasma is shown in Fig. 1. The poloidal confinement field (B_{ze}) consists of the externally applied magnetic field of an external coil (B_{z0}), and the self-generated magnetic field of the toroidal plasma current (I_θ : $I_\theta > 2B_{z0}/\mu_0$). The FRC consists of an axially symmetric magnetized plasma, a plasma liner and a simply connected configuration. Then, the beta value $\beta = p/\left(B_{ze}^2/2\mu_0\right)$, which is the ratio of confined plasma pressure (p) to the confinement magnetic field pressure ($B_{ze}^2/2\mu_0$), is extremely high. The system has a closed field line region in which the high temperature plasma is confined, and an open field line region which acts as a natural diverter.

A scrape-off layer is formed in the open field line region. Two singularities in the magnetic field, i.e., X-points, are formed at the intersections of the symmetric device axis with the separatrix ($B_{ze} = 0$). A null field surface ($B_z = 0$) is also formed in the closed separatrix region. The radius (R) of the null surface at midplane (minor radius) is $R = r_s/\sqrt{2}$ (r_s: radius of the separatrix at midplane) in the pressure equilibrium state. The separatrix length l_s is defined as the distance between the two X-points (Armstrong et al., 1981).

An FRC has three essential geometrical plasma parameters (S^*: radial size parameter; E: separatrix elongation; and X_s: normalized separatrix radius), which are related to the physical properties of the FRC plasma. $S^{*}=r_s/(c/\omega_{pi})$ is defined as the ratio of the separatrix radius to the ion skin depth (c/ω_{pi}). Here, r_s, c and ω_{pi} are the separatrix radius, speed of light and ion plasma frequency, respectively. Two other radial size parameters, $S \equiv r_s/\sqrt{2}\rho_i$ and $\bar{s} = \int_R^{r_s} r dr/r_s\rho_i$, are sometimes used, where ρ_i and ρ_{io} are the local ion Larmor-radius and the reference ion Larmor-radius, respectively, based on the external magnetic field B_{ze}. These parameters indicate the importance of the two-fluid (ion and electron fluid) and finite-Larmor-radius effects; for example, $S \approx 1.3S^{*}$ and $\bar{s} = X_s S/5$ under under $T_i \sim 2T_e$ (T_i: ion temperature; T_e: electron temperature) (Steinhauer, 2011). $E = l_s/2r_s$ is defined as the ratio of the separatrix radius to the diameter, and indicates the elongation of the separatrix, which is different from that of a tokamak system. It is known that this elongation affects the global plasma stability of an FRC. Oblate and prolate FRC plasmas are usually categorized as $0 < E < 1$ and $E > 1$ FRCs, respectively. $X_s = r_s/r_w$ is defined as the ratio of the separatrix to the confinement coil radius. This normalized radius has a strong relation to the poloidal flux (ϕ_p) of an elongated (prolate) FRC plasma, and to the FRC confinement time scaling. The average beta value $\langle\beta\rangle$ can be written as $\langle\beta\rangle = 1 - 0.5x_s^2$ (Armstrong et al., 1981).

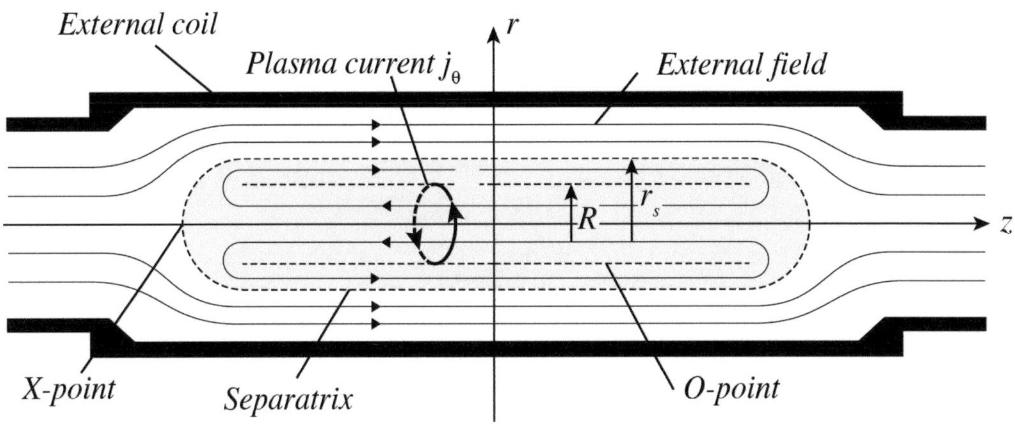

Fig. 1. A field-reversed configuration plasma formed by the field-reversed theta-pinch (FRTP) method.

The FRC topology is similar to an elongated, low-aspect-ratio, toroidal version of the Z-pinch, as shown in Fig. 1. Since the FRC plasma has no toroidal field, and no center conductor, theoretical studies predict that FRC plasma is unstable with respect to an MHD mode with low toroidal mode number. The principal instabilities of the FRC, predicted by magnetohydrodynamics (MHD) theory, are listed in Table 1 (Tuszewski, 1988; Slough & Hoffman, 1993). Here, the tokomak nomenclature has been adopted, with n and m being the toroial and poloidal mode numbers, respectively.

The plasma current of FRC, at just after formation, is primarily carried by electrons. On the other hand, ions are approximately at rest. However, in most of the FRC experiments, ions soon begin to rotate to a diamagnetic direction. The rotation speed often reaches to a supersonic level. Instabilities driven by the Rotational mode is appeared. The origin of rotation has not yet been completely understood. But, given rotation, the condition for instability has been fairly well understood. The stability threshold was expressed in terms of a parameter $\alpha=\Omega/\Omega_{Di}$, where Ω and Ω_{Di} are rotation angular frequency and ion diamagnetic drift frequency. The threshold for n=1 mode was α=1 and in n=2 mode could also grow for a greater than 1.2 -1.4, for zero bias limit (Freidberg & Pearlstein, 1978). For FRC plasma, a similar threshold is α~1.3-1.5 (Seyler, 1979).

	m (poloidal mode)	Mode Character	Mode Name	Experimental Observation
1. Local Ideal Mode				
∞	0		Interchange	Only lowest order (n=1,2,3)
∞	1,2	Axial or Radial	Co-interchange (Ballooning)	No
2. Global Mode				
No rotating				
0	1	Axial	Roman candle	No
1	0	Sausage Shift	Interchange	Often, occasional
1	1	Radial	Sideways shift (Tilt)	Seldom
>1	0	Flute	Interchange	Often, always (high-$\bar{s}$)
>1	1	Axial	Tilt	No
Rotating				
1	1	Radial	Wobble	Yes (occasional)
2	1	Radial	n=2	Yes (always)
>2	1	Radial	n>2	Yes (often, high-$\bar{s}$)
3. Resistive Mode				
0	2	Radial and Axial	Tearing	Yes (always?)

Table 1. FRC Stability: MHD Theory versus Experimental Observation

These two global ideal modes driven by rotation—the n = 1 wobble, of little concern since it saturates at low amplitude, and the n = 2 rotational instability, which destroys most FRCs—have been regularly observed experimentally. These rotational modes (n = 1, 2) have been controlled by applying a straight or helical multi-pole field (Ohi et al., 1983; Shimamura & Nogi, 1986; Fujimoto et al., 2002). Higher-order (n > 2) rotational modes have often been observed in large-$\bar{s}$ FRC experiments, with $\bar{s}$ in the region of $3 < \bar{s} < 8$ (Slough & Hoffman, 1993). FRC plasma with a higher $\bar{s}$ value behaves as a MHD plasma and with low $\bar{s}$ one becomes more kinematically. According to the several theoretical works, FRC plasmas have been predicted to be unstable because of a bad curvature of a closed confinement field. Various local and global non-rotating ideal MHD modes are listed in Table 1, and it is worth noting that stable FRC plasma is impregnable against these low n-modes. The m/n=1/1 tilt mode instability is thought to be most dangerous. The stability of prolate and oblate FRC plasma has been investigated experimentally on several FRTP devices (FRX-C/LSM (Tuszewski et al., 1990, 1991), LSX, (Slough & Hoffman, 1993), NUCTE-III (Kumashiro et al., 1993; Asai et al., 2006; Ikeyama et al., 2008), etc.), and on spheromak-merging facilities such as the TS-3 (Ono et al., 1993) and MRX (Gerhardt et al., 2006). In addition, the stability of prolate and oblate FRCs has been analyzed by means of visible and x-ray photography with an end-on camera (Slough & Hoffman, 1993; Tuszewski et al., 1991), computer tomography reconstruction of the visible emission profile (Asai et al., 2006), as well as mode analysis of the external B_θ-magnetic probe array (Mirnov coil array) (Slough & Hoffman, 1993; Tuszewski et al., 1990, 1991; Kumashiro et al., 1993; Ikeyama et al., 2008) and the internal magnetic probe array (Ono et al., 1993; Gerhardt et al., 2006).

In the following sections, the formation methods for FRC plasma (Section 2) and the stability of FRC plasma (Section 3) are described based on these experimental results and some theoretical studies.

2. Formation methods for FRC plasma

An FRC is formed through a violent formation process dominated by self-organization. In this process, plasma pressure is built up, coinciding with the formation of reversed magnetic structure and plasma current drive within the diffusion time. The beta value of the FRC at the magnetic axis is infinite, and the volume-averaged beta value $<\beta>$ is nearly equal to one.

The plasma current density j_θ(r), the confinement magnetic field B_{ze}, the coil magnetic field B_{z0}, and the formed plasma pressure $p(r)$, satisfy the following condition:

$$\begin{aligned} I_\vartheta &= \int_0^{r_s} j_\vartheta(r)dr \geq 2B_{z0}/\mu_0 \\ \nabla p(r) &= \frac{\partial p(r)}{\partial r} = j_\vartheta(r)B_z(r) \end{aligned} \quad . \tag{1}$$

FRC plasma is traditionally formed by the field-reversed theta-pinch (FRTP) method (Armstrong et al., 1981). Since the 1990's, a variety of alternative formation methods have emerged (Ono, Y et al., 1993; Gerhardt et al. 2008; Slough & Miller, 2000; Knight & Jones, 1990; A. Hoffman et al., 2002; Guo et al., 2007; Logan et al., 1976; Davis, 1976; Greenly et al., 1986; Schamiloglu et. al., 1993). At the same time, the FRTP-based formation method has been significantly improved (Slough et al., 1989; Hoffman et al., 1993; Pietrzyk et al., 1987; Pierce et al., 1995; Guo et al., 2004, 2005; Asai et al., 2000; Binderbauer et al., 2010; Guo, et al., 2011). Recently initiated new formation methods include (1) counter-helicity spheromak-merging (CHSM) (Yamada et al., 1990; Ono et al., 1993; Gerhardt et al. 2008), (2) rotating magnetic field (RMF) (Slough & Miller, 2000; Knight & Jones, 1990; A. Hoffman et al., 2002; Guo et al., 2007), (3) field-reversed mirror configuration (FRM) driven by neutral beam injection (NBI) (Logan et al., 1976), relativistic electron beam (REB) (Davis, 1976), and intense light ion beam (ILIB) injections (Greenly et al., 1986; Schamiloglu et. al., 1993).

Improved FRTP methods have also introduced low inductive voltage and program formation by FRTP (Slough et al., 1989; Hoffman et al., 1993), the coaxial slow source (CSS) (Pietrzyk et al., 1987; Pierce et al., 1995), translation-trapping formation by FRTP (Guo et al., 2004, 2005; Asai et al., 2000), and collision FRC merging by FRTP (Binderbauer et al., 2010; Guo et al., 2011). Through these innovative new methods and improved FRTP methods, the

FRC lifetime has been prolonged to the order of several ms, and the confinement properties have also been improved [34]. The plasma parameters and lifetimes of FRCs formed by the above methods are summarized in Table 2.

Experimental and theoretical studies of FRC stability have mainly focused on elongated ($E > 1$) and oblate ($0 < E < 1$) FRC plasmas, formed by the FRTP (Slough & Hoffman, (1993), Fujimoto et al., 2002; Tuszewski, et al., 1990; Tuszewski et al., 1991; Kumashiro et al.,1993; Asai et al., 2006) and CHSW methods (Yamada et al., 1990; Ono et al., 1993; Gerhardt et al. 2008), respectively. Details of the two methods are introduced in the next section.

Formation method	B_e (T)	r_s (m)	n_p $(\times 10^{20} m^{-3})$	T_i (keV)	Φ_p (mWb)	τ_{life} (ms)	E	$\overline{S}$
FRTP	0.3-2	0.05-0.24	5-500	0.1-15	0.4-12	0.03-0.4	2.5-10	0.5-5
FRTP+translation-trapping	0.01-0.06	0.12-0.5	0.4-5	0.2-0.5	1.5	0.2-0.4	5-8	0.8
FRTP+collision-merging	1.0-1.1	0.3-0.4	1	0.5-0.6	12	~1.0	5	1
RMF	0.006-0.025	0.03-0.45	0.007-0.17	0.02-0.2	1-10	1-2.5	1-3	-
Spheromak-merging	0.2-0.3	0.4-0.5	2	0.02-0.2	≤ 10	0.075-0.1	0.35-0.65	1-3
Spheromak-merging+CS	0.2-0.3	0.4-0.5	1	0.02-0.2	2-3	0.35-0.6	0.35-0.65	1-3

Table 2. FRC Plasma parameters for various formation methods

2.1 Field-reversed theta-pinch method (FRTP)

The schematic of a typical field-reversed theta-pinch device, NUCTE-III (Nihon University Compact Torus Experiment 3), is shown in Fig. 2 (Asai et al., 2006). A transparent fused silica glass discharge tube lies in a cylindrical one-turn coil. The tube is filled with a working gas (usually hydrogen or deuterium gas) by static filling or gas puffing, and then a *z*-discharge or inductive theta-discharge (θ-discharge) generates a pre-ionized plasma of the working gas, which is embedded in the reversed-bias field of 0.03-0.08 T, produced by 2 mF of the bias bank. A main bank of 67.5 μF rapidly reverses the magnetic field in the discharge tube (rising time of 4 μs). The circuit of the main bank is crowbarred on reaching the maximum current, and resistively decays with a decay time of 120 μs. A thin current sheet is initially formed around the inner wall of the discharge tube by an inductive electric field ($E_\theta = 0.5 r dB_z/dt$), and shields the plasma from the rising forward field. The rising field works as a 'magnetic piston' to implode the plasma radially. At both ends of the coil, the reversed-bias field is reconnected with the forward field, and a closed magnetic structure is created. The tension formed due to the magnetic curvature produces a shock-like axial contraction. Then the radial and axial dynamics rapidly dissipate within about 20 μs, and the FRC plasma reaches an equilibrium/quiescent phase. Figure 3 shows the separatrix and the equi-magnetic surface estimated by our improved excluded flux method, and the radial profile of bremsstrahlung (proportional to $n_e^2/T_e^{0.5}$). An FRC plasma with a separatrix radius of 0.055 m and a length of 0.8 m is formed at about 20 μs, and is isolated from the discharge tube (r_t = 0.13 m).

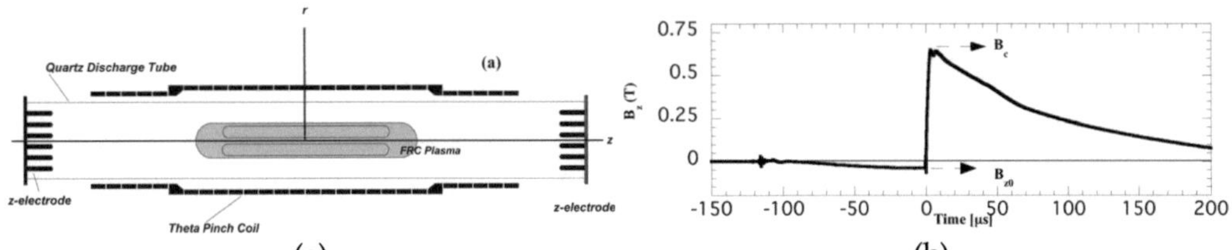

(a) (b)

Fig. 2. (a) Schematic diagram of NUCTE-III and (b) a typical waveform of magnetic field on FRTP method.method.

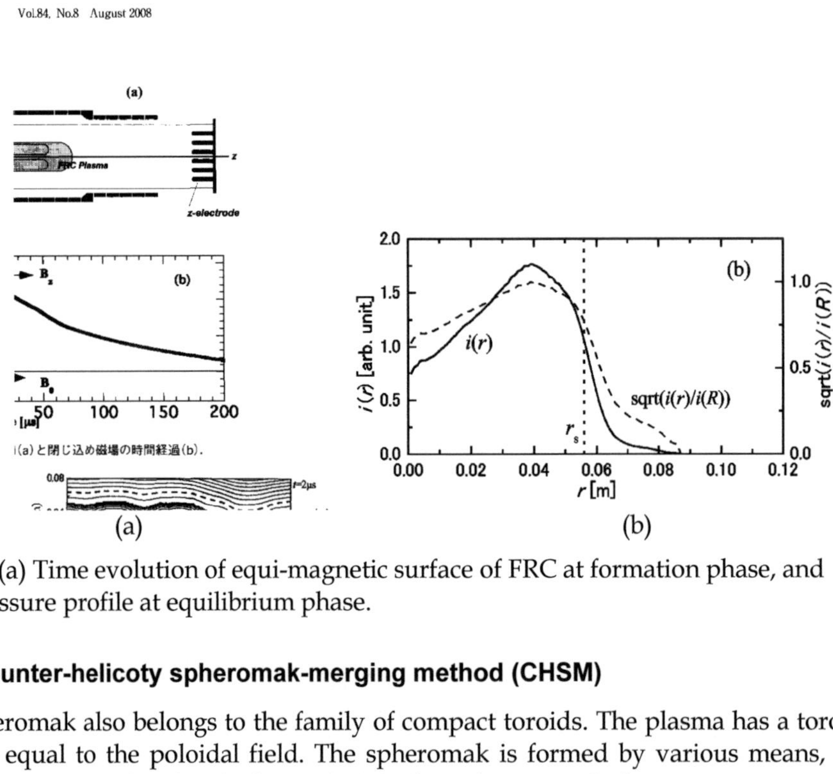

(a) (b)

Fig. 3. (a) Time evolution of equi-magnetic surface of FRC at formation phase, and (b) pressure profile at equilibrium phase.

2.2 Counter-helicoty spheromak-merging method (CHSM)

A spheromak also belongs to the family of compact toroids. The plasma has a toroidal field nearly equal to the poloidal field. The spheromak is formed by various means, such as a simultaneous axial and θ discharge (z-θ discharge), a coaxial plasma gun, or a toroidal flux core containing both toroidal and poloidal winding (Ono et al., 1993; Gerhardt et al., 2006; Yamada et al., 1990; Ono et al., 1993; Gerhardt et al., 2008).

Two spheromaks, with a common geometric axis and opposite helicity (opposite toroidal fields) of equal value, are separately formed, and merge to form the FRC. If the respective helicity values of the spheromaks differ significantly, the merged plasma remains a spheromak (Ono et al., 1993; Yamada et al., 1990). This formation method naturally forms an oblate FRC, in contrast to the prolate FRC formed by the FRTP method.

In the Tokyo Spheromak 3 (TS-3) experiment, merging spheromaks are formed by z-θ discharge (Ono et al., 1993, 1997). The Swarthmore Spheromak Experiment (SSX) utilizes the coaxial plasma gun method (Cohen et al., 2003). The Tokyo Sheromak 4 (TS-4) and the Magnetic Reconnection Experiment (MRX) employ the flux core method (Gerhardt et al., 2006). The TS-3 device is illustrated in Fig. 4 (Ono et al., 1997). The time evolutions of the magnetic surface of the poloidal and toroidal fields, the toroidal flow, and the ion temperature, are also shown in the figure.

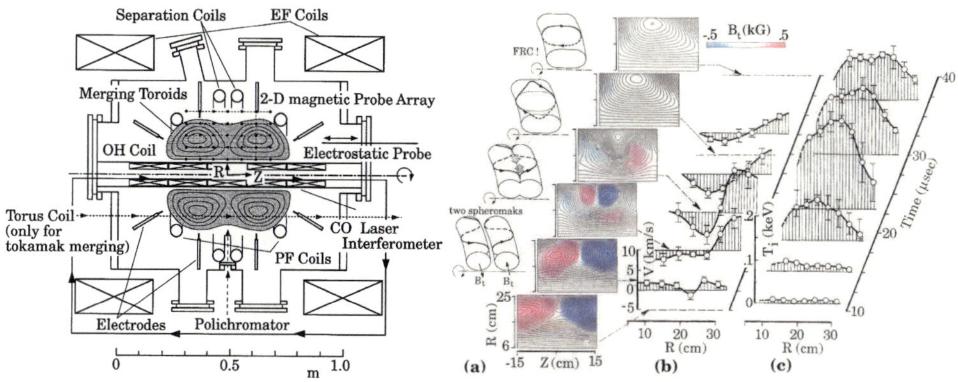

Fig. 4. (a) TS-3 merging device and 2D contours of poloidal flux surface and toroidal magnetic field on the R-Z plane; radial profiles of (b) ion toroidal velocity V, and (c) ion temperature (T_i), on the midplane, during the counter helicity merging of two spheromaks with equal but opposing B_t. The red and blue colors indicate the positive and negative amplitudes of B_t.

3. MHD behavior of FRCs

3.1 MHD behavior of prolate FRCs

3.1.1 General picture of prolate FRC MHD behavior

The global deformation of the internal structure of an FRC, and its time evolution, were investigated by means of an optical diagnostic system (Takahashi et al., 2004), combined with tomographic reconstruction, in the NUCTE facility (Asai et al., 2006). Fourier image transform was applied to the reconstructed image, and the correlation of global modes with n = 1 and 2 was investigated. The typical plasma parameters are separatrix radius of 0.06 m, separatrix length of 0.8 m, electron density of 2.5 x 10^{-20} m^{-3}, total temperature of 270 eV, particle confinement time of 80 μs, and $\bar{s}$ -value of 1.9. Figure 5 shows the time evolution of the 2D emissivity profile of bremsstrahlung of 550 nm. Here, the intensity of bremsstrahlung is proportional to $n_e^2/T_e^{0.5}$. Figure 5 (b) - (g) shows a reconstructed tomographic image of the cross-sectional structure at each phase indicated in the time history of line integrated electron density, measured along the y-axis (Fig. 5 (a)). Figure 5 (b) shows the emissivity structure 1 μs after application of the main compression field. We can see that the radial compression has started at the chamber wall. The following radial compression phase is shown in Fig. 5 (c). The circular boundary of the bright area indicates azimuthally uniform compression. After the formation phase, the equilibrium phase, with a circular cross-sectional structure, lasts approximately 20 μs (Fig. 5 (d)). The oscillation observed in the latter phase is caused by rotational instability with toroidal mode number of n = 2. At the first stage of deformation, illustrated in Fig. 5 (e), the reconstructed cross section is deformed into an oval shape.

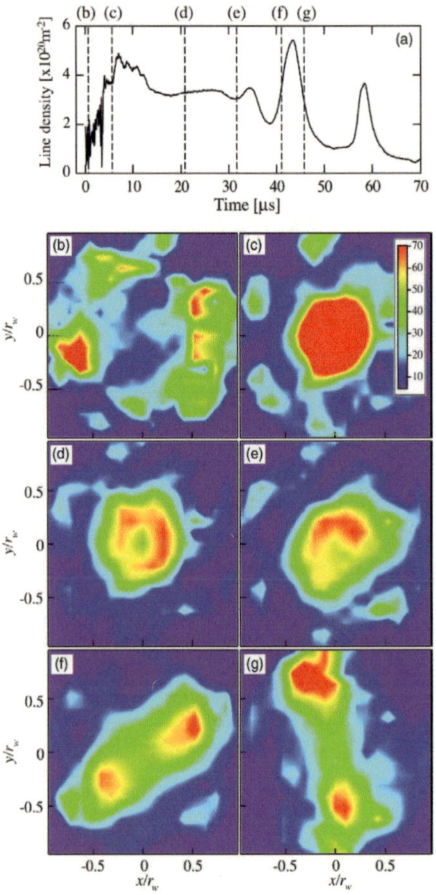

Fig. 5. (a) Time evolution of line integrated electron density, and (b) - (g) reconstructed cross-sectional structure of an FRC.

Generally, this deformation is called 'elliptical deformation.' However, the reconstructed image indicates an internal shift (n = 1) in an oval separatrix. In this rotational instability phase, deformation grows due to centrifugal distortion. In the early stage of the growth of instability, the structure of the FRC has a dumbbell-like shape, as shown in Fig. 5 (f). The internal structure at the final stage of the discharge is shown in Fig. 5 (g). The distribution of emissivity shows two clear peaks which orbit around the separatrix axis like binary stars. The analyzed time evolution of mode intensity and phase by the Fourier image transform are shown in Fig. 6. This result indicates that the n = 1 shift motion of the plasma column (wobble motion) increases prior to the growth of the n = 2 mode. The amplitude of n = 1 increases in the equilibrium phase of 20 – 30 μs. It is thus apparent that the dominant mode changes to n = 2 after a modest peak of n = 1. In the tomographic image of this transition region of n = 1 to 2 (Fig. 5 (e)), an internal shift of the bright area, which has a different rotational phase from that of the oval plasma boundary, is seen. More specifically, this $n = 1$ shift motion and the rotational torque are possible sources of the $n = 2$ mode deformation. This result suggests that the suppression of this shift motion in the formation and equilibrium phase might impede the growth of $n = 2$ mode deformation.

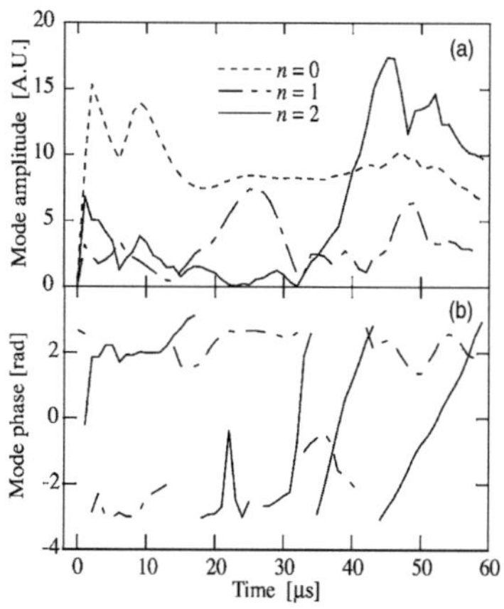

Fig. 6. Time evolution of (a) toroidal mode intensity, and (b) phase.

3.1.2 Stability of prolate FRCs with respect to rotational mode

These two global ideal modes driven by rotation—the $n = 1$ wobble and the $n = 2$ rotational instability—have been regularly observed in experiments. These rotational modes have been controlled by applying a straight or helical multipole field (Ohi et al., 1983; Shimamura & Nogi, 1986). The stability criterion (B_{sc} ($n = 2$)) of the straight multipole field of an m-pole for the $n = 2$ rotational instability has already been developed, on the basis of MHD equations, by Ishimura (Ishimura, 1984) as

$$B_{sc}(n=2) = \frac{1}{2}\sqrt{\frac{\mu_0 \rho}{m-1}} r_s |\Omega| , \tag{2}$$

where Ω, r, and μ_0 are, respectively, the rotational angular velocity of the $n = 2$ deformation (which is twice that of the angular velocity of the plasma column), the mass density of the plasma column, and the magnetic permeability of free space. On the other hand, the stability criterion for the $n = 1$ wobble motion (B_{sc} ($n = 1$)) has been derived from the experimental results of NUCTE-III (Fujimoto et al., 2002) as

$$B_{sc}(n=1) = \frac{1}{f}\sqrt{\frac{\mu_0 \rho}{2(m-1)}} r_s \omega_{n=1} , \tag{3}$$

where $\omega_{n=1}$ is the angular velocity of the $n = 1$ wobble motion, and f is an amplitude reduction coefficient which is defined by experiments with different pole numbers and is about 0.3 for the NUCTE FRCs. The ratio of B_{sc} ($n = 1$)/ B_{sc} ($n = 2$) suggests that the amplitude of $n = 1$ mode motion can always be maintained at a low level by the application of B_{sc} ($n = 2$), provided $\omega_{n=1}/f < \omega_{n=2}$.

The experimental results of the FIX (FRC injection experiment) with neutral beam injection into FRC plasma formed by FRTP, indicate that the global $n = 1$ mode motion was controlled by neutral beam injection (Asai, et al., 2003). The neutral beam was injected obliquely to the axial direction due to the limited poloidal flux. The stabilization effects of ion rings confined by mirror fields at each end have been noted. Improved confinement properties (e.g., prolonged decay time of plasma volume and increased electron temperature) have also been observed.

TCS (Translation, Confinement and Sustainment) experiments at Washington University indicate that the $n = 2$ mode rotational instability can be controlled by the self-generated toroidal field, which converts from a toroidal into a poloidal field during the capture process of translated FRCs (Guo et al., 2004, 2005). The stabilization effect of the toroidal fields was investigated using the modified energy principle with the magnetic shear effect (Milroy et al., 2008). The following analytic stability criterion was derived as

$$B_{SC} > 0.66\sqrt{\mu_0 \rho} r_s \Omega \, . \tag{4}$$

This stability criterion is very similar to the one for the multipole field. This formula indicates that a relatively modest toroidal field, which is about 12% in comparison to the external poloidal field, can stabilize the FRC in the case of TCS experiments.

To supply the modest toroidal field to the FRC plasma, a magnetized coaxial plasma gun (MCPG) has been employed in the NUCTE facility (Asai et al., 2010). The MCPG generates a spheromak-like plasmoid which can then travel axially to merge with a pre-existing FRC. Since the MCPG is mounted on-axis and generates a significant helicity, it provides the FRC-relevant version of coaxial helicity injection (CHI) that has been applied to both spheromaks and spherical tokamaks. When CHI is applied, the onset of elliptical deformation of the FRC cross section is delayed until 45 - 50 μs from FRC formation, compared to an onset time of 25 μs without CHI. Besides delaying instability, MCPG application reduces the toroidal rotation frequency from 67 kHz to 41 kHz. Moreover, the flux decay time is extended from 57 to 67 μs. These changes occur despite the quite modest flux content of the plasmoid: ~ 0.05 mWb of poloidal and 0.01 mWb of toroidal flux, compared with the 0.4 mWb of poloidal flux in the pre-formed FRC. The MCPG introduces a different stabilization mechanism, which may be the same as that observed in translated FRCs, because of the existence of modest toroidal flux. The observed global stabilization and confinement improvements suggest that the MCPG can actively control the rotational instability.

In STX experiments, stabilization effects due to RMF have been observed. The stabilization effects can be attributed to two-fluid effects produced by rotational and ponderomotive forces. In TCS experiments, the stabilization effects of RMF on the $n = 2$ interchange mode and rotational mode have been reported (Guo et al., 2005). The stability criterion of RMF field strength (B_ω) was derived as

$$B_\omega \geq 1.14\sqrt{\rho \mu_0} r_s \Omega \, . \tag{5}$$

The stability diagram for FRC formed and sustained by the RMF at a different frequency is indicated in Fig. 6 of reference of Guo et al., 2005.

3.1.3 Stability of prolate FRCs with respect to tilt mode

In the FRX-C/LSM, which is a conventional FRTP device using non-tearing reconnection, the stability of FRC plasmas with $1 < \bar{s} < 3.5$ and $3 < E < 9$ (highly kinetic and elongated) was investigated using a Mirnov loop array of 64 external B_θ pick-up loops and a soft X-ray end-on camera (Tuszewski et al., 1991). Tilt-like asymmetries (the $n = 1$ axial odd component of B_θ) were found, which strongly correlates with FRC confinement. Tilt and other instabilities also appeared with an increase in the bias magnetic field and/or filling pressure (i.e., higher $\bar{s}$ - value). An increase in the bias field and filling pressure also coincidently causes strong axial dynamics, which triggers confinement degradation. These experimental results suggest that the tilt-stability condition for kinetic and elongated FRCs is in the range of $s/e < 0.2 - 0.3$ ($S^*/E < 3$) (Fig. 7), and becomes $s/e \sim 1$ for MHD-like FRC (Fig. 8). Strong axial dynamics during FRC formation results in lesser elongation of the FRC. Therefore it eventually fosters the growth of tilt instability. For an FRC with $s/e \sim 1$, the tilt instability grows from small initial perturbations, and becomes large enough to cause major plasma disruptions after 10 - 20 μs, which is 3 - 4 times longer than the growth time of instability. In the case of low filling pressures, higher order ($n = 2$ and 3) axially odd asymmetries are also observed. However, the amplitude of these modes is much less than that of the $n = 1$ tilt components. In the case of higher filling pressures, higher order modes appear earlier and grow vigorously.

In the LSX, using an improved FRTP formation method with a programmed formation scheme, the correlation between plasma distortions and the confinement properties was investigated (Slough & Hoffman, 1993). A B_θ probe array and an end-on soft X-ray camera were employed to determine separatrix movement, which might indicate the existence of lower order modes, such as tilt mode. Experiments were conducted over a large range of $\bar{s}$ ($1 < \bar{s} < 8$) and no correlation was observed between the quality of confinement and the B_θ signal. In fact, the confinement quality correlates more with the shape of the equilibrium radial profile than with $\bar{s}$. Details of the experimental results are summarized in Table 3.

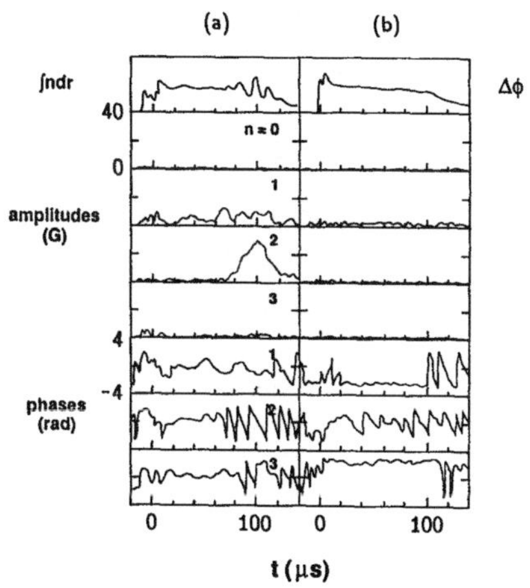

Fig. 7. Toroidal Fourier analysis of the B_θ for good confinement. The Fourier amplitudes and phases are shown as functions of time for (a) even and (b) odd components. The top trace of (a) is a line-integrated electron density and (b) diamagnetism.

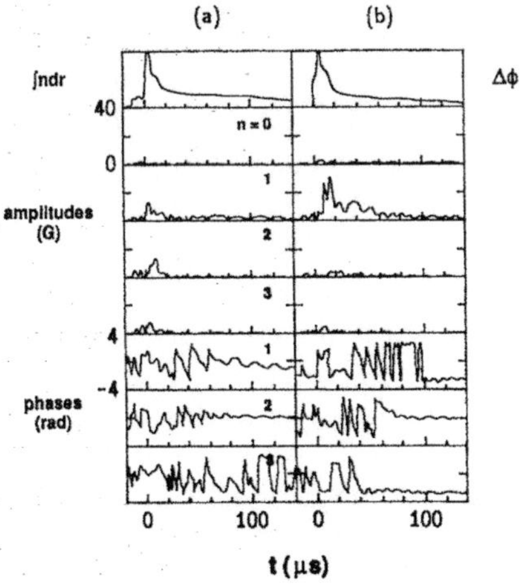

Fig. 8. Toroidal Fourier analysis of the B_θ for bad confinement. The Fourier amplitudes and phases are shown as functions of time for (a) even and (b) odd components The top trace of (a) is a line-integrated electron density and (b) diamagnetism.

Mode (*1)	Observed	Time	Confinement (degraded)	Loss of equilibrium
Co-interchange:				
N=1, M=1 (Tilt)	seldom	various	No	No
Interchange:				
N=1, M=0	occasional	equilibrium	No	No
N=1, M=0 (shift)	occasional	equilibrium	No	No
N=1, M=0 (wobble)	occasional	form/equilibrium	No	No
N>1, M=0 (flute)	seldom	formation	No	No
Rotation :				
N=2, M=1	always	late equilibrium	No (a)	Yes
N>2, M=1	seldom	late equilibrium	No (a)	Yes
Mode (*2)	Observed	Time	Confinement (degraded)	Loss of equilibrium
Co-interchange:				
N=1, M=1 (Tilt)	seldom	various	No	No
Interchange:				
N=1, M=0 (sausage)	often	equilibrium	No	No
N=1, M=0 (shift)	occasional	equilibrium	No	No
N=1, M=0 (wobble)	occasional	form/equilibrium	No	No
N>1, M=0 (flute)	often	formation	No (b)	No
Rotation :				
N=2, M=1	always	late equilibrium	Yes	Yes
N>2, M=1	often	late equilibrium	Yes	Yes
Mode (*3)	Observed	Time	Confinement (degraded)	Loss of equilibrium
Co-interchange:				
N=1, M=1 (Tilt)	seldom	various	No	No
Interchange:				
N=1, M=0 (sausage)	often	equilibrium	No	No
N=1, M=0 (shift)	occasional	equilibrium	No	No
N=1, M=0 (wobble)	occasional	form/equilibrium	No	No
N>1, M=0 (flute)	always	formation	Yes (c)	No (d)
Rotation :				
N=2, M=1	always	late equilibrium	Yes	Yes
N>2, M=1	often	late equilibrium	Yes	Yes

(*1) $1 \le \overline{s} \le 3$, (*2) $3 \le \overline{s} \le 5$, (*3) $5 \le \overline{s} \le 8$, (a) Confinement was not influenced until the mode amplitude was quite large, (b) Poor confinement correlated with non-optimal formation modes that resulted in large-amplitude flutes, (c) All very high-s discharges employed a non-optimal formation sequence, (d) A highly nonlinear flute destroyed the configuration formation

Table 3. Stability properties of FRC in LSX

3.1.4 Recent progress in theoretical understanding

To resolve this discrepancy between MHD predictions and experimental observation, significant progress in the theoretical understanding of FRC stability has been achieved. A host of stabilization effects—for example, the ion FLR effect, the effects of the Hall term and sheared ion flow, resonant particle effects, modern relaxation theory, and two-fluid flowing equilibrium—have been considered in the theoretical studies. Systematic studies of the stability properties of prolate and oblate FRC plasmas have also been presented in a series of Belova's works (Belova et al., 2000, 2001, 2003, 2004, 2006[a], 2006[b]).

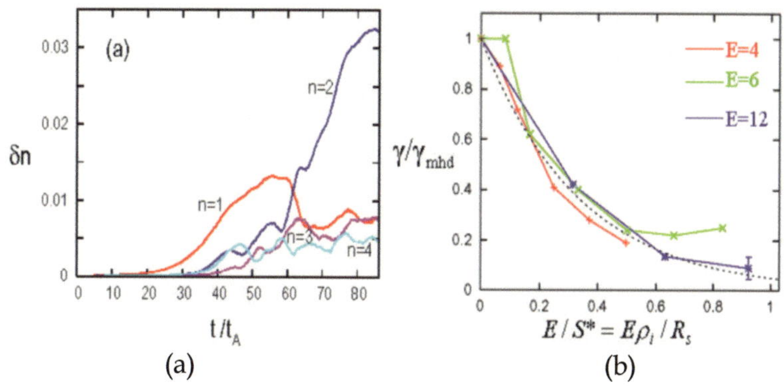

Fig. 9. (a) Time evolution of the amplitudes of different n-modes in prolate FRC with S*=20 and (b) Growth rate of n = 1 tilt instability for three elliptical FRC equilibria with E = 4, 6 and 12

The time evolution of the amplitudes (δn) of different n-modes in prolate FRC with S^* = 20 is shown in Fig. 9 (a) (Belova et al., 2004). Nonlinear saturation of the tilt modes, and growth of the n = 2 rotational mode due to ion toroidal spin-up, have been demonstrated. For oblate FRCs, the scaling of the linear growth rate of n = 1 internal tilt instability, with the parameter of S^*/E for elliptical FRC equilibria (E = 4, 6, 12), has also been investigated, and is shown in Fig. 9 (b) (Belova et al., 2006[a]). The growth rate of the n = 1 tilt mode is decreased in the range of S*/E < 3 ~ 4. For oblate FRC plasma, the stabilized region has been found for all n = 1 modes (the tilt mode, the radial shift, the interchange mode, and the co-interchange mode), with a closed conducting shell and neutral beam injection (Belova et al. 2006[a], 2006[b]).

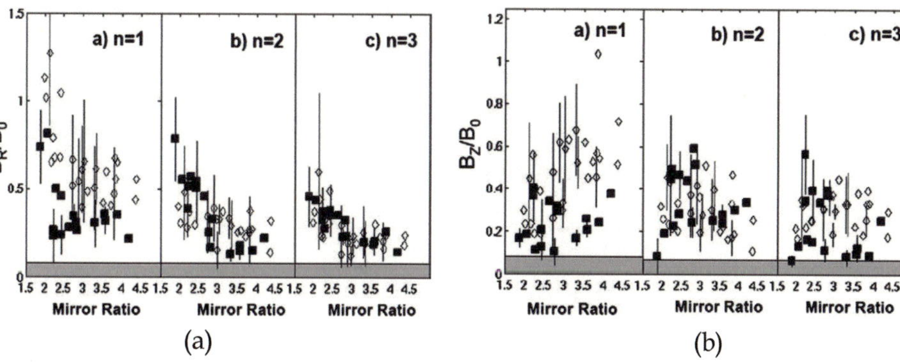

Fig. 10. Magnitude of (a) n = 1, (b) n = 2, (c) n = 3 perturbation in (a) B_r and (b) B_z. Hollow symbols are for cases without the center column, and solid symbols for cases with the center column.

3.2 MHD behavior and stability of oblate FRCs

In the MRX device, the stability of oblate FRC plasma was investigated. The plasma parameters are electron density of 0.5 - 2 x 10^{20} m^{-3}, ion temperature of ~18 eV, B_z of 300 - 200 G, E of 0.65 - 0.35, $\overline{S}$ of 1-3, and $\overline{s}/E$ of 2-5 (Gerhardt et al., 2006). Without a passive stabilizer (a hollow conducting center conductor), tilt and shift instability (n = 1 mode of Fig. 10 (a) and (b)) often appeared. The tilt mode limits the plasma lifetime. The tilt instability can be mitigated by either including a passive stabilizing conductor or forming highly oblate plasmas with a strong mirror field. Without the center column, the growth of the shift mode is reduced, apparently by the large magnetic fields on the outboard side of device. Large perturbations (n = 2 and 3) may still remain after passive stabilization is applied ((b) and (c) of Fig. 10 (a)). These perturbations have the characteristics of co-interchange modes, which have never been observed in conventional oblate FRCs. Such modes cause the early termination of the oblate FRC shape. These co-interchange modes can be stabilized in oblate plasma with a high mirror ratio, and this produces an FRC with maximum configuration lifetime.

4. Summary

In Table 4, the MHD stability properties of both prolate and oblate FRCs are summarized (Yamada et al., 2007). In oblate FRC plasma, the global mode (n = 1 external tilt and shift mode, co-interchange mode) is unstable. But these modes can be stabilized by employing a close-fitting conducting shell or shaping with a strong external magnetic field. In prolate plasma, the internal tilt mode and co-interchange mode are MHD-unstable. However, these can be stabilized by nonlinear effects, such as FLR, rotation and sheared flow. It is difficult to observe these modes clearly in the experiments. The most destructive rotational mode (n = 2), and wobble (n = 1) mode, are almost always observable in experiments. Active stabilization methods without degradation of confinement, such as CHI, need to be developed.

	Prolate [E>1]	**Oblate [E<1]**
Internal Tilt, n=1	MHD unstable, stabilized by FLR, rotation and nonlinear effects for S*<20, E$\geq$5.	MHD Stable
External Tilt and Radial Shift, n=1	MHD stable	MHD unstable, stabilized by conducting shell
Co-interchange, n>1	MHD unstable, stabilized by FLR	MHD unstable, stabilized by NBI + conducting shell
Interchange, n>1	MHD unstable, stabilized by compressional effects	MHD unstable, stabilized by compressional effects
Rotational, n=2	MHD unstable, stabilized by quadrupole field, RMF and conducting shell	MHD unstable, stabilized by quadrupole field, RMF and conducting shell

Table 4. Stability properties of prolate and oblate FRCs

5. References

Armstrong, W. T., et al., (1981), Field-reversed experiments (FRX) on compact toroids, *Physics of Fluids* Vol. 24, No.11, pp. 2068-2088

Asai, T., et al., (2000), Experimental evidence of improved confinement in a high-beta field-reversed configuration plasma by neutral beam injection, *Phys. Plasmas*, Vol. 7, No.6, pp. 2294-2297

Asai, T., et al., (2003), Stabilization of global movement on a field-reversed configuration due to fast neutral beam ions, *Physics of Plasmas*, Vol.10, No.9, pp. 3608-3613

Asai, T., et al., (2006), Tomographic reconstruction of deformed internal structure of a field-reversed configuration, *Physics of Plasmas* Vol. 13, 072508, pp. 1-6.

Asai, T., et al., (2010), Active Stability Control of a High-Beta Self-Organized Compact Torus, *23rd IAEA Fusion Energy*,ICC/P5/01, Available from http:// www-pub.iaea.org/mtcd/meetings/cn180_papers.asp

Belova, E. V., et al., (2000), Numerical study of tilt stability of prolate field-reversed configurations, *Physics of Plasmas*, Vol.7, No. 12, pp. 4996-5006

Belova, E. V., et al., (2001), Numerical study of global stability of oblate field-reversed configurations, *Physics of Plasmas*, Vol.8, No.4, pp. 1267-1277

Belova, E. V., et al., (2003), Kinetic effects on the stability properties of field-reversed configurations. I. Linear stability, *Physics of Plasmas*, Vol.10, No.11, pp. 2361-2371

Belova, E. V., et al., (2004), Kinetic effects on the stability properties of field-reversed configurations. II. Nonlinear évolution, *Physics of Plasmas*, Vol.11, pp. 2523-2531

Belova, E. V., et al., (2006), Advances in the numerical modeling of field-reversed configurations, *Physics of Plasmas,* Vol. 13, 056115, pp. 1-5

Belova, E. V., et al., (2006), Numerical study of the formation, ion spin-up and nonlinear stability properties of field-reversed configurations, *Nuclear Fusion,* Vol.46, pp. 162-170

Binderbauer, M. W., et al., (2010), Dynamic Formation of a Hot Field Reversed Configuration with Improved Confinement by Supersonic Merging of Two Colliding High-Compact Toroids, *Physical Review Letters,* Vol. 105, 045003, pp.1-4.

Cohen, C. D., et al., (2003), Spheromak merging and field reversed configuration formation at the Swarthmore Spheromak Experiment, *Physics of Plasmas*, Vol. 10, No.5 pp. 1748-1754

Davis, H. A., et al., (1976), Generation of Field-Reversing E Layers with Millisecond Lifetimes, *Physical Review Letters,* Vol.37, No.9, pp. 542-545

Freidberg, J. P. and Pearlstein, L. D., (1978), Rotational instabilities in a theta pinch, *Physics of Fluids,* Vol. 21, 1207-1217

Fujimoto, K., et al., (2002), Control of a global motion on field-reversed configuration, *Physics of Plasmas* Vol.9, No.1, pp. 171-176

Gerhardt, S. P. et al., (2006), Equilibrium and stability studies of oblate field-reversed configurations in the Magnetic Reconnection Experiment, *Physics of Plasmas* Vol. 13, 112508, pp.1-18

Gerhardt, S. P., et al., (2008), Field-reversed configuration formation scheme utilizing a spheromak and solenoid induction, *Physics of Plasmas*, Vol.15, 022503, pp.1-11

Greenly, J. B., et al., (1986), Efficient proton ring trapping in an ion ring experiment, *Physics of Fluids*, Vol. 29, No.4, pp.908-911

Guo, H. Y., et al., (2004), Flux Conversion and Evidence of Relaxation in a High- Plasma Formed by High-Speed Injection into a Mirror Confinement Structure, *Physical Review Letters*, Vol. 92, 245001, pp. 1-4

Guo, H. Y., et al., (2005), Observations of Improved Stability and Confinement in a High-β Self-Organized Spherical-Torus-Like Field-Reversed Configuration, *Physical Review Letters*. Vol. 94, 175001, pp. 1-4

Guo, H. Y., et al., (2005), Stabilization of Interchange Modes by Rotating Magnetic Fields, *Physical Review Letters* , Vol.94, 185001, pp. 1-4

Guo, H. Y., et al., (2007), Improved confinement and current drive of high temperature field reversed configurations in the new translation, confinement, and sustainment upgrade device, *Physics of Plasmas*, Vol.14, 112502, pp.1-7

Guo, H. Y., et al., (2011), Formation of a long-lived hot field reversed configuration by dynamically merging two colliding high- compact toroids, *Physics of Plasmas*, Vol. 18, 056110, pp. 1-10

Hoffman, A., et al., (1993), Field reversed configuration lifetime scaling based on measurements from the Large s Experiment, *Nuclear Fusion*, Vol.33, No.1, pp.27-38

Hoffman, A., et al., (2002), Formation and steady-state maintenance of field reversed configuration using rotating magnetic field current drive, *Physics of Plasmas*, Vol.9, No.1, pp. 185-200,

Ikeyama, T., et. al., (2008), Analysis of magnetic probe signals including effect of cylindrical conducting wall for field-reversed configuration experiment, *Review of Science Instruments* Vol.79, 063501, pp. 1-10.

Ishimura, T., (1984), Multipole-field stabilization of the rotational instability in a theta pinch. I, *Physics of Fluids*, Vol. 27, No.8, pp. 2139-2150

Knight, A. J. & Jones, I. R., (1990), A quantitative investigation of rotating magnetic field current drive in a field-reversed configuration, *Plasma Physics and Controlled Nuclear Fusion*, Vol. 32, No. 8, pp. 575-605

Kumashiro, S., et al., (1993), Sources of Fluctuating Field on Field-Reversed Configuration Plasma, *Journal of the Physical Society of Japan*, Vol.62, No.5, pp. 1539-1551

Logan, G. B., et al., (1976), High- β , Gas-Stabilized, Mirror-Confined Plasma, *Physical Review. Letters*, Vol. 37, No.22, pp. 1468-1471

Milroy, R. D., et al., (2008), Toroidal field stabilization of the rotational instability in field-reversed configurations, *Physics of Plasmas* Vol.15, 022508, pp. 1-8

Momota, H., et al., (1992), CONCEPTIONAL DESIGN OF THE D-3He REACTOR ARTEMIS, Fusion Technology Vol. 21, pp.2307-2323

Ohi, S., et al., (1983), Quadrupole stabilization of the n=2 rotational instability of a field-reversed theta-pinch plasma, *Physical Review Letters*, Vol.51, No.12, pp.1042-1045

Ono, Y., et al., (1993), Experimental investigation of three-dimensional magnetic reconnection by use of two colliding spheromaks, *Physics of Fluids B*, Vol.5, No. 10, pp. 3691-3701,

Ono, Y., et al., (1997), Experimental investigation of three-component magnetic reconnection by use of merging spheromaks and tokamaks, *Physics of Plasmas,* Vol. 4, No.5, pp. 1953-1963.

Pierce, W. F., et al., (1995), Stabilization and saturation of the ideal tilt mode in a driven annular field-reversed configuration, *Physics of Plasmas,* Vol. 2, No.3, pp. 846-858

Pietrzyk, Z. P., et al., (1987), Initial results from the Coaxial Slow Source FRC device, *Nuclear. Fusion*, Vol. 27, No.9, pp. 1478-1487

Schamiloglu, E., et. al., (1993), Ion ring propagation in a magnetized plasma, *Physics of Fluids B*, Vol.5, No8, pp. 3069-3078

Shimamura, S. & Nogi, Y. , (1986), Helical Quadrupole Field Stabilization of Field Reversed Configuration Plasma, *Fusion Technology*, Vol.9, No.1, pp. 69-74

Slough, J. T., et al., (1989), Formation studies of field-reversed configurations in a slow field-reversed theta pinch, *Physics of Fluids B,* Vol.1, No.4, pp. 840-850

Slough, J. T. & Hoffman, A. L., (1993), Stability of field-reversed configurations in the large s experiment (LSX), *Physics of Fluids B*, Vol.5, No.12, pp. 4366-4377

Slough, J. T., & Miller, K., (2000), Flux generation and sustainment of a field reversed configuration with rotating magnetic field current drive, *Physics of Plasmas*. Vol.7, No.5, pp.1945-1950

Slough,J. T., et al., (2007), Compression, Heating and Fusion of Colliding Plasmoids by a Z-theta Driven Plasma Linear, *Journal of Fusion Energy*, Vol. 26, No. 1-2, pp. 191-197

Slough, J. T., et al., (2007), The Pulsed High Density Experiment: Concept, Design, and Initial Results, *Journal of Fusion Energy,* Vol. 26, Numbers 1-2, pp. 199-205

Steinhauer, L. C., (2011), Review of field-reversed configurations , *Physics of Plasmas, Vol.* 18, 070501, pp. 1-38

Seyler, C. E., (1979), Vlasov–fluid stability of a rigidly rotating theta pinch, *Physics of Fluids Vol.* 22, pp. 2324-2330

Taccetti, J. M., et al., (2003), FRX-L: A field-reversed configuration plasma injector for magnetized target fusion, *Review of Science Instruments, Vol 74, No.10,* pp.4314-4323

Takahashi, T., et al., (2004), Multichannel optical diagnostic system for field-reversed configuration plasmas, *Review of Science. Instruments*, Vol. 75, No.12, pp. 5205-5212.

Tuszewski, M., (1988), Field reversed configurations, *Nuclear Fusion,* Vol. 28, No.11, pp. 2033-2092

Tuszewski, M., et al., (1990), The n=1 rotational instability in field-reversed configurations, *Physics of Fluids B Vol. 2,* No.11 pp. 2541-2543

Tuszewski, M., et al., (1991), Axial dynamics in field-reversed theta pinches II, Stability, *Physics of Fluids B* Vol.3, No.3, pp. 2856-2870

Yamada, M., et al., (1990), Magnetic reconnection of plasma toroids with cohelicity and counterhelicity, *Physical. Review Letters*, Vol. 65, No.6, pp. 721-724

Yamada, M., et al., (2007), A Self-Organized Plasma with Induction, Reconnection, and Injection Techniques: the SPIRT Concept for Field Reversed Configuration Research, *Plasma and Fusion Research*, Vol. 2, 004, pp. 1-14

6

Review of the Magnetohydrodynamic Waves and Their Stability in Solar Spicules and X-Ray Jets

Ivan Zhelyazkov
Faculty of Physics, Sofia University
Bulgaria

1. Introduction

One of the most enduring mysteries in solar physics is why the Sun's outer atmosphere, or corona, is millions of kelvins hotter than its surface. Among suggested theories for coronal heating is one that considers the role of spicules – narrow jets of plasma shooting up from just above the Sun's surface – in that process (Athay & Holzer, 1982; Athay, 2000). For decades, it was thought that spicules might be sending heat into the corona. However, following observational research in the 1980s, it was found that spicule plasma did not reach coronal temperatures, and so this line of study largely fell out of vogue. Kukhianidze et al. (Kukhianidze et al., 2006) were first to report the observation of kink waves in solar spicules – the wavelength was found to be $\sim$3500 km, and the period of waves has been estimated to be in the range of 35–70 s. The authors argue that these waves may carry photospheric energy into the corona and therefore can be of importance in coronal heating. Zaqarashvili et al. (Zaqarashvili et al., 2007) analyzed consecutive height series of Hα spectra in solar limb spicules at the heights of 3800–8700 km above the photosphere and detected Doppler-shift oscillations with periods of 20–25 and 75–110 s. According to authors, the oscillations can be caused by waves' propagation in thin magnetic flux tubes anchored in the photosphere. Moreover, observed waves can be used as a tool for spicule seismology, and the magnetic filed induction in spicules at the height of $\sim$6000 km above the photosphere is estimated as 12–15 G. De Pontieu et al. (De Pontieu et al., 2007) identified a new class of spicules (see Fig. 1) that moved much faster and were shorter lived than the traditional spicules, which have speeds of between 20 and 40 km s^{-1} and lifespans of 3 to 7 minutes. These Type II spicules, observed in Ca II 854.2 nm and Hα lines (Sterling et al., 2010), are much more dynamic: they form rapidly (in $\sim$10 s), are very thin ($\leqslant$200 km wide), have lifetimes of 10 to 150 s (at any one height), and shoot upwards at high speeds, often in excess of 100–150 km s^{-1}, before disappearing. The rapid disappearance of these jets had suggested that the plasma they carried might get very hot, but direct observational evidence of this process was missing. Both types of spicules are observed to carry Alfvén waves with significant amplitudes of order 20 km s^{-1}. In a recent paper, De Pontieu et al. (De Pontieu et al., 2011) used new observations from the Atmospheric Imaging Assembly on NASA's recently launched *Solar Dynamics Observatory* and its Focal Plane Package for the Solar Optical Telescope (SOT) on the Japanese *Hinode* satellite. Their observations reveal "a ubiquitous coronal mass supply in which chromospheric plasma in fountainlike jets or spicules (see Fig. 2) is accelerated upward into the corona, with much of the plasma heated to temperatures between $\sim$0.02 and 0.1 million kelvin (MK) and a small but sufficient fraction to temperatures above 1 MK. These observations provide constraints

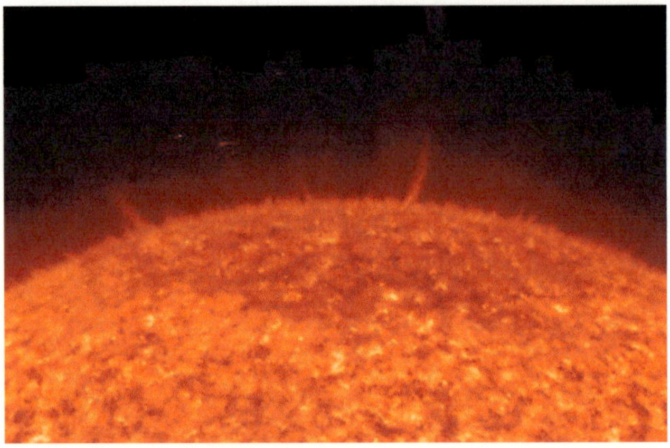

Fig. 1. Solar spicules on the Sun recorded on August 3, 2007. Credit: NASA/*STEREO*.

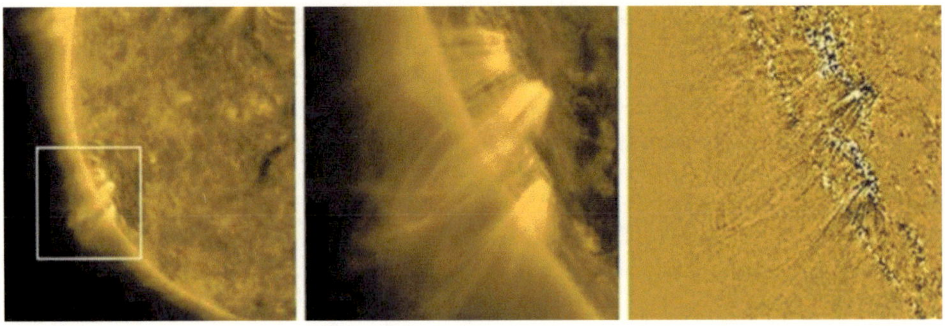

Fig. 2. Solar spicules recorded by the *Solar Dynamics Observatory* on April 25, 2010. Credit: NASA/*SDO*.

on the coronal heating mechanism(s) and highlight the importance of the interface region between photosphere and corona." Nevertheless, Moore et al. (Moore et al., 2011) from *Hinode* observations of solar X-ray jets, Type II spicules, and granule-size emerging bipolar magnetic fields in quiet regions and coronal holes, advocate a scenario for powering coronal heating and the solar wind. In this scenario, Type II spicules and Alfvén waves are generated by the granule-size emerging bipoles in the manner of the generation of X-ray jets by larger magnetic bipoles. From observations and this scenario, the authors estimate that Type II spicules and their co-generated Alfvén waves carry into the corona an area-average flux of mechanical energy of $\sim 7 \times 10^5$ erg s^{-1} cm^{-2}. This is enough to power the corona and solar wind in quiet regions and coronal holes, hence indicates that the granule-size emerging bipoles are the main engines that generate and sustain the entire heliosphere. The upward propagation of high- and low-frequency Alfvén waves along spicules detected from SOT's observations on *Hinode* was also reported by He et al. (He et al, 1999) and Tavabi et al. (Tavabi et al., 2011). He et al. found in four cases that the spicules are modulated by high-frequency ($\geqslant$0.02 Hz) transverse fluctuations. These fluctuations are suggested to be Alfvén waves that propagate upwards along the spicules with phase speed ranges from 50 to 150 km s^{-1}. Three of the modulated spicules show clear wave-like shapes with short wavelengths less than 8 Mm. We note that at the same time, Kudoh & Shibata (Kudoh & Shibata, 1999) presented a torsional Alfvén-wave model of spicules (actually the classical Type I spicules) and discussed the possibility for wave coronal heating – the energy flux transported into corona was estimated to be of about 3×10^5 erg s^{-1} cm^{-2}, i.e., roughly half of the flux carried by the Alfvén waves running on Type II spicules (Moore et al., 2011). Tavabi et al. (Tavabi et al., 2011), performed a statistical analysis of the SOT/*Hinode* observations of solar spicules and their wave-like behavior, and argued that there is a possible upward propagation of Alfvén waves inside a doublet spicule with a typical wave's period of 110 s.

No less effective in coronal heating are the so called X-ray jets. We recall, however, that whilst the classical spicules were first discovered in 1870's by the Jesuit astronomer Pietro Angelo Secchi (Secchi, 1877) and named as "spicules" by Roberts (Roberts, 1945), the X-ray jets are relatively a new discovered phenomenon. They, the jets, were extensively observed with the Soft X-ray Telescope on *Yohkoh* (Shibata et al., 1992; Shimojo et al., 1996), and their structure and dynamics have been better resolved by the X-Ray Telescope (XRT) on *Hinode*, in movies having 1 arc sec pixels and ~1-minute cadence (Cirtain et al., 2007) – see Fig. 3. According

Fig. 3. Three X-ray jets recorded by the *Hinode* spacecraft on January 10, 2007. Credit: SAO/NASA/JAXA/NAOJ.

to Cirtain et al. (Cirtain et al., 2007), "coronal magnetic fields are dynamic, and field lines may misalign, reassemble, and release energy by means of magnetic reconnection. Giant releases may generate solar flares and coronal mass ejections and, on a smaller scale, produce X-ray jets. *Hinode* observations of polar coronal holes reveal that X-ray jets have two distinct velocities: one near the Alfvén speed (~800 kilometers per second) and another near the sound speed (200 kilometers per second). The X-ray jets are from 2×10^3 to 2×10^4 kilometers wide and 1×10^5 kilometers long and last from 100 to 2500 seconds. The large number of events, coupled with the high velocities of the apparent outflows, indicates that the jets may contribute to the high-speed solar wind." The more recent observations (Madjarska, 2011; Shimojo & Shibata, 2000) yield that the temperature of X-ray jets is from 1.3 to 12 MK (i.e., the jets are hotter than the ambient corona) and the electron/ion number density is of about $(0.7–4) \times 10^9$ cm^{-3} with average of 1.7×10^9 cm^{-3}. The X-ray jets can have velocities above 10^3 km s^{-1}, reach heights of a solar radius or more, and have kinetic energies of the order of 10^{29} erg.

Since both spicules and X-ray jets support Alfvén (or more generally magnetohydrodynamic) waves' propagation it is of great importance to determine their dispersion characteristics and more specifically their stability/instability status. If while propagating along the jets MHD waves become unstable and the expected instability is of the Kelvin–Helmholtz type, that instability can trigger the onset of wave turbulence leading to an effective plasma jet heating and the acceleration of the charged particles. We note that the Alfvénic turbulence is considered to be the most promising source of heating in the chromosphere and extended corona (van Ballegooijen et al., 2011). In this study, we investigate these travelling wave properties for a realistic, cylindrical geometry of the spicules and X-ray jets considering appropriate values for the basic plasma jet parameters (mass density, magnetic fields, sound, Alfvén, and jet speeds), as well as those of the surrounding medium. For detailed reviews of the oscillations and waves in magnetically structured solar spicules we refer the reader to (Zaqarashvili & Erdélyi, 2009) and (Zaqarashvili, 2011). Our research concerns the dispersion curves of kink and sausage modes for the MHD waves travelling primarily along the Type II spicules and X-ray jets for various values of the jet speed. In studying wave propagation characteristics, we assume that the axial wave number k_z ($\hat{z}$ is the direction of the embedded constant magnetic fields in the two media) is real, while the angular wave frequency, ω, is complex. The imaginary part of that complex frequency is the wave growth rate when a given mode becomes unstable. All of our analysis is based on a linearized set of equations for the adopted form of magnetohydrodynamics. We show that the stability/instability status of the travelling waves depends entirely on the magnitudes of the flow velocities and the values of two important control parameters, namely the so-called density contrast (the ratio of the mass density inside to that outside the flux tube) and the ratio of the background magnetic field of the environment to that of the spicules and X-ray jets.

2. Geometry and basic magnetohydrodynamic equations

The simplest model of spicules is a straight vertical cylinder (see Fig. 4) with radius a

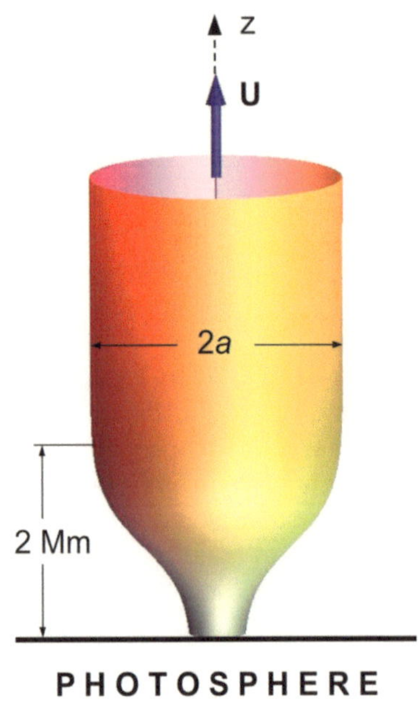

Fig. 4. Geometry of a spicule flux tube containing flowing plasma with velocity **U**.

filled with ideal compressible plasma of density $\rho_i \sim 3 \times 10^{-13}$ g cm^{-3} (Sterling, 2000) and immersed in a constant magnetic field $\mathbf{B}_i$ directed along the z axis. Such a cylinder is usually termed *magnetic flux tube* or simply 'flux tube.' The most natural discontinuity, which occurs at the surface binding the cylinder, is the tangential one because it is the discontinuity that ensures an equilibrium total pressure balance. Moreover, it is worth noting that the jet is non-rotating and without twist – otherwise the centrifugal and the magnetic tension forces should be taken into account. Due to the specific form of the real flux tube which models a spicule, that part of the whole flux tube having a constant radius actually starts at the height of 2 Mm from the tube footpoint. The flow velocity, $\mathbf{U}_i$, like the ambient magnetic field, is directed along the z axis. The mass density of the environment, ρ_e, is much, say 50–100 times, less than that of the spicule, while the magnetic field induction B_e might be of the order or less than $B_i \sim$ 10–15 G. Both the magnetic field, $\mathbf{B}_e$, and flow velocity, $\mathbf{U}_e$ (if any), are also in the $\hat{z}$-direction. We note that while the parameters of classical Type I spicules are well-documented (Beckers, 1968; 1972) those of Type II spicules are generally disputed; Centeno et al. (Centeno et al., 2010), for example, on using a novel inversion code for Stokes profiles caused by the joint action of atomic level polarization and the Hanle and Zeeman effects to interpret the observations, claim that magnetic fields as strong as ~50 G were detected in a very localized area of the slit, which might represent a lower field strength of organized network spicules.

The flux tube modelling of the X-ray jets is actually the same as that for spicules, however, with different magnitudes of the mass densities, flow velocities, and background magnetic fields. When studying waves' propagation and their stability/instability status for a given solar structure (spicule or X-ray jet), the values of the basic parameters will be additionally specified. Now let us see what are the basic magnetohydrodynamic equations governing the motions in a flowing solar plasma.

2.1 Basic equations of ideal magnetohydrodynamics

Magnetohydrodynamics (MHD) studies the dynamics of electrically conducting fluids. Examples of such fluids include plasmas and liquid metals. The field of MHD was initiated in 1942 by the Swedish physicist Hannes Alfvén (1908–1995), who received the Nobel Prize in Physics (1970) for "fundamental work and discoveries in magnetohydrodynamics with fruitful applications in different parts of plasma physics." The fundamental concept behind MHD is that magnetic fields can induce currents in a moving conductive fluid, which in turn creates forces on the fluid and also changes the magnetic field itself. The set of equations, which describe MHD are a combination of the equations of motion of fluid dynamics (Navier–Stokes equations) and Maxwell's equations of electromagnetism. These partial differential equations have to be solved simultaneously, either analytically or numerically.

Magnetohydrodynamics is a macroscopic theory. Its equations can in principle be derived from the kinetic Boltzmann's equation assuming space and time scales to be larger than all inherent scale-lengths such as the Debye length or the gyro-radii of the charged particles (Chen, 1995). It is, however, more convenient to obtain the MHD equations in a phenomenological way as an electromagnetic extension of the hydrodynamic equations of ordinary fluids, where the main approximation is to neglect the displacement current $\propto \partial \mathbf{E}/\partial t$ in Ampère's law.

In the standard nonrelativistic form the MHD equations consist of the basic conservation laws of mass, momentum, and energy together with the induction equation for the magnetic field. Thus, the MHD equations of our magnetized quasineutral plasma with singly charged ions (and electrons) are

$$\frac{\partial \rho}{\partial t} + \nabla \cdot \rho \mathbf{v} = 0, \tag{1}$$

where ρ is the mass density and $\mathbf{v}$ is the bulk fluid velocity. Equation (1) is the so called *continuity equation* in our basis set of equations.

The momentum equation is

$$\frac{\partial(\rho \mathbf{v})}{\partial t} + \rho(\mathbf{v} \cdot \nabla)\mathbf{v} = \mathbf{j} \times \mathbf{B} - \nabla p + \rho \mathbf{g}, \tag{2}$$

where $\mathbf{j} \times \mathbf{B}$ (with $\mathbf{j}$ being the current density and $\mathbf{B}$ magnetic field induction) is the *Lorentz force* term, $-\nabla p$ is the pressure-gradient term, and $\rho \mathbf{g}$ is the gravity force.

Faraday's law reads

$$\frac{\partial \mathbf{B}}{\partial t} = -\nabla \times \mathbf{E}, \tag{3}$$

where $\mathbf{E}$ is the electric field. The ideal Ohm's law for a plasma, which yields a useful relation between electric and magnetic fields, is

$$\mathbf{E} + \mathbf{v} \times \mathbf{B} = 0. \tag{4}$$

The low-frequency Ampère's law, which neglects the displacement current, is given by

$$\mu_0 \mathbf{j} = \nabla \times \mathbf{B}, \tag{5}$$

where μ_0 is the permeability of free space.

The magnetic divergency constraint is

$$\nabla \cdot \mathbf{B} = 0. \tag{6}$$

By determining the current density $\mathbf{j}$ from Ampère's Eq. (5), the expression of the Lorentz force can be presented in the form

$$\mathbf{j} \times \mathbf{B} = \frac{1}{\mu_0}(\mathbf{B} \cdot \nabla)\mathbf{B} - \nabla \left(\frac{B^2}{2\mu_0} \right),$$

where the first term on the right hand side is the magnetic tension force and the second term is the magnetic pressure force. Thus, momentum Eq. (2) can be rewritten in a more convenient form, notably

$$\frac{\partial(\rho \mathbf{v})}{\partial t} + \rho(\mathbf{v} \cdot \nabla)\mathbf{v} = -\nabla \left(p + \frac{B^2}{2\mu_0} \right) + \frac{1}{\mu_0}(\mathbf{B} \cdot \nabla)\mathbf{B} + \rho \mathbf{g}. \tag{7}$$

On the other hand, on using Ohm's law (4) the Faraday's law (or induction equation) takes the form

$$\frac{\partial \mathbf{B}}{\partial t} = -\nabla \times (\mathbf{v} \times \mathbf{B}). \tag{8}$$

Finally, the equation of the thermal energy is given by

$$\frac{\mathrm{d}}{\mathrm{d}t}\frac{p}{\rho^{\gamma}} = 0,$$

where $\gamma = 5/3$ is the ratio of specific heats for an adiabatic equation of state. This equation usually is written as an equation for the pressure, p,

$$\frac{\partial p}{\partial t} + \mathbf{v}\cdot\nabla p + \gamma p \nabla\cdot\mathbf{v} = 0. \quad (9)$$

Equation (9) implies that the equation of state of the ideal fully ionized gas has the form

$$p = 2(\rho/m_{\mathrm{i}})k_{\mathrm{B}}T,$$

where T is the temperature, m_{i} the ion mass, k_{B} is the Boltzmann constant, and the factor 2 arises because ions and electrons contribute equally.

In total the ideal MHD equations thus consist of two vector equations, (7) and (8), and two scalar equations, (1) and (9), respectively. Occasionally, when studying wave propagation in magnetized plasmas, one might also be necessary to use Eq. (6). We note that the basic variables of the ideal MHD are the mass density, ρ, the fluid bulk velocity, $\mathbf{v}$, the pressure, p, and the magnetic induction, $\mathbf{B}$; the electric field, $\mathbf{E}$, has been excluded via Ohm's law.

In MHD there is a few dimensionless numbers, which are widely used in studying various phenomena in magnetized plasmas. Such an important dimensionless number in MHD theory is the plasma beta, β, defined as the ratio of gas pressure, p, to the magnetic pressure,

$$\beta = \frac{p}{B^2/2\mu_0}.$$

When the magnetic field dominates in the fluid, $\beta \ll 1$, the fluid is forced to move along with the field. In the opposite case, when the field is weak, $\beta \gg 1$, the field is swirled along by the fluid.

We finish our short introduction to MHD recalling that in ideal MHD Lenz's law dictates that the fluid is in a sense tied to the magnetic field lines, or, equivalently, magnetic filed lines are *frozen into the fluid*. To explain, in ideal MHD a small rope-like volume of fluid surrounding a field line will continue to lie along a magnetic field line, even as it is twisted and distorted by fluid flows in the system. The connection between magnetic field lines and fluid in ideal MHD fixes the topology of the magnetic field in the fluid.

3. Wave dispersion relations

It is well-known that in infinite magnetized plasmas there exist three types of MHD waves (Chen, 1995), namely the Alfvén wave and the fast and slow magnetoacoustic waves. Alfvén wave (Alfvén, 1942; Gekelman et al., 2011), is a transverse wave propagating at speed $v_{\mathrm{A}} = B_0/(\mu_0\rho_0)^{1/2}$, where B_0 and ρ_0 are the equilibrium (not perturbed) magnetic field and mass density, respectively. The propagation characteristics of magnetoacoustic waves depend upon their plasma beta environment. In particular, in high-beta plasmas ($\beta \gg 1$) the fast magnetoacoustic wave behaves like a sound wave travelling at sound speed $c_{\mathrm{s}} = (\gamma p_0/\rho_0)^{1/2}$, while in low-beta plasmas ($\beta \ll 1$) it propagates roughly isotropically and across the magnetic field lines at Alfvén speed, v_A. The slow magnetoacoustic wave in high-beta plasmas is guided along the magnetic field $\mathbf{B}_0$ at Alfvén speed, v_A – in the opposite case of low-beta plasmas it is a longitudinally propagating along $\mathbf{B}_0$ wave at sound speed, c_s. A question that immediately raises is how these waves will change when the magnetized plasma is spatially bounded (or magnetically structured) as in our case of spicules or X-ray jets. The answer to that question is not trivial – we actually have to derive the *normal modes* supported by the flux tube, which models the jets.

As we will study a linear wave propagation, the basic MHD variables can be presented in the form

$$\rho = \rho_0 + \delta\rho, \quad p = p_0 + \delta p, \quad \mathbf{v} = \mathbf{U} + \delta\mathbf{v}, \quad \text{and} \quad \mathbf{B} = \mathbf{B}_0 + \delta\mathbf{B},$$

where ρ_0, p_0, and $\mathbf{B}_0$ are the equilibrium values in either medium, $\mathbf{U}_\mathrm{i}$ and $\mathbf{U}_\mathrm{e}$ are the flow velocities inside and outside the flux tube, $\delta\rho$, δp, $\delta\mathbf{v}$, and $\delta\mathbf{B}$ being the small perturbations of the basic MHD variables. For convenience, we chose the frame of reference to be attached to the ambient medium. In that case

$$\mathbf{U}^\mathrm{rel} = \mathbf{U}_\mathrm{i} - \mathbf{U}_\mathrm{e}$$

is the *relative* flow velocity whose magnitude is a non-zero number inside the jet, and zero in the surrounding medium. For spicules, $U_\mathrm{e} \approx 0$; which is why the relative flow velocity is indeed the jet velocity, which we later denote as simply $\mathbf{U}$.

With the above assumptions, the basic set of MHD equations for the perturbations of the mass density, pressure, fluid velocity, and magnetic field become

$$\frac{\partial}{\partial t}\delta\rho + (\mathbf{U}\cdot\nabla)\delta\rho + \rho_0\nabla\delta\mathbf{v} = 0, \tag{10}$$

$$\rho_0\frac{\partial}{\partial t}\delta\mathbf{v} + \rho_0\left(\mathbf{U}\cdot\nabla\right)\delta\mathbf{v} + \nabla\left(\delta p + \frac{1}{\mu_0}\mathbf{B}_0\cdot\delta\mathbf{B}\right) - \frac{1}{\mu_0}(\mathbf{B}_0\cdot\nabla)\delta\mathbf{B} = 0, \tag{11}$$

$$\frac{\partial}{\partial t}\delta\mathbf{B} + (\mathbf{U}\cdot\nabla)\delta\mathbf{B} - (\mathbf{B}_0\cdot\nabla)\delta\mathbf{v} + \mathbf{B}_0\nabla\cdot\delta\mathbf{v} = 0, \tag{12}$$

$$\frac{\partial}{\partial t}\delta p + (\mathbf{U}\cdot\nabla)\delta p + \gamma p_0\nabla\cdot\delta\mathbf{v} = 0, \tag{13}$$

$$\nabla\cdot\delta\mathbf{B} = 0. \tag{14}$$

We note that the gravity force term in momentum Eq. (11) has been omitted because one assumes that the mass density of the jet does not change appreciably in the limits of the spicule's length of order 10–11 Mm.

From Eq. (10) we obtain that

$$\nabla\cdot\delta\mathbf{v} = -\frac{1}{\rho_0}\left[\frac{\partial}{\partial t}\delta\rho + (\mathbf{U}\cdot\nabla)\delta\rho\right]. \tag{15}$$

Inserting this expression into Eq. (13) we get

$$\left[\frac{\partial}{\partial t} + (\mathbf{U}\cdot\nabla)\right]\delta p - c_\mathrm{s}^2\left[\frac{\partial}{\partial t} + (\mathbf{U}\cdot\nabla)\right]\delta\rho = 0,$$

which means that the pressure's and density's perturbations are related via the expression

$$\delta p = c_\mathrm{s}^2\delta\rho, \qquad \text{where} \qquad c_\mathrm{s} = \left(\gamma p_0/\rho_0\right)^{1/2}. \tag{16}$$

Assuming that each perturbation is presented as a plain wave $g(r)\exp\left[\mathrm{i}\left(-\omega t + m\varphi + k_z z\right)\right]$ with its amplitude $g(r)$ being just a function of r, and that in cylindrical coordinates the nabla operator has the form

$$\nabla \equiv \frac{\partial}{\partial r}\hat{r} + \frac{1}{r}\frac{\partial}{\partial \varphi}\hat{\varphi} + \frac{\partial}{\partial z}\hat{z},$$

Eq. (11) reads

$$-\mathrm{i}\rho_0(\omega - \mathbf{k}\cdot\mathbf{U})\delta v_r + \frac{\mathrm{d}}{\mathrm{d}r}\left(\delta p + \frac{1}{\mu_0}B_0\delta B_z\right) - \mathrm{i}k_z\frac{1}{\mu_0}B_0\delta B_r = 0, \tag{17}$$

$$-\rho_0(\omega - \mathbf{k}\cdot\mathbf{U})\delta v_\varphi + \frac{m}{r}\left(\delta p + \frac{1}{\mu_0}B_0\delta B_z\right) - k_z\frac{1}{\mu_0}B_0\delta B_\varphi = 0, \tag{18}$$

$$-\rho_0(\omega - \mathbf{k}\cdot\mathbf{U})\delta v_z + k_z\left(\delta p + \frac{1}{\mu_0}B_0\delta B_z\right) - k_z\frac{1}{\mu_0}B_0\delta B_z = 0. \tag{19}$$

Accordingly Eq. (13) yields

$$-\mathrm{i}(\omega - \mathbf{k}\cdot\mathbf{U})\delta p + \gamma p_0\nabla\cdot\delta\mathbf{v} = 0. \tag{20}$$

Induction Eq. (12) gives

$$(\omega - \mathbf{k}\cdot\mathbf{U})\delta B_r - k_z B_0\delta v_r = 0, \tag{21}$$

$$(\omega - \mathbf{k}\cdot\mathbf{U})\delta B_\varphi - k_z B_0\delta v_\varphi = 0, \tag{22}$$

$$-\mathrm{i}(\omega - \mathbf{k}\cdot\mathbf{U})\delta B_z - \mathrm{i}k_z B_0\delta v_z + B_0\nabla\cdot\delta\mathbf{v} = 0. \tag{23}$$

Finally Eq. (14) yields

$$\frac{\mathrm{d}}{\mathrm{d}r}\delta B_r + \frac{1}{r}\delta B_r + \mathrm{i}\frac{m}{r}\delta B_\varphi + \mathrm{i}k_z\delta B_z = 0. \tag{24}$$

From Eq. (19) we obtain

$$\delta v_z = \frac{k_z}{\omega - \mathbf{k}\cdot\mathbf{U}}\frac{1}{\rho_0}\delta p \quad \text{or} \quad \delta p = \rho_0\frac{\omega - \mathbf{k}\cdot\mathbf{U}}{k_z}\delta v_z, \tag{25}$$

while Eq. (20) gives

$$\delta p = -\mathrm{i}\frac{1}{\omega - \mathbf{k}\cdot\mathbf{U}}\gamma p_0\nabla\cdot\delta\mathbf{v},$$

which means that

$$\delta v_z = -\mathrm{i}\frac{k_z}{(\omega - \mathbf{k}\cdot\mathbf{U})^2}\frac{\gamma p_0}{\rho_0}\left(\frac{\mathrm{d}}{\mathrm{d}r}\delta v_r + \frac{1}{r}\delta v_r + \mathrm{i}\frac{m}{r}\delta v_\varphi + \mathrm{i}k_z\delta v_z\right).$$

After some rearranging this expression can be rewritten in the form

$$\frac{\mathrm{d}}{\mathrm{d}r}\delta v_r + \frac{1}{r}\delta v_r + \mathrm{i}\frac{m}{r}\delta v_\varphi = \mathrm{i}\frac{(\omega - \mathbf{k}\cdot\mathbf{U})^2 - k_z^2 c_\mathrm{s}^2}{k_z c_\mathrm{s}^2}\delta v_z \tag{26}$$

Let us now differentiate Eq. (17) with respect to r:

$$-\mathrm{i}\rho_0(\omega - \mathbf{k}\cdot\mathbf{U})\frac{\mathrm{d}}{\mathrm{d}r}\delta v_r + \frac{\mathrm{d}^2}{\mathrm{d}r^2}\left(\delta p + \frac{1}{\mu_0}B_0\delta B_z\right) - \frac{1}{\mu_0}B_0\mathrm{i}k_z\frac{\mathrm{d}}{\mathrm{d}r}\delta B_r = 0. \tag{27}$$

But according to Eqs. (26) and (24)

$$\frac{\mathrm{d}}{\mathrm{d}r}\delta v_r = -\frac{1}{r}\delta v_r - \mathrm{i}\frac{m}{r}\delta v_\varphi - \mathrm{i}k_z\left[1-\frac{(\omega-\mathbf{k}\cdot\mathbf{U})^2}{k_z^2c_\mathrm{s}^2}\right]\delta v_z,$$

$$\frac{\mathrm{d}}{\mathrm{d}r}\delta B_r = -\frac{1}{r}\delta B_r - \mathrm{i}\frac{m}{r}\delta B_\varphi - \mathrm{i}k_z\delta B_z.$$

Then Eq. (27) becomes

$$\mathrm{i}\rho_0(\omega-\mathbf{k}\cdot\mathbf{U})\frac{1}{r}\delta v_r - \rho_0(\omega-\mathbf{k}\cdot\mathbf{U})\frac{m}{r}\delta v_\varphi - \rho_0(\omega-\mathbf{k}\cdot\mathbf{U})\left[1-\frac{(\omega-\mathbf{k}\cdot\mathbf{U})^2}{k_z^2c_\mathrm{s}^2}\right]k_z\delta v_z$$

$$+\frac{\mathrm{d}^2}{\mathrm{d}r^2}\left(\delta p + \frac{1}{\mu_0}B_0\delta B_z\right) + \frac{1}{\mu_0}B_0\mathrm{i}k_z\frac{1}{r}\delta B_r - \frac{1}{\mu_0}B_0k_z\frac{m}{r}\delta B_\varphi - \frac{1}{\mu_0}B_0k_z^2\delta B_z = 0 \quad (28)$$

In order to simplify notation we introduce a new variable, namely the perturbation of the total pressure, $\delta p_\mathrm{tot} = \delta p + \frac{1}{\mu_0}B_0\delta B_z$. From Eqs. (17) to (19) one can get that

$$\frac{1}{\mu_0}B_0\mathrm{i}k_z\frac{1}{r}\delta B_r = -\mathrm{i}\rho_0(\omega-\mathbf{k}\cdot\mathbf{U})\frac{1}{r}\delta v_r + \frac{1}{r}\frac{\mathrm{d}}{\mathrm{d}r}\delta p_\mathrm{tot},$$

$$-\frac{1}{\mu_0}B_0\mathrm{i}k_z\frac{m}{r}\delta B_\varphi = \rho_0(\omega-\mathbf{k}\cdot\mathbf{U})\frac{m}{r}\delta v_\varphi - \frac{m^2}{r^2}\delta p_\mathrm{tot},$$

$$-\frac{1}{\mu_0}B_0\mathrm{i}k_z^2\delta B_z = \rho_0(\omega-\mathbf{k}\cdot\mathbf{U})k_z\delta v_z - k_z^2\delta p_\mathrm{tot}.$$

Inserting these expressions into Eq. (28), we obtain

$$\left[\frac{\mathrm{d}^2}{\mathrm{d}r^2} + \frac{1}{r}\frac{\mathrm{d}}{\mathrm{d}r} - \left(k_z^2 + \frac{m^2}{r^2}\right)\right]\delta p_\mathrm{tot} + \rho_0\frac{(\omega-\mathbf{k}\cdot\mathbf{U})^3}{k_zc_\mathrm{s}^2}\delta v_z = 0. \quad (29)$$

Bearing in mind that according to Eq. (15)

$$\nabla\cdot\delta\mathbf{v} = \mathrm{i}(\omega-\mathbf{k}\cdot\mathbf{U})\frac{\delta\rho}{\rho_0},$$

from Eq. (23) we get

$$(\omega-\mathbf{k}\cdot\mathbf{U})\delta B_z - k_zB_0\delta v_z + B_0(\omega-\mathbf{k}\cdot\mathbf{U})\frac{\delta\rho}{\rho_0} = 0.$$

On using Eq. (16) we express $\delta\rho$ in the above equation as $\delta p/c_\mathrm{s}^2$, multiply it by

$$-\frac{1}{\mu_0}B_0\frac{1}{\omega-\mathbf{k}\cdot\mathbf{U}}$$

to get after some algebra that

$$\delta p + \frac{1}{\mu_0}B_0\delta B_z = -\frac{k_z\rho_0}{\omega-\mathbf{k}\cdot\mathbf{U}}\frac{B_0^2}{\mu_0\rho_0}\delta v_z + \delta p\left(1+\frac{v_\mathrm{A}^2}{c_\mathrm{s}^2}\right),$$

where, we remember, $v_A = B_0/(\mu_0\rho_0)^{1/2}$ is the Alfvén speed. After inserting in the above equation δp expressed in terms of δv_z – see Eq. (25) – and performing some straightforward algebra we obtain that

$$\delta v_z = -\frac{1}{\rho_0}\frac{\omega - \mathbf{k}\cdot\mathbf{U}}{k_z}\frac{k_z^2 c_s^2}{k_z^2 c_s^2 v_A^2 - (\omega - \mathbf{k}\cdot\mathbf{U})^2\left(c_s^2 + v_A^2\right)}\delta p_{\rm tot}.$$

Next step is to insert above expression of δv_z into Eq. (29) and combine the $-k_z^2\,\delta p$-term with the last member in the same equation to get a new form of Eq. (29), notably

$$\left[\frac{d^2}{dr^2} + \frac{1}{r}\frac{d}{dr} - \left(\kappa^2 + \frac{m^2}{r^2}\right)\right]\delta p_{\rm tot} = 0. \qquad (30)$$

Here, κ^2 is given by the expression

$$\kappa^2 = -\frac{\left[(\omega - \mathbf{k}\cdot\mathbf{U})^2 - k_z^2 c_s^2\right]\left[(\omega - \mathbf{k}\cdot\mathbf{U})^2 - k_z^2 v_A^2\right]}{\left(c_s^2 + v_A^2\right)\left[(\omega - \mathbf{k}\cdot\mathbf{U})^2 - k_z^2 c_T^2\right]}, \qquad (31)$$

where

$$c_T = \frac{c_s v_A}{\left(c_s^2 + v_A^2\right)^{1/2}} \qquad (32)$$

is the so-called *tube velocity* (Edwin & Roberts, 1983). It is important to notice that both κ^2 (respectively κ) and the tube velocity, c_T, *have different values inside and outside the jet* due to the different sound and Alfvén speeds, which characterize correspondingly the jet and its surrounding medium.

As can be seen, Eq. (30) is the equation for the modified Bessel functions I_m and K_m and, accordingly, its solutions in both media (the jet and its environment) are:

$$\delta p_{\rm tot}(r) = \begin{cases} A_i I_m(\kappa_i r) & \text{for } r \leqslant a, \\ A_e K_m(\kappa_e r) & \text{for } r \geqslant a. \end{cases}$$

From Eq. (17) one can obtain an expression of δv_r and inserting it in the expression of δB_r deduced from Eq. (21) one gets a formula relating δv_r with the first derivative (with respect to r) of $\delta p_{\rm tot}$

$$\delta v_r = -\frac{i}{\rho_0}\frac{\omega - \mathbf{k}\cdot\mathbf{U}}{(\omega - \mathbf{k}\cdot\mathbf{U})^2 - k_z^2 v_A^2}\frac{d}{dr}\delta p_{\rm tot}. \qquad (33)$$

It is clear that we have two different expressions of δv_r, which, bearing in mind the solutions to the ordinary second order differential Eq. (30), read

$$\delta v_r(r \leqslant a) = -\frac{i}{\rho_i}\frac{\omega - \mathbf{k}\cdot\mathbf{U}}{(\omega - \mathbf{k}\cdot\mathbf{U})^2 - k_z^2 v_{Ai}^2}\kappa_i A_i I_m'(\kappa_i r)$$

and

$$\delta v_r(r \geqslant a) = -\frac{i}{\rho_e}\frac{\omega}{\omega^2 - k_z^2 v_{Ae}^2}\kappa_e A_e K_m'(\kappa_e r),$$

respectively. Now it is time to apply some boundary conditions, which link the solutions of total pressure and fluid velocity perturbations at the interface $r = a$. The appropriate boundary conditions are:

- $\delta p_{\rm tot}$ has to be continuous across the interface,
- the perturbed interface, $\dfrac{\delta v_r}{\omega - \mathbf{k}\cdot\mathbf{U}}$, has also to be continuous (Chandrasekhar, 1961).

After applying the boundary conditions (we recall that for the ambient medium $U = 0$) finally we arrive at the required dispersion relation of the normal MHD modes propagating along the jet (Nakariakov, 2007; Terra-Homen et al., 2003)

$$\frac{\rho_{\rm e}}{\rho_{\rm i}}\left(\omega^2 - k_z^2 v_{\rm Ae}^2\right)\kappa_{\rm i}\frac{I_m'(\kappa_{\rm i}a)}{I_m(\kappa_{\rm i}a)} - \left[(\omega - \mathbf{k}\cdot\mathbf{U})^2 - k_z^2 v_{\rm Ai}^2\right]\kappa_{\rm e}\frac{K_m'(\kappa_{\rm e}a)}{K_m(\kappa_{\rm e}a)} = 0. \tag{34}$$

For the azimuthal mode number $m = 0$ the above equation describes the propagation of so called *sausage* waves, while with $m = 1$ it governs the propagation of the *kink* waves (Edwin & Roberts, 1983). As we have already seen, the wave frequency, ω, is Doppler-shifted inside the jet. The two quantities $\kappa_{\rm i}$ and $\kappa_{\rm e}$, whose squared magnitudes are given by Eq. (31) are termed *wave attenuation coefficients*. They characterize how quickly the wave amplitude having its maximal value at the interface, $r = a$, decreases as we go away in both directions. Depending on the specific sound and Alfvén speeds in a given medium, as well as on the density contrast, $\eta = \rho_{\rm e}/\rho_{\rm e}$, and the ratio of the embedded magnetic fields, $b = B_{\rm e}/B_{\rm e}$, the attenuation coefficients can be real or imaginary quantities. In the case when both $\kappa_{\rm i}$ and $\kappa_{\rm e}$ are real, we have a *pure surface wave*. The case $\kappa_{\rm i}$ imaginary and $\kappa_{\rm e}$ real corresponds to *pseudosurface waves* (or *body waves* according to Edwin & Roberts terminology (Edwin & Roberts, 1983)). In that case the modified Bessel function inside the jet, I_0, becomes the spatially periodic Bessel function J_0. In the opposite situation the wave energy is carried away from the flux tube – then the wave is called *leaky wave* (Cally, 1986). The waves, which propagate in spicules and X-ray jets, are generally pseudosurface waves, that can however, at some flow speeds become pure surface modes.

For the kink waves one defines the *kink speed* (Edwin & Roberts, 1983)

$$c_{\rm k} = \left(\frac{\rho_{\rm i}v_{\rm Ai}^2 + \rho_{\rm e}v_{\rm Ae}^2}{\rho_{\rm i} + \rho_{\rm e}}\right)^{1/2} = \left(\frac{v_{\rm Ai}^2 + (\rho_{\rm e}/\rho_{\rm i})v_{\rm Ae}^2}{1 + \rho_{\rm e}/\rho_{\rm i}}\right)^{1/2}, \tag{35}$$

which is independent of sound speeds and characterizes the propagation of transverse perturbations.

Our study of the dispersion characteristics of kink and sausage waves, as well as their stability status will be performed in two steps. First, at given sound and Alfvén speeds inside the jet and its environment and a fixed flow speed U, we solve the transcendental dispersion Eq. (34) assuming that the wave angular frequency, ω, and the wave number, k_z, are real quantities. In the next step, when studying their stability/instability status, we assume that the wave frequency and correspondingly the wave phase velocity, $v_{\rm ph} = \omega/k_z$, become complex. Then, as the imaginary part of the complex frequency/phase velocity at a given wave number, k_z, and a critical jet speed, $U_{\rm crt}$, has some non-zero positive value, one says that the wave becomes unstable – its amplitude begins to grow with time. In this case, the linear theory is no longer applicable and one ought to investigate the further wave propagation by means of a nonlinear theory. Our linear approach can determine just the instability threshold only.

In the next two section we numerically derive the dispersion curves of kink and sausage waves running along spicules and X-ray jets, respectively.

4. Dispersion diagrams of MHD surface waves in spicules

Before starting solving the wave dispersion relation (34), we have to specify some input parameters, characterizing both media (the jet and its surrounding). Bearing in mind, as we have already mention in the beginning of Sec. 2, the mass density of the environment is much less (50–100 times); thus we take the density contrast – the ratio of equilibrium plasma density outside to that inside of spicule – to be $\eta = 0.02$. Our choice of the sound and Alfvén speeds in the jet is $c_{si} = 10$ km s^{-1} and $v_{Ai} = 80$ km s^{-1}, respectively, while those speeds in the environment are correspondingly $c_{se} \cong 488$ km s^{-1} and $v_{Ae} = 200$ km s^{-1}. All these values are in agreement with the condition for the balance of total pressures at the flux tube interface – that condition can be expressed in the form

$$p_i + \frac{B_i^2}{2\mu} = p_e + \frac{B_e^2}{2\mu},$$

which yields (Edwin & Roberts (1983)

$$\frac{\rho_e}{\rho_i} = \frac{c_{si}^2 + \frac{\gamma}{2}v_{Ai}^2}{c_{se}^2 + \frac{\gamma}{2}v_{Ae}^2}. \tag{36}$$

The two tube speeds (look at Eq. (32)) are $c_{Ti} = 9.9$ km s^{-1} and $c_{Te} = 185$ km s^{-1}. The kink speed, associated with the kink waves, in our case (see Eq. (35)) is 84 km s^{-1}.

It is obvious that dispersion Eq. (34) of either mode can be solved only numerically. Before starting that job, we normalize all velocities to the Alfvén speed v_{Ai} inside the jet thus defining the dimensionless phase velocity $V_{ph} = v_{ph}/v_{Ai}$ and the *Alfvén–Mach number* $M_A = U/v_{Ai}$. The wavelength is normalized to the tube radius a, which means that the dimensionless wave number is $K = k_z a$. The calculation of wave attenuation coefficients requires the introduction of three numbers, notably the two ratios $\bar{\beta} = c_s^2/v_A^2$ correspondingly in the jet and its environment, and the ratio of the background magnetic field outside to that inside the flow, $b = B_e/B_i$, in addition to the density contrast, η. We recall that the two $\bar{\beta}$s are 1.2 times smaller than the corresponding plasma betas in both media – the latter are given by the expressions

$$\beta_{i,e} = 2\bar{\beta}_{i,e}/\gamma.$$

Thus, the input parameters in the numerical procedure are

$$\eta = 0.02, \quad \bar{\beta}_i \cong 0.016, \quad \bar{\beta}_e \cong 5.96, \quad b \cong 0.35, \quad \text{and} \quad M_A.$$

The value of the Alfvén–Mach number, M_A, naturally depends on the value of the streaming velocity, U. Our choice of this value is 100 km s^{-1} that yields $M_A = 1.25$. With these input values, we calculate the dispersion curves of first kink waves and then sausage ones.

4.1 Kink waves in spicules

We start by calculating the dispersion curves of kink waves assuming that the angular wave frequency, ω, is real. As a reference, we first assume that the plasma in the flux tube is static, i.e., $M_A = 0$. The dispersion curves, which present the dependence of the normalized wave phase velocity on the normalized wave number, are in this case shown in Fig. 5. One can recognize three types of waves: a sub-Alfvénic slow magnetoacoustic wave (in

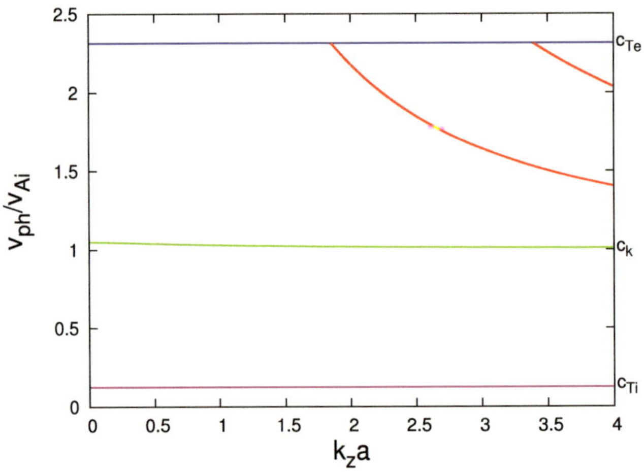

Fig. 5. Dispersion curves of kink waves propagating along the flux tube at $M_{\mathrm{A}} = 0$.

magenta colour) labelled $\mathrm{c_{Ti}}$ (which is actually the normalized value of c_{Ti} to v_{Ai}), an almost Alfvén wave labelled $\mathrm{c_k}$ (the green curve), and a family of super-Alfvénic waves (the red dispersion curves). We note that one can get by numerically solving Eq. (34) the mirror images (with respect to the zeroth line) of the $\mathrm{c_k}$-labelled dispersion curve, as well as of the fast super-Alfvénic waves – both being backward propagating modes that are not plotted in Fig. 5. The next Fig. 6 shows how all these dispersion curves change when the plasma inside the

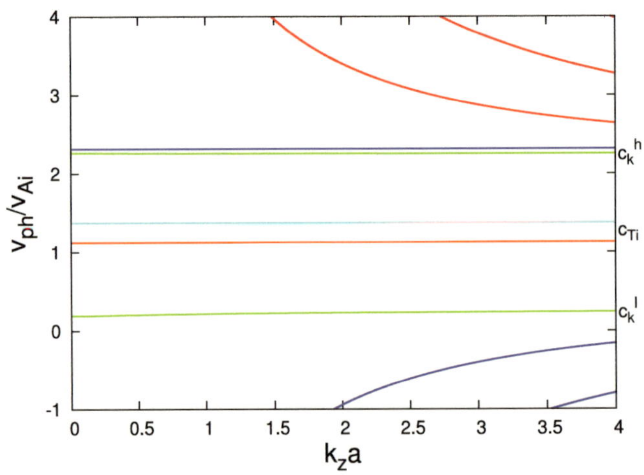

Fig. 6. Dispersion curves of kink waves propagating along the flux tube at $M_{\mathrm{A}} = 1.25$.

tube flows. One sees that the flow first shifts upwards the almost Alfvén wave now labelled $\mathrm{c_k^h}$, as well as high-harmonic super-Alfvénic waves. Second, the slow magnetoacoustic wave ($\mathrm{c_{Ti}}$ in Fig. 5) is replaced by two, now, super-Alfvénic waves, whose dispersion curves (in orange and cyan colours) are collectively labelled $\mathrm{c_{Ti}}$. These two waves have practically constant normalized phase velocities equal to 1.126 and 1.374, respectively, which are the $(M_{\mathrm{A}} \mp c^0_{\mathrm{Ti}})$-values, where c^0_{Ti} is the normalized magnitude of the slow magnetoacoustic wave at $M_{\mathrm{A}} = 0$. Unsurprisingly, one gets a $\mathrm{c_k^l}$-labelled curve, which is the mirror image of the $\mathrm{c_k^h}$-labelled curve. That is why this curve is plotted in green and, as can be seen, it is now a forward propagating wave that has, however, a lower normalized phase velocity than that of its sister c_k^h-labelled dispersion curve. Moreover, there appears to be a family of generally backward propagating waves (below the c_k^l-labelled curve) plotted in blue colour that can similarly be considered as a mirror image of the high-harmonic super-Alfvénic waves.

The most interesting waves especially for the Type II spicules seems to be the waves labelled c_k. It would be interesting to see whether these modes can become unstable at some, say critical, value of the Alfvén–Mach number, M_A. To study this, we have to assume that the wave frequency is complex, i.e., $\omega \to \omega + i\bar{\gamma}$, where $\bar{\gamma}$ is the expected instability growth rate. Thus, the dispersion equation becomes complex (complex wave phase velocity and real wave number) and the solving a transcendental complex equation is generally a difficult task (Acton, 1990).

Before starting to derive a numerical solution to the complex version of Eq. (34), we can simplify that equation. Bearing in mind that the plasma beta inside the jet is very small ($\beta_i \cong 0.02$) and that of the surrounding medium quite high (of order 7), we can treat the jet as a cool plasma and the environment as a hot incompressible fluid. We point out that according to the numerical simulation of spicules by Matsumoto & Shibata (Matsumoto & Shibata, 2010) the plasma beta at heights greater than 2 Mm is of that order (0.03–0.04) – look at Fig. 4 in their paper. For cool plasma, $c_s \to 0$; hence the normalized wave attenuation coefficient $\kappa_i a = \left[1 - (V_{ph} - M_A)^2\right]^{1/2} K$, while for the incompressible environment $c_s \to \infty$ and the corresponding attenuation coefficient is simply equal to k_z, i.e., $\kappa_e a = K$. Under these circumstances the simplified dispersion equation of kink waves takes the form

$$\left(V_{ph}^2\eta - b^2\right)\left[1 - \left(V_{ph} - M_A\right)^2\right]^{1/2}\frac{I_1'(\kappa_i a)}{I_1(\kappa_i a)} - \left[\left(V_{ph} - M_A\right)^2 - 1\right]\frac{K_1'(K)}{K_1(K)} = 0, \qquad (37)$$

where, we recall, that $\kappa_i a = \left[1 - (V_{ph} - M_A)^2\right]^{1/2} K$, and the normalized wave phase velocity, V_{ph}, is a complex number. We note that this simplified version of the dispersion equation of the kink waves closely reproduces the dispersion curves labelled c_k in Figs. 5 and 6.

To investigate the stability/instability status of kink waves we numerically solve Eq. (37) using the Müller method (Muller, 1956) for finding the complex roots at fixed input parameters $\eta = 0.02$ and $b = 0.35$ and varying the Alfvén–Mach number, M_A, from zero to some reasonable numbers. Before starting any numerical procedure for solving the aforementioned dispersion equation, we note that for each input value of M_A one can get two c_k-dispersion curves one of which (for relatively small magnitudes of M_A) has normalized phase velocity roughly equal to $M_A - 1$ and a second dispersion curve associated with dimensionless phase velocity equal to $M_A + 1$. These curves are similar to the dispersion curves labelled c_k^l and c_k^h in Fig. 6. The results of the numerical solving Eq. (37) are shown in Fig. 7. For $M_A = 0$, except for the dispersion curve with normalized phase velocity approximately equal to 1, one can find a dispersion curve with normalized phase velocity close to -1 – that curve is not plotted in Fig. 7. Similarly, for $M_A = 2$ one obtains a curve at $V_{ph} = 1$ and another at $V_{ph} = 3$, and so on. With increasing the magnitude of the Alfvén–Mach number kink waves change their structure – for small numbers being pseudosurface (body) waves and for $M_A \geqslant 4$ becoming pure surface modes. Another effect associated with the increase in M_A, is that, for instance at $M_A \geqslant 6$, the shapes of pairs of dispersion curves begin to visibly change as can be seen in Figs. 7 and 8. The most interesting observation is that for $M_A \geqslant 8$ both curves begin to

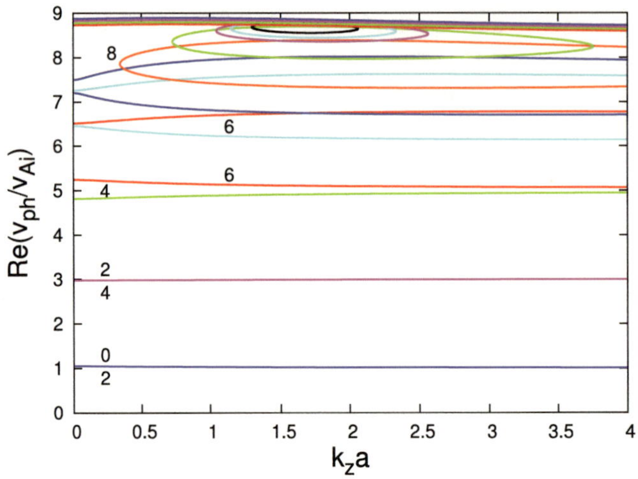

Fig. 7. Dispersion curves of kink waves propagating along the flux tube at various values of M_A.

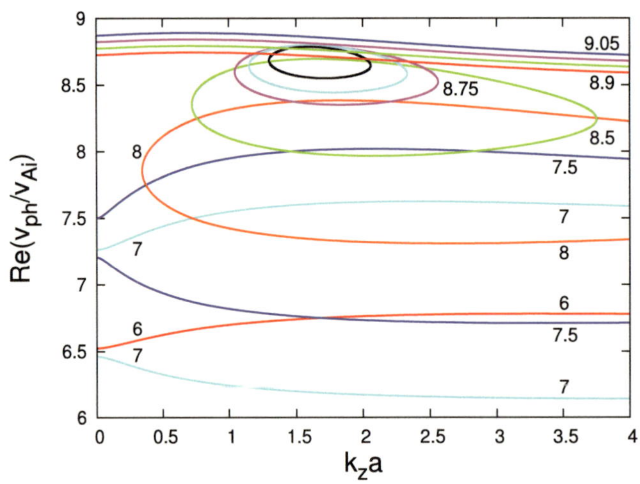

Fig. 8. Dispersion curves of kink waves propagating along the flux tube for relatively large values of M_A.

merge and at $M_A = 8.5$ they form a closed dispersion curve. The ever increasing of M_A yields yet smaller closed dispersion curves – the two non-labelled ones depicted in Fig. 8 correspond to $M_A = 8.8$ and 8.85, respectively. All these dispersion curves present stable propagation of the kink waves. However, for $M_A \geqslant 8.9$ we obtain a new family of wave dispersion curves that correspond to an unstable wave propagation. We plot in Fig. 8 four curves of that kind that have been calculated for $M_A = 8.9, 8.95, 9,$ and 9.05, respectively. The growth rates of the unstable waves are shown in Fig. 9. The instability that arises is of the Kelvin–Helmholtz type. We recall that the Kelvin–Helmholtz instability, which is named after Lord Kelvin and Hermann von Helmholtz, can occur when velocity shear is present within a continuous fluid, or when there is a sufficient velocity difference across the interface between two fluids (Chandrasekhar, 1961). In our case, we have the second option and the relative jet

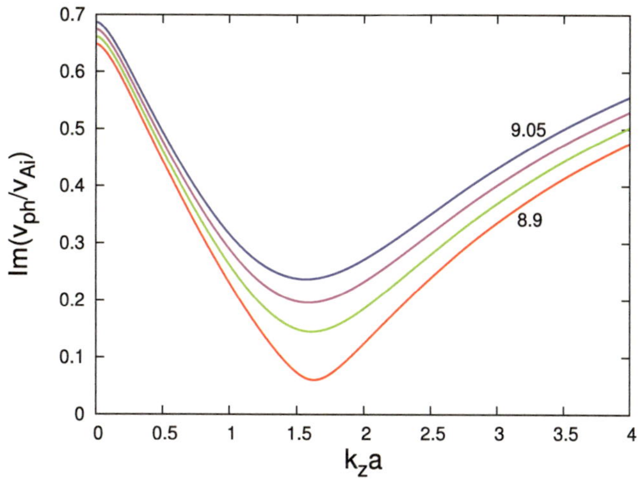

Fig. 9. Growth rates of unstable kink waves propagating along the flux tube at values of M_A equal to 8.9, 8.95, 9, and 9.05, respectively.

velocity, U, plays the role of the necessary velocity difference across the interface between the spicule and its environment.

The big question that immediately springs to mind is whether one can really observe such an instability in spicules. The answer to that question is obviously negative – to register the onset of a Kelvin–Helmholtz instability of kink waves travelling on a Type II spicule one would need to observe jet velocities of the order of or higher than 712 km s^{-1}! If we assume that the density contrast, η, possesses the greater value of 0.01 (which means that the jet mass density is 100 times larger than that of the ambient medium) and the ratio of the background magnetic fields, b, is equal to 0.36 (which may be obtained from a slightly different set of characteristic sound and Alfvén speeds in both media), the critical Alfvén–Mach number at which the instability starts is even much higher (equal to 12.6) – in that case the corresponding jet speed is $U_{\rm crt} = 882$ km s^{-1} – too high to be registered in a spicule. The value of 882 was computed under the assumption that the Alfvén speed inside the jet is 70 km s^{-1}.

We note that very similar dispersion curves and growth rates of unstable kink waves like those shown in Figs. 8 and 9 were obtained for cylindrical jets when both media were treated as incompressible fluids. In that case, dispersion Eq. (37) becomes a quadratic equation

$$\left(V_{\rm ph}^2\eta - b^2\right)\frac{I'_m(K)}{I_m(K)} - \left[\left(V_{\rm ph} - M_{\rm A}\right)^2 - 1\right]\frac{K'_m(K)}{K_m(K)} = 0, \tag{38}$$

that provides solutions for the real and imaginary parts of the normalized wave phase velocity in closed forms, notably (Zhelyazkov, 2010; 2011)

$$V_{\rm ph} = \frac{-M_{\rm A}B \pm \sqrt{D}}{\eta A - B},$$

where

$$A = I'_m(K)/I_m(K), \qquad B = K'_m(K)/K_m(K),$$

and the discriminant D is

$$D = M_A^2 B^2 - (\eta A - B)\left[\left(1 - M_A^2\right)B - Ab^2\right].$$

Obviously, if $D \geqslant 0$, then

$$\mathrm{Re}(V_{\mathrm{ph}}) = \frac{-M_A B \pm \sqrt{D}}{\eta A - B}, \qquad \mathrm{Im}(V_{\mathrm{ph}}) = 0,$$

else

$$\mathrm{Re}(V_{\mathrm{ph}}) = -\frac{M_A B}{\eta A - B}, \qquad \mathrm{Im}(V_{\mathrm{ph}}) = \frac{\sqrt{D}}{\eta A - B}.$$

We note that our choice of the sign of $\sqrt{D}$ in the expression of $\mathrm{Im}(V_{\mathrm{ph}})$ is plus although, in principal, it might also be minus – in that case, due to the arising instability, the wave's energy is transferred to the jet.

It is interesting to note that for our jet with $b = 0.35$ and $\eta = 0.02$ the quadratic dispersion Eq. (38) yields a critical Alfvén–Mach number for the onset of a Kelvin–Helmholtz instability equal to 8.87, which is lower than its magnitude obtained from Eq. (37). With this new critical Alfvén–Mach number, the required jet speed for the instability onset is $\cong$710 km s^{-1}. The most astonishing result, however, is the observation that the dispersion curves and the corresponding growth rates, when kink waves become unstable, – look at Figs. 10 and 11 – are very similar to those shown in Figs. 8 and 9. It is worth mentioning that for the

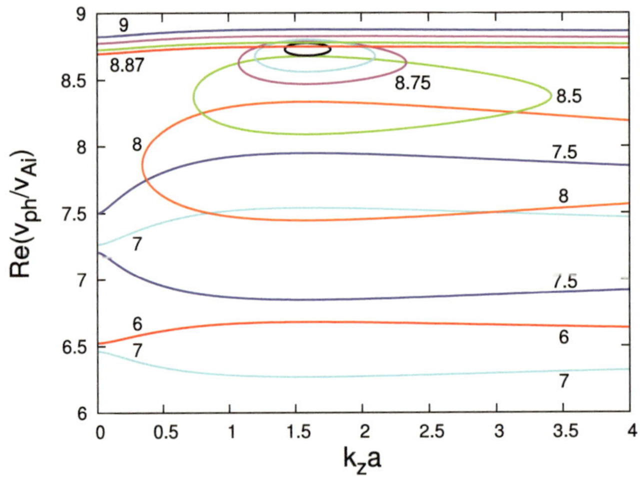

Fig. 10. Dispersion curves of kink waves derived from Eq. (38) for relatively large values of M_A.

same $\eta = 0.02$, but for $b = 1$ (equal background magnetic fields), the quadratic equation yields a much higher critical Alfvén–Mach number (=11.09), which means that the critical jet speed grows up to 887 km s^{-1}. This consideration shows that both the density contrast, η, and the ratio of the constant magnetic fields, b, are equally important in determining the critical Alfvén–Mach number. Moreover, since Eq. (37) and its simplified form as quadratic Eq. (38) yield almost similar results (both for dispersion curves and growth rates when kink waves become unstable) firmly corroborates the correctness of the numerical solutions to the complex dispersion Eq. (37).

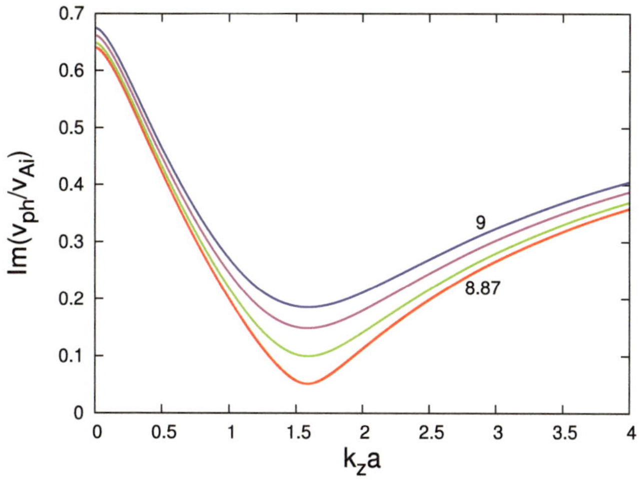

Fig. 11. Growth rates of unstable kink waves calculated from Eq. (38) at values of M_A equal to 8.87, 8.9, 8.95, and 9, respectively.

4.2 Sausage waves in spicules

The dispersion curves of sausage waves both in a static and in a flowing plasma shown in Figs. 12 and 13 are very similar to those of kink waves (compare with Figs. 5 and 6). The latter curves were calculated from dispersion Eq. (34) with azimuthal mode number $m = 0$ for the

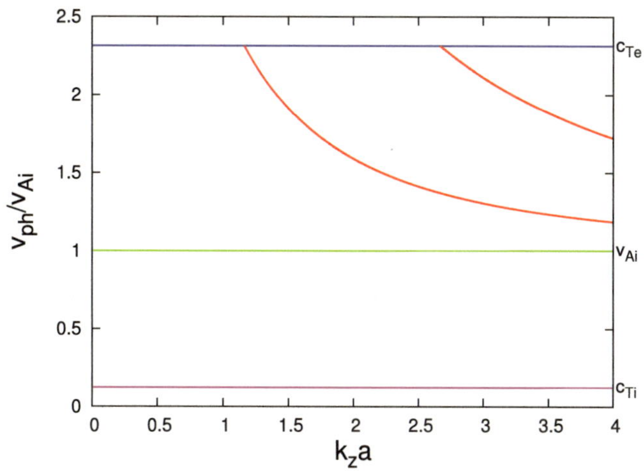

Fig. 12. Dispersion curves of sausage waves propagating along the flux tube at $M_A = 0$.

same input parameters as in the case of kink waves. The main difference is that the c_k-labelled green dispersion curve is replaced by a curve corresponding to the Alfvén wave inside the jet. We note that the dispersion curve in Fig. 13 corresponding to a normalized phase velocity 0.25 is labelled v_{Ai}^{l} because it can be considered as the one dispersion curve of the (1.25 ∓ 1)-curves that can be derived from the dispersion equation. As in the case of kink waves, the dispersion curve corresponding to the higher speed has the label v_{Ai}^{h}. Here we also get the two almost dispersionless curves collectively labelled c_{Ti} (in the same colours, orange and cyan, as in Fig. 6) with normalized wave phase velocities equal to 1.126 and 1.374.

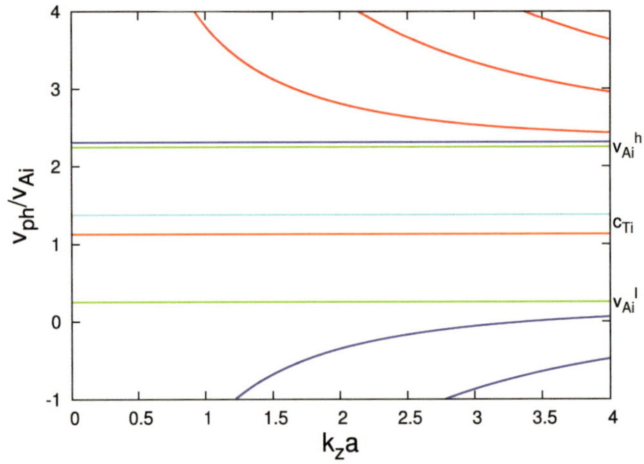

Fig. 13. Dispersion curves of sausage waves propagating along the flux tube at $M_A = 1.25$.

When examining the stability properties of sausage waves as a function of the Alfvén–Mach number, M_A, we use the same Eq. (37) while changing the order of the modified Bessel functions from 1 to 0. As in the case of kink waves, we are interested primarily in the behaviour of the waves whose phase velocities are multiples of the Alfvén speed. The results of numerical calculations of the complex dispersion equation are shown in Fig. 14. It turns out that for all reasonable Alfvénic Mach numbers the waves are stable. This is unsurprising

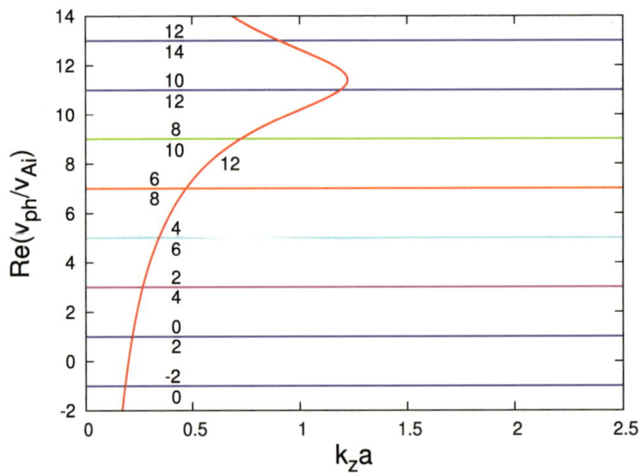

Fig. 14. Dispersion curves of sausage waves propagating along the flux tube at various values of M_A.

because the same conclusion was drawn by solving precisely the complex dispersion equation governing the propagation of sausage waves in incompressible flowing cylindrical plasmas (Zhelyazkov, 2010; 2011). In Fig. 14 almost all dispersion curves have two labels: one for the ($M_A - 1$)-labelled curve at given M_A (the label is below the curve), and second for the ($M'_A + 1$)-labelled curve associated with the corresponding ($M'_A = M_A - 2$)-value (the label is above the curve). This labelling is quite complex because for all M_A we find dispersion curves that overlap: for instance, the higher-speed dispersion curve (i.e., that associated with the $(M_A + 1)$-value) for $M_A = 0$ coincides with the lower-speed dispersion curve (i.e., that associated with the $(M_A - 1)$-value) for $M_A = 2$. In contrast to the kink waves, which for $M_A \geqslant 4$ are pure surface modes, the sausage waves can be both pseudosurface and pure surface modes, or one of the pair can be a surface mode while the other is a pseudosurface one. For example, all dispersion curves for $M_A = 0$ and 8 correspond to the pseudosurface waves while the curves' pair associated with $M_A = 4$ describes the dispersion properties of pure surface waves. For the other Alfvén–Mach numbers, one of the wave is a pseudosurface and the other is a pure surface. However, there is a 'rule': if, for instance, the higher-speed wave with $M_A = 10$ is a pseudosurface mode, the lower-speed wave for $M_A = 12$ is a pure surface wave. We finish the discussion of sausage waves with the following conclusion: with increasing the Alfvén–Mach number M_A the initially independent high-harmonic waves and their mirroring counterparts begin to merge – this is clearly seen in Fig. 14 for $M_A = 12$ – the resulting dispersion curve is in red colour. A similar dispersion curve can be obtained, for example, for $M_A = 10$; the merging point of the corresponding two high-harmonic dispersion curves moves, however, to the right – it lies at $k_z a = 1.943$. It is also evident that in the long wavelength limit the bottom part of the red-coloured dispersion curve describes a backward propagating sausage pseudosurface wave. Another peculiarity of the same dispersion curve is the circumstance that for the range of dimensionless wave numbers between 0.7 and 1.23, one can have two different wave phase velocities. Which one is detected, the theory cannot predict.

5. Dispersion diagrams of MHD surface waves in soft X-ray jets

The geometry model of solar X-ray jets is the same as for the spicules – straight cylinder with radius a. Before starting the numerical calculations, we have to specify, as before, the input parameters. The sound and Alfvén speed that are typical for X-ray jets and their environment are correspondingly $c_{si} = 200$ km s^{-1}, $v_{Ai} = 800$ km s^{-1}, $c_{se} = 120$ km s^{-1}, and $v_{Ae} = 2300$ km s^{-1}. With these speeds the density contrast is $\eta = 0.13$. The same η (calculated from a slightly different set of sound and Alfvén speeds) Vasheghani Farahani et al. (Vasheghani Farahani et al., 2009) used in studying the propagation of transfer waves in soft X-ray coronal jets. Their analysis, however, is restricted to the long-wavelength limit, $|k|a \ll 1$ in their notation, while our approach considers the solving the exact dispersion relation without any limitations for the wavelength – such a treating is necessary bearing in mind that the wavelengths of the propagating along the jets fast magnetoacoustic waves might be of the order of X-ray jets radii. We remember that the soft X-ray coronal jets are much ticker than the Type II spicules.

With our choice of sound and Alfvén speeds, the tube velocities in both media (look at Eq. (32)), respectively, are $c_{Ti} \cong 194$ km s^{-1} and $c_{Te} = 119.8$ km s^{-1}. The kink speed (see Eq. (35)) turns out to be rather high, namely $\cong$1078 km s^{-1}. To compare our result of the critical jet speed for triggering the Kelvin–Helmholtz instability with that found by Vasheghani Farahani et al. (Vasheghani Farahani et al., 2009), we take the same jet speed as theirs, notably $U = 580$ km s^{-1}, which yields Alfvén–Mach number equal to 0.725. (For simplicity we assume that the ambient medium is static, i.e., $U_e = 0$.) Thus, our input parameters for the numerical computations are

$$\eta = 0.13, \quad \bar{\beta}_i \cong 0.06, \quad \bar{\beta}_e \cong 0.003, \quad b = 1.035, \quad \text{and} \quad M_A = 0.725.$$

We note that $b = 1.035$ means that the equilibrium magnetic fields inside and outside the X-ray coronal jet are almost identical. Moreover, due to the relatively small plasma betas, $\beta_e = 0.0033$ and $\beta_i = 0.075$, respectively, the magnetic pressure dominates the gas one in both media and the propagating waves along X-ray jets should accordingly be predominantly transverse.

5.1 Kink waves in soft X-ray coronal jets

The dispersion diagrams of kink waves propagating along a static-plasma ($U = 0$) flux tube are shown in Fig. 15. They, the dispersion curves, have been obtained by numerically finding

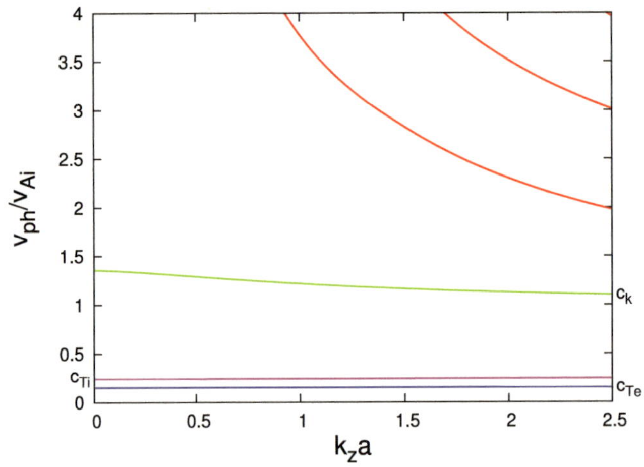

Fig. 15. Dispersion curves of kink waves propagating along a flux tube modelling X-ray jet at $M_A = 0$.

the solutions to dispersion Eq. (34) with mode number $m = 1$ and input data listed in the introductory part of this section with $M_A = 0$. The dispersion curves are very similar to those for spicules (look at Fig. 5). Here, there is, however, one distinctive difference: the c_{Te}-labelled dispersion curve (blue color) lies below the curve corresponding to the tube velocity inside the jet (magenta coloured line labelled c_{Ti}). The dispersion curves of the high-harmonic super-Alfvénic waves (red colour) lye, as usual, above the green curve associated with the kink speed. What actually does the flow change when is taken into account? The answer to this question is given in Fig. 16. As in the case with spicules, the flow duplicates the c_{Ti}-labelled dispersion curve in Fig. 15. The two, again collectively labelled c_{Ti} dispersion curves, are sub-Alfvénic waves having normalized phase velocities equal to 0.482 and 0.968 in correspondence to the ($M_A \mp c^0_{Ti}$)-rule. All the rest curves have the same behaviour and notation as in Fig. 6. The only difference here is the circumstance that the lower-speed c_k-curve lies below the zero line, i.e., it describes a *backward* propagating kink pseudosurface wave. This is because the Alfvén–Mach number now is less than one. We note also that the c_{Te}-labelled dispersion curve is unaffected by the presence of flow.

The most intriguing question is whether the c_k-labelled wave can become unstable at any reasonable flow velocity. Before answering that question, we have, as before, to simplify dispersion Eq. (34). Since the two plasma betas, as we have already mentioned, are much less that one, we can treat both media (the X-ray jet and its environment) as cool plasmas. In

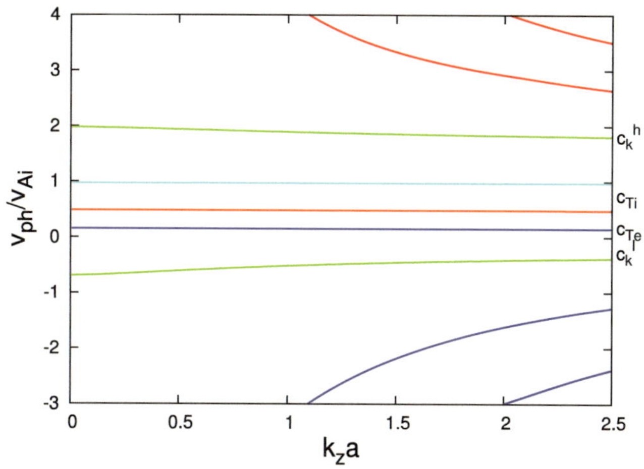

Fig. 16. Dispersion curves of kink waves propagating along a flux tube modelling X-ray jet at $M_A = 0.725$.

this case, the simplified dispersion equation of kink waves (in complex variables!) takes the form

$$\left(V_{\mathrm{ph}}^2\eta - b^2\right)\left[1 - \left(V_{\mathrm{ph}} - M_{\mathrm{A}}\right)^2\right]^{1/2} \frac{I_1'(\kappa_{\mathrm{i}}a)}{I_1(\kappa_{\mathrm{i}}a)}$$

$$-\left[\left(V_{\mathrm{ph}} - M_{\mathrm{A}}\right)^2 - 1\right]\left(1 - V_{\mathrm{ph}}^2\eta\right)\frac{K_1'(\kappa_{\mathrm{e}}a)}{K_1(\kappa_{\mathrm{e}}a)} = 0, \tag{39}$$

where

$$\kappa_{\mathrm{i}}a = \left[1 - \left(V_{\mathrm{ph}} - M_{\mathrm{A}}\right)^2\right]^{1/2} K \quad \text{and} \quad \kappa_{\mathrm{e}}a = \left(1 - V_{\mathrm{ph}}^2\eta\right)^{1/2} K.$$

We numerically solve this equation by varying the magnitude of the Alfvén–Mach number, M_{A}, using as before the Müller method and the dispersion curves of both stable and unstable kink waves are shown in Fig. 17. In this figure, we display only the most interesting, upper, part of the dispersion diagram, where one can observe the changes in the shape of the dispersion curves related to the corresponding c_{k}-speeds. First and foremost, the shape of the merging dispersion curves (labelled 4, 4.1, 4.2, and 4.23 in Fig. 17) is distinctly different from that of the similar curves in Fig. 8. Here, the curves, which are close to the dispersion curves corresponding to an unstable wave propagation (the first one is with label 4.25) are *semi-closed* in contrast to the closed curves in Fig. 8. The wave growth rates corresponding to Alfvén–Mach numbers 4.25, 4.3, 4.35, and 4.4 are shown in Fig. 18. As can be seen, the shape of those curves is completely different to that of the wave growth rates shown in Figs. 9 and 11. We note that all dispersion curves for $M_{\mathrm{A}} \geqslant 4$ correspond to pure surface kink waves.

It is clear from Figs. 17 and 18 that the critical Alfvén–Mach number, which determines the onset of a Kelvin–Helmholtz instability of the kink waves, is equal to 4.25 – the corresponding flow speed is 3400 km s^{-1}, that is much higher than the value we have used for calculating the dispersion curves in Fig. 16. The critical Alfvén–Mach number evaluated by Vasheghani Farahani et al. (Vasheghani Farahani et al., 2009), is 4.47, that means the corresponding flow speed must be at least 3576 km s^{-1}. If we use our Eq. (39) with the same $\rho_{\mathrm{e}}/\rho_{\mathrm{i}} = 0.13$, but

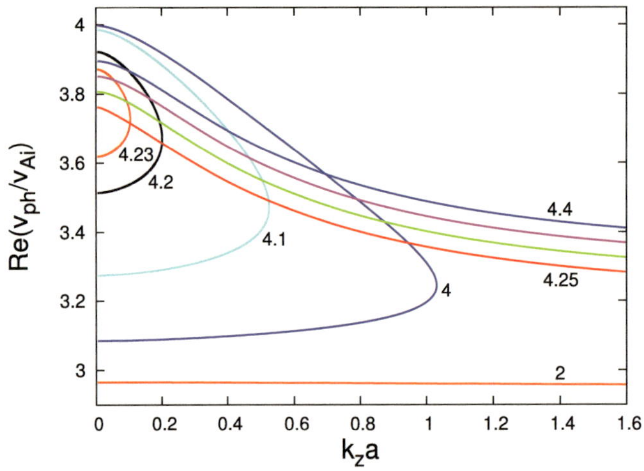

Fig. 17. Dispersion curves of kink waves propagating along a flux tube modelling X-ray jets for relatively large values of M_A.

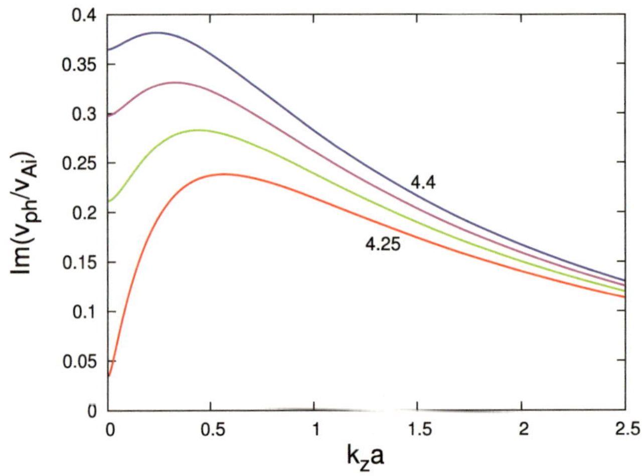

Fig. 18. Growth rates of unstable kink waves propagating along a flux tube modelling X-ray jets at values of M_A equal to 4.25, 4.3, 4.35, and 4.4, respectively.

with a little bit higher $B_e/B_i = 1.1132$, we get that the critical jet speed for triggering the Kelvin–Helmholtz instability in a soft X-ray coronal get would be $4.41 v_{Ai} = 3528$ km s^{-1}. It is necessary, however, to point out that the correct density contrast that can be calculated from Eq. (36) with $c_{si} = 360$ km s^{-1}, $v_{Ai} = 800$ km s^{-1}, $c_{se} = 120$ km s^{-1}, and $v_{Ae} = 2400$ km s^{-1} (the basic speeds in Vasheghani Farahani et al. paper) is $\rho_e/\rho_i = 0.137698$, which is closer to 0.14 rather than to 0.13. The solving Eq. (39) with the exact value of the density contrast (=0.1377) and the same B_e/B_i as before (=1.1132) yields a critical flow speed equal to $4.31 v_{Ai} = 3448$ km s^{-1}. All these calculations show that even small variations in the two ratios ρ_e/ρ_i and B_e/B_i lead to visibly different critical Alfvén–Mach numbers – our choice of the sound and Alfvén speeds gives the smallest value of the critical M_A. According to the more recent observations (Madjarska, 2011; Shimojo & Shibata, 2000), the soft X-ray coronal jets can have velocities above 10^3 km s^{-1} and it remains to be seen whether a speed of 3400 km s^{-1} can trigger the onset of a Kelvin–Helmholtz instability of the kink surface waves running along the jets.

5.2 Sausage waves in soft X-ray coronal jets

The dispersion diagram of sausage waves in a static-plasma flux tube should be more or less similar to that of the kink waves under the same circumstances. Here, however, the green curve in Fig. 15, associated with the kink speed c_{k}, is now replaced by a dispersionless line

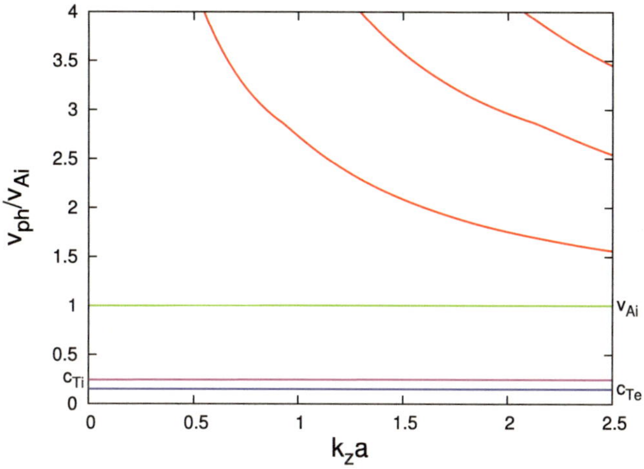

Fig. 19. Dispersion curves of sausage waves propagating along a flux tube modelling X-ray jet at $M_{\mathrm{A}} = 0$.

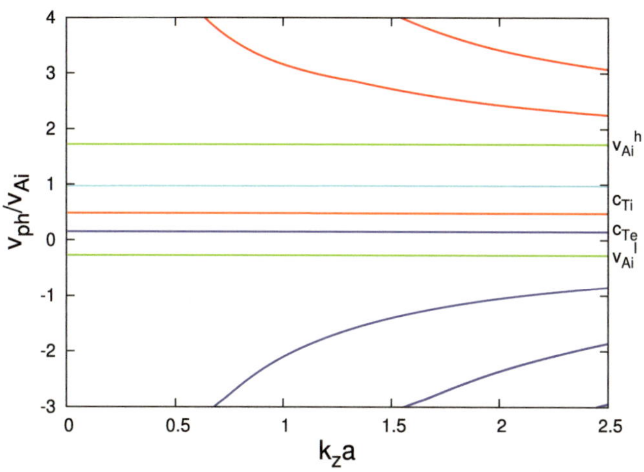

Fig. 20. Dispersion curves of sausage waves propagating along a flux tube modelling X-ray jet at $M_{\mathrm{A}} = 0.725$.

related to the Alfvén speed – see the green curve in Fig. 19. Another difference is the number of the red-coloured high-harmonic super-Alfvénic waves – here it is 3 against 2 in Fig. 15. The dispersion diagram of the same mode in a flow with $M_{\mathrm{A}} = 0.725$ ($U = 580$ km s^{-1}) is also predictable – the presence of the flow is the reason for splitting the green v_{Ai}-labelled curve in Fig. 19 into two sister curves labelled, respectively, v^l_{Ai} and v^h_{Ai} – look at Fig. 20. Observe that the normalized speeds of those two waves are, as expected, equal to $M_A \mp 1$ – in our case the lower-speed Alfvén wave is a backward propagating one. The two sub-Alfvénic waves, whose dispersion curves are in orange and cyan colours and collectively labelled c_{Ti} have practically the same normalized phase velocities as the corresponding curves in Fig. 16. We note that one of the aforementioned curve is slightly decreasing (the orange curve) whilst the other, cyan-coloured, curve is slightly increasing when the normalized wave number $k_z a$ becomes larger – the same holds for the analogous waves in spicules. One can see in Fig. 20 a symmetry between the upper and bottom parts of the dispersion diagram – the 'mirror line' lies somewhere between the orange and cyan dispersion curves.

The 'evolution' of the green v_{Ai}-labelled curves in Fig. 20 with the increase in the Alfvén–Mach number is illustrated in Fig. 21. It is unsurprising that the sausage surface waves in soft X-ray

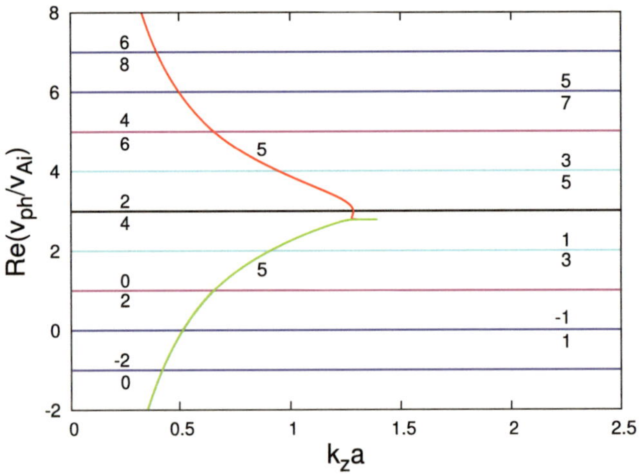

Fig. 21. Dispersion curves of sausage waves propagating along a flux tube modelling X-ray jets at various values of M_A.

coronal jets (like in spicules) are unaffected by the Kelvin–Helmholtz instability. Similarly as in Fig. 14, we have an overlapping of the dispersion curves associated with different Alfvén–Mach numbers. The labelling of dispersion curves in Fig. 21 is according to the previously discussed in Sec. 4.2 rule, namely each horizontal dispersion curve possesses two labels: one for the $(M_A - 1)$-curve at given M_A (the label is below the curve), and second for the $(M'_A + 1)$-curve associated with the corresponding $(M'_A = M_A - 2)$-value (the label is above the curve). Interestingly, even for the relatively low $M_A = 1$ both the lower- and the high-speed curves describe pure surface sausage waves. The same is also valid for the dispersion lines corresponding to Alfvén–Mach numbers equal to 4 and 5. The lower-speed Alfvénic curve at $M_A = 6$ is related to a pseudosurface sausage wave while the higher-speed one (with normalized phase velocity equal to 7) corresponds to a pure surface mode. (With $M_A = 2$ we have just the opposite situation.) At $M_A \geqslant 7$ all waves are pseudosurface ones. Each choice of the Alfvén–Mach number indeed requires separate studying of the wave's proper mode. Apart from Alfvénic waves and the pair of sub-Alfvénic modes (orange and cyan curves in Fig. 20), there appear to be families of high-speed harmonic waves (with red and blue colours of their dispersion curves), which also change with the increase of M_A. Initially being independent, with the growing of the Alfvén–Mach number, they change

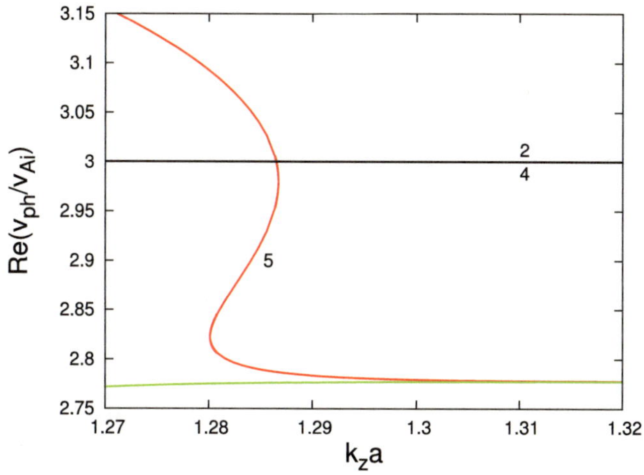

Fig. 22. Zoomed part of the dispersion diagram in Fig. 21 where two dispersion curves of super-Alfvénic sausage waves (at $M_{\rm A} = 5$) are touching each other.

their shapes and one may occur to observe the merging of, for instance, the first curves of each family as this has been shown in Fig. 14 (see the red curve there). Here, however, the situation is über-complicated – instead of merging we encounter a new phenomenon, notably the *touching* of two dispersion curves – see the green and red curves in Fig. 21 (calculated for $M_{\rm A} = 5$), and with more details in Fig. 22. The tip of the horizontal spike lies at $k_z a = 1.395$. Another peculiarity of this complex curve is the inverted-s shape of the red curve between the dimensionless wave numbers 1.28 and 1.29. Across that range, at a fixed $k_z a$, one can 'detect' four different normalized wave phase velocities. Similar sophisticated dispersion curves might also be obtained for $M_{\rm A} = 4$ or $M_{\rm A} = 6$. Maybe nowadays the sausage mode is not too interesting for the spacecrafts' observers but, who knows, it can sometime become important in interpretation observational data.

6. Conclusion

We now summarize the main findings of our chapter. We have studied the dispersion properties and the stability of the MHD normal modes running along the length of Type II spicules and soft X-ray coronal jets. Both have been modelled as straight cylindrical jets of ideal cool plasma surrounded by a warm/hot fully ionized medium (for spicules) or as flux tubes of almost cool plasma surrounded by a cool medium (for the X-ray jets). The wave propagation has been investigated in the context of standard magnetohydrodynamics by using linearized equations for the perturbations of the basic quantities: mass density, pressure, fluid velocity, and wave magnetic field. The derived dispersion equations describe the well-known kink and sausage mode influenced by the presence of spicules' or X-ray jets' moving plasma. The streaming plasma is characterized by its velocity $\mathbf{U}$, which is directed along the background magnetic fields $\mathbf{B}_{\rm i}$ and $\mathbf{B}_{\rm e}$ inside the jet and in its environment. An alternative and more convenient way of specifying the jet is by defining the Alfvén–Mach number: the ratio of jet speed to the Alfvén speed inside the jet, $M_{\rm A} = U/v_{\rm Ai}$. The key parameters controlling the dispersion properties of the waves are the so-called density contrast, $\eta = \rho_{\rm e}/\rho_{\rm i}$, the ratio of the two background magnetic fields, $b = B_{\rm e}/B_{\rm i}$, and the two ratios of the squared sound and Alfvén speeds, $\bar{\beta}_{\rm e} = c_{\rm se}^2/v_{\rm Ae}^2$ and $\bar{\beta}_{\rm i} = c_{\rm si}^2/v_{\rm Ai}^2$. How does the jet change the dispersion curves of both modes (kink and sausage waves) in a static-plasma flux tube? The answers to that question are as follows:

- The flow shifts upwards the specific dispersion curves, the kink-speed curve for kink waves and Alfvén-speed curve for sausage waves, as well as the high-harmonic fast waves of both modes. The sub-Alfvénic tube speed inside the jet, c_{Ti}, belongs to two waves with normalized phase velocities equal to $M_{\mathrm{A}} \mp c_{\mathrm{Ti}}/v_{\mathrm{Ai}}$. One also observes such a duplication of the c_{k}- or v_{Ai}-speed curve of kink or sausage waves. Below the lower-speed c_{k}- or v_{Ai}-curve there appears to be a set of dispersion curves, which are a mirror image of the high-harmonic fast waves. We note that the flow does not affect the c_{Te}-speed dispersion curve associated with the tube velocity in the environment.
- For a typical set of characteristic sound and Alfvén speeds in both media (the jet and its environment) at relatively small Alfvén–Mach numbers both modes are pseudosurface waves. With increasing M_{A}, some of them become pure surface waves. For kink waves, this finding is valid for $M_{\mathrm{A}} \geqslant 4$.
- The kink waves running along the jet can become unstable when the Alfvén–Mach number, M_{A}, exceeds some critical value. That critical value depends upon the two input parameters, η and b; the increase in the density contrast, $\rho_{\mathrm{e}}/\rho_{\mathrm{i}}$, decreases the magnitude of the critical Alfvén–Mach number, whilst the increase in the background magnetic fields ratio, $B_{\mathrm{e}}/B_{\mathrm{i}}$, leads to an increase in the critical M_{A}. For our choice of parameters for Type II spicules ($\eta = 0.02$ and $b = 0.35$) the value of the critical M_{A} is 8.9. This means that the speed of the jet must be at least 712 km s^{-1} for the onset of the Kelvin–Helmholtz instability of the propagating kink waves. Such high speeds of Type II spicules have not yet been detected. For the soft X-ray coronal jets, due to the greater density contrast ($\eta = 0.13$) and almost equal background magnetic fields ($b = 1.035$), the critical Alfvén–Mach number is approximately twice smaller (=4.25), but since the jet Alfvén speed is 10 times larger than that of spicules, the critical flow speed, U_{crt}, is much higher, namely 3400 km s^{-1}. Such high jet speeds can be in principal registered in soft X-ray coronal jets.

  A rough criterion for the appearance of the Kelvin–Helmholtz instability of kink waves is the satisfaction of an inequality suggested by Andries & Goossens (Andies & Goossens, 2001), which in our notation reads

  $$M_{\mathrm{A}} > 1 + b/\sqrt{\eta}.$$

  This criterion provides more reliable predictions for the critical M_{A} when $b \approx 1$ (Zhelyazkov, 2010). In particular, for a X-ray jet with $\eta = 0.13$ and $b = 1.035$ the above criterion yields $M_{\mathrm{A}} > 3.87$, which is lower than the numerically found value of 4.25.
- The onset of the Kelvin–Helmholtz instability for kink surface waves running along a cylindrical jet, modelling a Type II spicule, is preceded by a substantial reorganization of wave dispersion curves. As we increase the Alfvén–Mach number, the pairs of high- and low-speed curves (look at Fig. 8) begin to merge transforming into closed dispersion curves. After a further increase in M_{A}, these closed dispersion curves become smaller – this is an indication that we have reached the critical M_{A} at which the kink waves are subjected to the Kelvin–Helmholtz instability – the unstable waves propagate across the entire $k_z a$-range having growth rates depending upon the value of the current M_{A}. We note that this behaviour has been observed for kink waves travelling on flowing solar-wind plasma (Zhelyazkov, 2010; 2011).

For the X-ray jets, the dispersion curves' reorganization, because the environment has been considered as a cool plasma, is different – now, at high enough flow speeds, the merging lower- and higher-speed c_k-dispersion curves take the form of semi-closed loops (see Fig. 17). As we increase the flow speed (or equivalently the Alfvén–Mach number), the semi-closed loops shrink and at some critical flow speed the kink wave becomes unstable and the instability is of the Kelvin–Helmholtz type. We note that the shapes of the waves' growth rates of kink waves in spicules and soft X-ray coronal jets are distinctly different – compare Figs. 9 and 18.

- We have found that the sausage waves are unaffected by the Kelvin–Helmholtz instability. This conclusion was also previously drawn for the sausage modes in flowing solar-wind plasma (Zhelyazkov, 2010; 2011).

As we have seen, very high jet speeds are required to ensure that the Kelvin–Helmholtz instability occurs for kink waves propagating in Type II spicules associated with a subsequent triggering of Alfvén-wave turbulence, hence the possibility that this mechanism is responsible for chromospheric/coronal heating has to be excluded. However, a twist in the magnetic field of the flux tube or its environment may have the effect of lowering the instability threshold (Bennett et al., 1999; Zaqarashvili et al., 2010) and eventually lead to the triggering of the Kelvin–Helmholtz instability. According to Antolin & Shibata (Antolin & Shibata, 2010), a promising way to ensure spicules'/coronal heating is by means of the mode conversion and parametric decay of Alfvén waves generated by magnetic reconnection or driven by the magneto-convection at the photosphere. However, spicules can be considered as Alfvén wave resonant cavities (Holweg, 1981; Leroy, 1981) and as Matsumoto & Shibata (Matsumoto & Shibata, 2010) claim, the waves of the period around 100–500 s can transport a large amount of wave energy to the corona. Zahariev & Mishonov (Zahariev & Mishonov, 2011) state that the corona may be heated through a self-induced opacity of high-frequency Alfvén waves propagating in the transition region between the chromosphere and the corona owing to a considerable spectral density of the Alfvén waves in the photosphere. Another trend in explaining the mechanism of coronal heating is the dissipation of Alfvén waves' energy by strong wave damping due to the collisions between ions and neutrals (Song & Vasyliūnas, 2011; Tsap et al., 2011). In particular, Song & Vasyliūnas, by analytically solving a self-consistent one-dimensional model of the plasma–neutral–electromagnetic system, show that the damping is extremely strong for weaker magnetic field and less strong for strong field. Under either condition, the high-frequency portion of the source power spectrum is strongly damped at the lower altitudes, depositing heat there, whereas the lower-frequency perturbations are nearly undamped and can be observed in the corona and above when the field is strong.

The idea that Alfvén waves propagating in the transition region can contribute to the coronal heating was firmly supported by the observational data recorded on April 25, 2010 by NASA's *Solar Dynamics Observatory* (see Fig. 2). As McIntosh et al. (McIntosh et al., 2011) claim, "*SDO* has amazing resolution, so you can actually see individual waves. Previous observations of Alfvénic waves in the corona revealed amplitudes far too small (0.5 km s^{-1}) to supply the energy flux (100–200 W m^{-2}) required to drive the fast solar wind or balance the radiative losses of the quiet corona. Here we report observations of the transition region (between the chromosphere and the corona) and of the corona that reveal how Alfvénic motions permeate the dynamic and finely structured outer solar atmosphere. The ubiquitous outward-propagating Alfvénic motions observed have amplitudes of the order of 20 km s^{-1}

and periods of the order of 100–500 s throughout the quiescent atmosphere (compatible with recent investigations), and are energetic enough to accelerate the fast solar wind and heat the quiet corona."

Notwithstanding, as we have already mentioned in the end of Sec. 5.1, the possibility for the onset of a Kelvin–Helmholtz instability of kink waves running along soft X-ray coronal jets should not be excluded – at high enough flow speeds, which in principal are reachable, one can expect a dramatic change in the waves' behaviour associated with an emerging instability, and subsequently, with an Alfvén-wave-turbulence heating.

In all cases, the question of whether large coronal spicules can reach coronal temperatures remains open – for a discussion from an observational point of view we refer to the paper by Madjarska et al. (Madjarska et al., 2011).

7. References

Acton, F. S. (1990). *Numerical Methods That (Usually) Work*, Mathematical Association of America, ISBN: 0-88385-450-3, Washington, D.C., ch. 14.

Alfvén, H. (1942). Existence of Electromagnetic-Hydrodynamic Waves. *Nature*, Vol. 150, No. 3805, October 3 1942, 405–406.

Andries, J. & Goossens, M. (2001). Kelvin–Helmholtz instabilities and resonant flow instabilities for a coronal plume model with plasma pressure. *Astron. Astrophys.*, Vol. 368, No. 3, March IV 2001, 1083–1094.

Antolin, P. & Shibata, K. (2010). The Role of Torsional Alfvén Waves in Coronal Heating. *Astrophys. J.*, Vol. 712, No. 1, March 20 2010, 494–510.

Athay, R. G. & Holzer, T. E. (1982). Role of spicules in heating the solar atmosphere. *Astrophys. J.*, Vol. 255, April 15 1982, 743–752.

Athay, R. G. (2000). Are spicules related to coronal heating? *Solar Phys.*, Vol. 197, No. 1, November 2000, 31–42.

Beckers, J. M. (1968). Solar Spicules. *Solar Phys.*, Vol. 3, No. 3, March 1968, 367–433.

Beckers, J. M. (1972). Solar Spicules. *Ann. Rev. Astron. Astrophys.*, Vol. 10, September 1972, 73–100.

Bennett, K.; Roberts, B. & Narain, U. (1999). Waves in Twisted Magnetic Flux Tubes. *Solar Phys.*, Vol. 185, No. 1, March 1999, 41–59.

Cally, P. S. (1986). Leaky and non-leaky oscillations in magnetic flux tubes. *Solar Phys.*, Vol. 103, No. 2, February 1986, 277–298.

Centeno, R.; Bueno, J. T. & Ramos, A. A. (2010). On the Magnetic Field of Off-limb Spicules. *Astrophys. J.*, Vol. 708, No. 2, January 10 2010, 1579–1584.

Chandrasekhar, S. (1961). *Hydrodynamic and Hydromagnetic Stability*, Clarendon Press, ISBN: 0-486-64071-X, Oxford, ch. 11.

Chen, F. F. (1995). *Introduction to Plasma Physics*, Springer, ISBN: 0306307553, Berlin.

Cirtain, J. W.; Golub, L.; Lundquist, L. et al. (2007). Evidence for Alfvén Waves in Solar X-ray Jets. *Nature*, Vol. 318, No. 5856, December 7 2007, 1580–1582.

de Pontieu, B.; McIntosh, S.; Hansteen, V. H. et al. (2007). A Tale of Two Spicules: The Impact of Spicules on the Magnetic Chromosphere. *Publ. Astron. Soc. Japan*, Vol. 59, No. SP3, S655–S662.

De Pontieu, B.; McIntosh, S. W.; Carlsson, M. et al. (2011). The Origins of Hot Plasma in the Solar Corona. *Science*, Vol. 331, No. 6013, January 7 2011, 55–58.

Edwin, P. M. & Roberts, B. (1983). Wave Propagation in a Magnetic Cylinder. *Solar Phys.*, Vol. 88, Nos. 1–2, October 1983, 179–191.

Gekelman, W.; Vincena, S.; Van Compernolle, B.; Morales, G. J.; Maggs, J. E.; Pribyl, P. & Carter, T. A. (2011). The many faces of shear Alfvén waves. *Phys. Plasmas*, Vol. 18, No. 5, May 2011, 055501 (26 pages), doi: 10.1063/1.3592210.

He, J.-S.; Tu, C.-Y.; Marsch, E. et al. (1999). Upward propagating high-frequency Alfvén waves as identified from dynamic wave-like spicules observed by SOT on *Hinode*. *Astron. Astrophys.*, Vol. 497, No. 2, April II 2009, 525–535.

Hollweg, J. V. (1981). Alfvén waves in the solar atmosphere. *Solar Phys.*, Vol. 70, No. 1, March 1981, 25–66.

Kudoh, T. & Shibata, K. (1999). Alfvén Wave Model of Spicules and Coronal Heating. *Astrophys. J.*, Vol. 514, No. 1, March 20 1999, 493–505.

Kukhianidze, V.; Zaqarashvili, T. V. & Khutsishvili, E. (2006). Observation of kink waves in solar spicules. *Astron. Astrophys.*, Vol. 449, No. 2, April II 2006, L35–L38.

Leroy, B. (1981). Propagation of waves in an atmosphere in the presence of a magnetic field. III – Alfvén waves in the solar atmosphere. *Astron. Astrophys.*, Vol. 97, No. 2, April 1981, 245–250.

Madjarska, M. S. (2011). Dynamics and plasma properties of an X-ray jet from SUMER, EIS, XRT, and EUVI A & B simultaneous observations. *Astron. Astrophys.*, Vol. 526, February 2011, A19 (24 pages), doi: 10.1051/0004-6361/201015269.

Madjarska, M. S.; Vanninathan, K. & Doyle, J. D. (2011). Can coronal hole spicules reach coronal temperatures? *Astron. Astrophys.*, Vol. 532, August 2011, L1 (4 pages), doi: 10.1051/0004-6361/201116735.

Matsumoto, T. & Shibata, K. (2010). Nonlinear Propagation of Alfvén Waves Driven by Observed Photospheric Motion: Application to the Coronal Heating and Spicule Formation. *Astrophys. J.*, Vol. 710, No. 2, February 20 2010, 1857–1867.

McIntosh, S. W.; De Pontieu, B.; Carlsson, M.; Hansteen, V.; Boerner, P. & Goossens, M. (2011). Alfvénic waves with sufficient energy to power the quiet solar corona and fast solar wind. *Nature*, Vol. 475, No. 7357, July 28 2011, 477–480.

Moore, R. L.; Sterling, A. C.; Cirtain, J. W. & Falconer, D. A. (2011). Solar X-ray Jets, Type-II Spicules, Granule-size Emerging Bipoles, and the Genesis of the Heliosphere. *Astrophys. J.*, Vol. 731, No. 1, April 10 2011, L18 (5 pages), doi:10.1088/2041-8205/731/1/L18.

Muller, D. A. (1956). A Method for Solving Algebraic Equations Using an Automatic Computer. *Math. Tables Other Aids Comput.*, Vol. 10, No. 56, October 1956, 208–215.

Nakariakov, V. M. (2007). MHD oscillations in solar and stellar coronae: Current results and perspectives. *Adv. Space Res.*, Vol. 39, No. 12, December 2007, 1804–1813.

Roberts, W. O. (1945). A Preliminary Report on Chromospheric Spicules in Extremely Short Livetime. *Astrophys. J.*, Vol. 101, March 1945, 136R–140R.

Secchi, P. A. (1877). *Le Soleil*, 2nd edition, Part II, Gauthiers-Villars, Paris.

Shibata, K.; Ishido, Y.; Acton, L. W. et al. (1992). Observations of X-Ray Jets with the *Yohkoh* Soft X-Ray Telescope. *Publ. Astron. Soc. Japan*, Vol. 44, No. 5, October 1992, L173–L179.

Shimojo, M.; Hashimoto, S.; Shibata, K.; Hirayama, T.; Hudson, H. S. & Acton, L. W. (1996). Statistical Study of Solar X-Ray Jets Observed with the *Yohkoh* Soft X-Ray Telescope. *Publ. Astron. Soc. Japan*, Vol. 48, No. Sp1, February 1996, S123–S136.

Shimojo, M. & Shibata, K. (2000). Physical Parameters of Solar X-Ray Jets. *Astrophys. J.*, Vol. 542, No. 2, October 20 2000, 1100–1108.

Song, P. & Vasyliūnas, V. M. (2011). Heating of the solar atmosphere by strong damping of Alfvén waves. *J. Geophys. Res.*, Vol. 116, September 27 2011, A09104 (17 pages), doi: 10.1029/2011JA016679.

Sterling, A. C. (2000). Solar Spicules: A Review of Recent Models and Targets for Future Observations – (Invited Review). *Solar Phys.*, Vol. 196, No. 1, September 2000, 79–111.

Sterling, A. C.; Moore, R. L. & DeForest, C. E. (2010). *Hinode* Solar Optical Telescope observations of the source regions and evolution of "Type II" spicules at the solar limb. *Astrophys. J.*, Vol. 714, No. 1, May 1 2010, L1–L6.

Tavabi, E.; Koutchmy, S. & Ajabshirizadeh, A. (2011). A Statistical Analysis of the SOT-*Hinode* Observations of Solar Spicules and their Wave-like Behavior. *New Astron.*, Vol. 16, No. 4, July 2011, 296–305.

Terra-Homen, M.; Erdélyi, R. & Ballai, I. (2003). Linear and non-linear MHD wave propagation in steady-state magnetic cylinders. *Solar Phys.*, Vol. 217, No. 2, November 2003, 199–223.

Tsap, Y. T.; Stepanov, A. V. & Kopylova, Y. G. (2011). Energy Flux of Alfvén Waves in Weakly Ionized Plasma and Coronal Heating of the Sun. *Solar Phys.*, Vol. 270, No. 1, May 2011, 205–211.

van Ballegooijen, A. A.; Asgari-Targhi, M.; Cranmer, S. R. & DeLuca, A. A. (2011). Heating of the Solar Chromosphere and Corona by Alfvén Wave Turbulence. *Astrophys. J.*, Vol. 736, No. 1. July 20 2011, 3 (27 pages), doi: 10.1088/0004-637X/736/1/3.

Vasheghani Farahani, S.; Van Doorsselaere, T.; Verwichte, E. & Nakariakov, V. M. (2009). Propagating transverse waves in soft X-ray coronal jets. *Astron. Astrophys.*, Vol. 498, No. 2, May I 2009, L29–L32.

Zahariev, N. I. & Mishonov, T. M. (2011). Heating of the Solar Corona by Alfvén Waves: Self-Induced Opacity. *Proceedings of 3rd School and Workshop on Space Plasma Physics: AIP Conf. Proc.*, Vol. 1356, pp. 123–137, ISBN: 978-0-7354-0914-9, Kiten (Bulgaria), September 1–12, 2010, American Institute of Physics, Melville, NY 11747.

Zaqarashvili, T. V.; Khutsishvili, E.; Kukhianidze, V. & Ramishvili, G. (2007). Doppler-shift oscillations in solar spicules. *Astron. Astrophys.*, Vol. 474, No. 2, November I 2007, 627–632.

Zaqarashvili, T. V. & Erdélyi, R. (2009). Oscillations and Waves in Solar Spicules. *Space Sci. Rev.*, Vol. 149, Nos. 1–4, December 2009, 355–388.

Zaqarashvili, T. V.; Días, A. J.; Oliver, R. & Ballester, J. L. (2010). Instability of twisted magnetic tubes with axial mass flows. *Astron. Astrophys.*, Vol. 516, June–July 2010, A84 (8 pages), doi: 10.1051/0004-6361/200913874.

Zaqarashvili, T. V. (2011). Solar Spicules: Recent Challenges in Observations and Theory, *Proceedings of 3rd School and Workshop on Space Plasma Physics: AIP Conf. Proc.*, Vol. 1356, pp. 106–116, ISBN: 978-0-7354-0914-9, Kiten (Bulgaria), September 1–12, 2010, American Institute of Physics, Melville, NY 11747.

Zhelyazkov, I. (2010). Hall-magnetohydrodynamic waves in flowing ideal incompressible solar-wind plasmas. *Plasma Phys. Control. Fusion*, Vol. 52, No. 6, June 2010, 065008 (16 pages), doi: 10.1088/0741-3335/52/6/065008

Zhelyazkov, I. (2011). Hall-Magnetohydrodynamic Waves and their Stability in Flowing Solar-Wind Plasmas. *Proceedings of 3rd School and Workshop on Space Plasma Physics: AIP Conf. Proc.*, Vol. 1356, pp. 138–155, ISBN: 978-0-7354-0914-9, Kiten (Bulgaria), September 1–12, 2010, American Institute of Physics, Melville, NY 11747.

7

Hamiltonian Representation of Magnetohydrodynamics for Boundary Energy Controls

Gou Nishida[1] and Noboru Sakamoto[2]
[1]*RIKEN*
[2]*Nagoya University*
Japan

1. Introduction

1.1 Brief summary of this chapter

This chapter shows that basic boundary control strategies for *magnetohydrodynamics* (MHD) can be derived from a formal system representation, called a *port-Hamiltonian system* (Van der Schaft and Maschke, 2002). The port-Hamiltonian formulation clarifies collocated input/output pairs used for stabilizing and assigning a global stable point. The controls called *passivity-based controls* (Arimoto, 1996; Ortega et al., 1998; Van der Schaft, 2000; Duindam et al., 2009) are simple and robust to disturbances. Moreover, port-Hamiltonian systems can be connected while keeping their consistency with respect to energy flows. Finally, we show that port-Hamiltonian systems can be used for boundary controls. In the future, this theory might be specialized, for instance, in order to control disruptions of Tokamak plasmas (Wesson, 2004; Pironti and Walker, 2005; Ariola and Pironti, 2008). This chapter emphasizes the versatility of control system representations.

1.2 Background and motivation

Control theory significantly progressed during the last two decades of the 20th century. *Linear control theory* (Zhou et al., 1996) was developed for systems whose states are limited to a neighborhood around stable points. The theory was extended to include particular classes of distributed parameter systems and nonlinear systems (Khalil, 2001; Isidori, 1995). However, dispite this progress, simpler and more intuitive methods like *PID controls* (Brogliato et al., 2006) are still in the mainstream of practical control designs. One reason for this trend is that advanced methods do not always remarkably produce significant improvements to the performance of controlled systems despite their theoretical complexity; rather, they are prone to modeling errors. The other reason is that simple methods are understandable and adjustable online, although the resulting performance is not exactly optimal.

On the other hand, actual controlled systems can be regarded as distributed parameter systems from a macroscopic viewpoint, e.g., as elastic continuums, and as discrete nonlinear systems from a microscopic viewpoint, e.g., as molecular dynamics systems. Moreover, their stable points are not always unique and vary according to the environment. Multi-physics and multi-scaling models are becoming increasingly significant in science and engineering because of rapid advances in computational devices and micromachining technology. However, such complexities have tended to be ignored in system modeling of conventional control designs, because controllers have to be simple enough to be integrated with other mechanisms and be quickly adjustable. Moreover, numerical analyses using more detailed models can be executed off-line by trial and error and in circumstance where there are no physical size limitations on the computational devices. Hence, it would be desirable to have a new framework of simple control designs like PID controls, but for complex systems. The port-Hamiltonian system, which is introduced in this chapter, is one of the most promising frameworks for this purpose. This chapter addresses the issue of how to derive simple and versatile controls for partial differential equations (PDEs), especially, those of MHD, from considerations about the storage and dissipation of energy in port-Hamiltonian systems.

1.3 History of topic and relevant research

Port-Hamiltonian systems are a framework for passivity-based controls. *Passivity* (Van der Schaft, 2000) is a property by which the energy supplied from the outside of systems through input/output variables can be expressed as a function of the stored energy. The storage function is equivalent to a Hamiltonian in dynamical systems. The collocated input/output variable pairs, called *port variables*, are defined systematically in terms of port-Hamiltonian systems, and they are used as controls and for making observations. Passivity-based controls consist of *shaping Hamiltonians* and *damping assignments*. The Hamiltonians of these systems can be changed by "connecting" them to other port-Hamiltonian systems by means of the port variables. The Hamiltonian of controlled systems is equal to the sum of those of the original system and controllers. Thus, if we can design such a changed Hamiltonian beforehand, the connections give the Hamiltonian of the original system "shaping". Such connected port-Hamiltonian systems with a shaped Hamiltonian can be stabilized to the minimum of the storage function by adding dissipating elements to the port variables.

The energy preserving properties of port-Hamiltonian systems can be described in terms of a *Dirac structure* (Van der Schaft, 2000; Courant, 1990), which is the generalization of symplectic and Poisson structures (Arnold, 1989). Dirac structures enable us to model complex systems as port-Hamiltonian representations, e.g., distributed parameter systems with nonlinearity (Van der Schaft and Maschke, 2002), systems with higher order derivatives (Le Gorrec et al., 2005; Nishida, 2004), thermodynamical systems (Eberard et al., 2007), discretized distributed systems (Golo et al., 2004; Voss and Scherpen, 2011) and their coupled systems. This chapter mainly uses the port-Hamiltonian representation of PDEs for boundary controls based on passivity, i.e., the DPH system. The boundary integrability of DPH systems is derived from a *Stokes-Dirac structure* (Van der Schaft and Maschke, 2002), which is an extended Dirac structure in the sense of Stokes theorem. Because of this boundary integrability, the change in the internal energy of DPH systems is equal to the energy supplied through port variables defined on the boundary of the system domain. Hence, passivity-based controls for distributed parameter systems can be considered to be boundary energy controls.

1.4 Construction of this chapter

In Section 2, we derive the geometric formulation of MHD defined by using differential forms (Flanders, 1963; Morita, 2001). After that, we rewrite the model in terms of DPH systems. The modeling procedure is systematically determined by a given Hamiltonian. Next, we explain passivity-based controls that can be applied to the DPH system of MHD, and their energy flows by means of the bond graph (Karnopp et al., 2006). Finally, we show that the boundary power balance equation of the DPH system is the extended energy principle of MHD (Wesson, 2004) in the sense of dynamical systems and boundary controls.

In Section 3, we extend the DPH model of MHD to include non-Hamiltonian subsystems corresponding to external force terms in Euler-Lagrange equations. Actual controlled systems represented by MHD might be affected by model perturbations, e.g., disturbances or other controllers, or model improvements. Such variations cannot always be modeled in terms of Hamiltonian systems. Some systems of PDEs can be decomposed into a Hamiltonian subsystem, which we call an *exact subsystem*, and a non-Hamiltonian subsystem, which we call a *dual-exact subsystem* (Nishida et al., 2007a). Through this decomposition, a PDE system can be described as a coupled system consisting of a port-Hamiltonian subsystem determined by a *pseudo potential* and other subsystems representing, e.g., external forces, dissipations and distributed controls.

In Section 4, we derive a boundary observer for detecting symmetry breaking (Nishida et al., 2009) from the DPH system of conservation laws associated with MHD. For example, Hamiltonian systems can be regarded as the conservation law with a symmetry that is the invariance of energy with respect to the time evolution. If a symmetry is broken, the associated conservation law becomes invalid. Symmetry breaking can be detected by checking whether quantities are conserved with the boundary port variables of the DPH system. Furthermore, we present a basic strategy for detecting the topological transitions of the domain of DPH systems. The formulation using differential forms defined on Riemannian manifolds can describe systems affected by such transitions. We use a general decomposition of differential forms on Riemannian manifolds and of vector fields on three-dimensional Riemannian manifolds and derive the boundary controls for creating a desired topological energy flow from this decomposition.

The last section is devoted to a brief introduction of future work on this topic.

2. Port-Hamiltonian systems and passivity-based controls for MHD

2.1 Ideal magnetohydrodynamical equations

Magnetohydrodynamics (MHD) is a discipline involving modeling magnetically confined plasmas (Wesson, 2004; Pironti and Walker, 2005; Ariola and Pironti, 2008). The ideal MHD system is a coupled system consisting of a single fluid and an electromagnetic field with certain constitutive relations.

The fluid is described by the two equations in three dimensions. The first is the mass conservation law,

$$\frac{\partial \rho}{\partial t} + \nabla \cdot (\rho v) = 0, \tag{1}$$

where $\rho(t,x) \in \mathbb{R}$ is the local mass density at time $t \in \mathbb{R}$ at the spatial position $x = (x^1, x^2, x^3) \in \mathbb{R}^3$, and $v(t,x) \in \mathbb{R}^3$ is the fluid (Eulerian) velocity at t and x. The second is Newton's law applied to an infinitesimal plasma element with an electromagnetic coupling,

$$\rho \frac{\partial v}{\partial t} = -\rho v \cdot \nabla v - \nabla p + \boldsymbol{J} \times \boldsymbol{B}, \tag{2}$$

where $p(t,x) \in \mathbb{R}$ is the kinetic pressure in plasma, $\boldsymbol{J}(t,x) \in \mathbb{R}^3$ is the free current density, $\boldsymbol{B}(t,x) \in \mathbb{R}^3$ is the magnetic field induction, and the Lorentz force term $\boldsymbol{J} \times \boldsymbol{B}$ means the coupling.

The electromagnetic field satisfies the Maxwell's equations consisting of Ampere's law, Faraday's law, and Gauss's law for the magnetic induction field:

$$-\frac{\partial \boldsymbol{D}}{\partial t} = -\nabla \times \boldsymbol{H} + \boldsymbol{J}, \quad -\frac{\partial \boldsymbol{B}}{\partial t} = \nabla \times \boldsymbol{E}, \quad \nabla \cdot \boldsymbol{B} = 0, \tag{3}$$

where the time derivative of the electric field induction $\boldsymbol{D} \in \mathbb{R}^3$ is neglected in MHD.

The constitutive relations are given by

$$\boldsymbol{B} = \mu \boldsymbol{H}, \quad \boldsymbol{E} + v \times \boldsymbol{B} = \eta \boldsymbol{J}, \tag{4}$$

where μ is the magnetic permeability and η is the resistance coefficient that is assumed to be zero in an ideal MHD system.

2.2 Geometric formulation of MHD

The main framework of this chapter is the port-Hamiltonian system for PDEs called a *distributed port-Hamiltonian (DPH) system* (Van der Schaft and Maschke, 2002). DPH systems are expressed in terms of differential forms (Flanders, 1963; Morita, 2001). Moreover, a formulation using differential forms defined on Riemannian manifolds can describe the relation between the vector fields of systems and the topological properties of system domains (see *Section 4*). Thus, we shall rewrite the equations of MHD by using differential forms to derive the DPH representation of MHD.

Let Y be an n-dimensional smooth Riemannian manifold. Let Z be an n-dimensional smooth Riemannian submanifold of Y with a smooth boundary ∂Z. We assume that the time coordinate $t \in \mathbb{R}$ is split from the spatial coordinates $x = (x^1, \cdots, x^n) \in Z$ in the local chart of Z. We denote the space of differential k-forms on Z by $\Omega^k(Z)$ for $0 \leq k \leq n$. We denote the infinite-dimensional vector space of all smooth vector fields in Z by $\mathfrak{X}(Z)$. We identify the 1-from v with the vector field $v^\sharp \in \mathfrak{X}(Z)$. The fluid equations (1) and (2) can be rewritten as follows:

$$\begin{cases} \dfrac{\partial \rho}{\partial t} = -de_v, \quad \dfrac{\partial v}{\partial t} = -de_\rho + g_1 + g_2, \\ e_v = \boldsymbol{i}_{v^\sharp}\rho, \quad e_\rho = \dfrac{1}{2}\langle v^\sharp, v^\sharp \rangle + w(*\rho), \\ g_1 = -(*\rho)^{-1} * (*dv \wedge *e_v), \quad g_2 = (*\rho)^{-1} * (*\boldsymbol{J} \wedge *\boldsymbol{B}), \end{cases} \tag{5}$$

where $n = 3$, $\rho \in \Omega^3(Z)$ is the mass density, $v \in \Omega^1(Z)$ is the fluid velocity, $J \in \Omega^2(Z)$ is the free current density, $B \in \Omega^2(Z)$ is the magnetic field induction, $\langle v^\sharp, v^\sharp \rangle = \|v^\sharp\|^2$ is the inner product with respect to $v^\sharp$, and we have introduced the following operators:

- $d\colon \Omega^k(Z) \to \Omega^{k+1}(Z) \cdots$ The exterior differential operator d on Z is defined as

$$d\omega = \sum_{j=1}^{n} \frac{\partial f_{i_1\cdots i_k}}{\partial x^j} dx^j \wedge dx^{i_1} \wedge \cdots \wedge dx^{i_k} \tag{6}$$

  for $\omega = f_{i_1\cdots i_k}(x)\, dx^{i_1} \wedge \cdots \wedge dx^{i_k} \in \Omega^k(Z)$, where $i_1 \cdots i_k$ is the combination of k different integers selected from 1 to n, and $j \neq i_1 \neq \cdots \neq i_k$.

- $*: \Omega^k(Z) \to \Omega^{n-k}(Z) \cdots$ The Hodge star operator $*$ induced in terms of a Riemannian metric on Z is defined as

$$*\omega = \sum_{i_1<\cdots<i_k} \operatorname{sgn}(I,J) f_{i_1\cdots i_k}\, dx^{j_1} \wedge \cdots \wedge dx^{j_{n-k}} \in \Omega^{n-k}(Z) \tag{7}$$

  for $\omega = \sum_{i_1<\cdots<i_k} f_{i_1\cdots i_k}(x)\, dx^{i_1} \wedge \cdots \wedge dx^{i_k} \in \Omega^k(Z)$, where $j_1 < \cdots < j_{n-k}$ is the rearrangement of the complement of $i_1 < \cdots < i_k$ in the set $\{1, \cdots, n\}$ in ascending order, and $\operatorname{sgn}(I,J)$ is the sign of the permutation of $i_1, \cdots, i_k, j_1, \cdots, j_{n-k}$ generated by interchanging of the basic forms dx^i (if we interchange dx^i and dx^j in ω for arbitrary i and j, the sign of ω changes, i.e., it is alternating).

- $i_{v^\sharp}: \Omega^k(Z) \to \Omega^{k-1}(Z) \cdots$ The interior product $i_{v^\sharp}$ with respect to $v^\sharp$ is defined as

$$i_{v^\sharp}\omega = \begin{cases} (-1)^{m-1} f_{i_1\cdots i_k} g_{i_m} dx^{i_1} \wedge \cdots \wedge dx^{i_{m-1}} \wedge dx^{i_{m+1}} \wedge \cdots \wedge dx^{i_k} & \text{if } j = i_m, \\ 0 & \text{if } j \neq i_m \end{cases} \tag{8}$$

  for $v^\sharp = g_j(x)(\partial/\partial x^j)$ and $\omega = f_{i_1\cdots i_k}(x) dx^{i_1} \wedge \cdots \wedge dx^{i_k}$.

In (5), we used the formula $(v \cdot \nabla)v = (1/2)\nabla(v \cdot v) + \mathrm{Curl}\, v \times v$, and the enthalpy $w(*\rho) = (\partial/\partial *\rho)(*\rho\, U(*\rho))$ is related to the pressure $p(*\rho)$ by $(*\rho)^{-1} dp(*\rho) = dw(*\rho)$, where $U(\rho)$ is the internal energy function of the fluid satisfying $p(*\rho) = w(*\rho)*\rho - U(*\rho)*\rho$.

Next, Maxwell's equations are defined as follows:

$$-\frac{\partial D}{\partial t} = -dH + J, \quad -\frac{\partial B}{\partial t} = dE, \quad dB = 0, \quad dD = \varrho, \tag{9}$$

where $D \in \Omega^2(Z)$ is the electric field induction, $H \in \Omega^1(Z)$ is the magnetic field intensity, $E \in \Omega^1(Z)$ is the electric field intensity, and $\varrho \in \Omega^3(Z)$ is the free charge density.

The constitutive relations are written as follows:

$$B = \mu * H, \quad *(E + i_{v^\sharp} B) = \eta J. \tag{10}$$

2.3 Definition of port-Hamiltonian system

Let us recall the definition of DPH systems. The advantage of these systems will be explained from the viewpoint of passivity and boundary controls in later sections.

The inner product of k-forms can be defined on Z as

$$\langle \omega, \eta \rangle = \omega \wedge *\eta, \quad \langle \omega, \eta \rangle_Z = \int_Z \langle \omega, \eta \rangle \tag{11}$$

for $\omega, \eta \in \Omega^k(Z)$. Moreover, we can identify the 1-from v with the vector field $v^\sharp \in \mathfrak{X}(Z)$; therefore, (11) can be defined as the inner product of vector fields, as in (5). DPH systems are defined by Stokes-Dirac structures (Van der Schaft and Maschke, 2002; Courant, 1990) with respect to the inner product (11).

Definition 2.1. *Let*

$$\begin{cases} (f^p, f^q, f^b) \in \Omega^p(Z) \times \Omega^q(Z) \times \Omega^{n-p}(\partial Z), \\ (e^p, e^q, e^b) \in \Omega^{n-p}(Z) \times \Omega^{n-q}(Z) \times \Omega^{n-q}(\partial Z), \\ (f_d^p, f_d^q) \in \Omega^p(Z) \times \Omega^q(Z), \\ (e_d^p, e_d^q) \in \Omega^{n-p}(Z) \times \Omega^{n-q}(Z), \end{cases} \tag{12}$$

where all f^i and e^i for $i \in \{p, q, b\}$ and all f_d^i and e_d^i for $i \in \{p, q\}$ constitute the pairs with respect to the inner product $\langle \cdot, \cdot \rangle_Z$. The Stokes-Dirac structure is defined as follows:

$$\begin{bmatrix} f^p \\ f^q \end{bmatrix} = \begin{bmatrix} 0 & (-1)^r d \\ d & 0 \end{bmatrix} \begin{bmatrix} e^p \\ e^q \end{bmatrix} - \begin{bmatrix} f_d^p \\ f_d^q \end{bmatrix}, \quad \begin{bmatrix} e_d^p \\ e_d^q \end{bmatrix} = \begin{bmatrix} e^p \\ e^q \end{bmatrix}, \quad \begin{bmatrix} f^b \\ e^b \end{bmatrix} = \begin{bmatrix} e^p|_{\partial Z} \\ (-1)^p e^q|_{\partial Z} \end{bmatrix}, \tag{13}$$

where $r = pq + 1$, $p + q = n + 1$, $|_{\partial Z}$ is the restriction of differential forms to ∂Z, $df_d^p \neq 0$, and $df_d^q \neq 0$.

A DPH system is formed by substituting the following variables obtained from a Hamiltonian density in the above Stokes-Dirac structure.

Definition 2.2. *Let $\mathcal{H}(\alpha^p, \alpha^q) \in \Omega^n(Z)$ be a Hamiltonian density, where $\alpha^i \in \Omega^i(Z)$ for $i \in \{p, q\}$. A DPH system is defined by substituting*

$$f^p = -\frac{\partial \alpha^p}{\partial t}, \quad f^q = -\frac{\partial \alpha^q}{\partial t}, \quad e^p = \frac{\partial \mathcal{H}}{\partial \alpha^p}, \quad e^q = \frac{\partial \mathcal{H}}{\partial \alpha^q} \tag{14}$$

into (13), where $\partial/\partial \alpha^i$ means the variational derivative with respect to α^i. The variables f_d^p and f_d^q cannot be derived from any Hamiltonian.

DPH systems satisfy the following boundary integrable relation that comes from Stokes theorem (Flanders, 1963; Morita, 2001).

Proposition 2.1 (Van der Schaft and Maschke (2002)). *A DPH system satisfies the following power balance:*

$$\int_Z (e^p \wedge f^p + e^q \wedge f^q) + \int_Z \left(e_d^p \wedge f_d^p + e_d^q \wedge f_d^q \right) + \int_{\partial Z} e^b \wedge f^b = 0. \tag{15}$$

where each term $e^i \wedge f^i$ for $i \in \{p, q, b\}$ has the dimension of power.

In DPH systems, each f^i and e^i for $i \in \{p, q\}$ are called *port variables*, and f^b and e^b are called *boundary port variables* that are a pair of boundary inputs and outputs. We call

$e^b \wedge f^b$ a *boundary energy flow*. On the other hand, the terms $e_d^p \wedge f_d^p$ and $e_d^q \wedge f_d^q$ are non-boundary-integrable; therefore, we cannot detect changes in them from the boundary energy flows. We call $e_d^p \wedge f_d^p$ and $e_d^q \wedge f_d^q$ *distributed energy flows*.

2.4 Passivity and boundary integrability of energy flows

The advantages of DPH systems are grounded in the following stability.

Definition 2.3. *Consider a system with an input vector $u(t)$ and an output vector $y(t)$. The system is called* passive *if there exists a C^0 class non-negative function $V(x)$ such that $V(0) = 0$ and*

$$V(x(t_1)) - V(x(t_0)) \leq \int_{t_0}^{t_1} u^\top(s)y(s)\,ds \tag{16}$$

for all inputs $u(t)$ and an initial value $x(t_0)$, where $t_0 \leq t_1$ and $\top$ means the transpose of vectors.

$V(x)$ can be regarded as the internal energy of the systems, which is an extended Lyapunov function. The inequality in (16) means that the energy always decreases; therefore, the system is stable in the sense of Lyapunov. Controls using the relation (16) are called *passivity-based controls*. Standard control systems with pairs of inputs/outputs satisfying (16) are called *port-Hamiltonian systems*. In this case, $V(x)$ corresponds with the Hamiltonian of the system. Hence, in (15), all port variables e_j^i and f_j^i for $i \in \{p, q, b\}$ and $j \in \{f, e\}$ might be inputs and outputs for passivity-based controls.

The boundary port variables f_j^b and e_j^b in (15) can be used as passivity-based boundary controls (Van der Schaft, 2000; Duindam et al., 2009) (details are given in *Section 2.6*). In (15), the first integral means the time variation of Hamiltonian; i.e., it is calculated by taking the interior product between a possible variational vector field and the variational derivative of Hamiltonian: $\boldsymbol{i}_{\mathfrak{X}_i} d\mathcal{H}_i$ for $i \in \{f, e\}$, where $\mathfrak{X}_i = \sum_j (\partial\alpha_j/\partial t)(\partial/\partial\alpha_j)$ is the variational vector field and α_j is the variational variable. The power of the first integral can be transformed into that of the third integral by appealing to boundary integrability of Stokes theorem. The second integral means non-boundary-integrable energy flows. Hence, if the second integral is zero, we can detect the variation of energies distributed on system domains from the variation on the boundary. In this sense, the power balance (15) is the principle of passivity-based boundary controls.

2.5 Port-Hamiltonian representation of MHD

In this section, we derive the DPH representation of MHD from the geometric formulation presented in *Section 2.2*, which has been partially treated as Maxwell's equations and as an ideal fluid in (Van der Schaft and Maschke, 2002).

Let $n = 3$. The DPH representation can be systematically constructed in terms of the Hamiltonian densities of the fluid and the electromagnetic field

$$\mathcal{H}_f = \int_Z \frac{1}{2}\langle v^\sharp, v^\sharp \rangle \rho + U(*\rho)\rho, \quad \mathcal{H}_e = \int_Z \frac{1}{2}\left(\boldsymbol{E} \wedge \boldsymbol{D} + \boldsymbol{H} \wedge \boldsymbol{B}\right) \tag{17}$$

under constraints defined by the system equations (5), (9) and (10). Indeed, the DPH system of MHD can be constructed as

$$\begin{cases} \begin{bmatrix} -\boldsymbol{\rho}_t \\ -\boldsymbol{v}_t \end{bmatrix} = \begin{bmatrix} 0 & d \\ d & 0 \end{bmatrix} \begin{bmatrix} e_\rho \\ e_v \end{bmatrix} - \begin{bmatrix} 0 \\ g_1 + g_2 \end{bmatrix}, & \begin{bmatrix} f_f^b \\ e_f^b \end{bmatrix} = \begin{bmatrix} e_\rho|_{\partial Z} \\ -e_v|_{\partial Z} \end{bmatrix}, \\ \begin{bmatrix} -\boldsymbol{D}_t \\ -\boldsymbol{B}_t \end{bmatrix} = \begin{bmatrix} 0 & -d \\ d & 0 \end{bmatrix} \begin{bmatrix} \boldsymbol{E} \\ \boldsymbol{H} \end{bmatrix} + \begin{bmatrix} \boldsymbol{J} \\ 0 \end{bmatrix}, & \begin{bmatrix} f_e^b \\ e_e^b \end{bmatrix} = \begin{bmatrix} \boldsymbol{E}|_{\partial Z} \\ \boldsymbol{H}|_{\partial Z} \end{bmatrix}, \end{cases} \tag{18}$$

where the subscript t means the partial derivative with respect to t, and we have defined

$$\begin{cases} e_v = \boldsymbol{i}_{\boldsymbol{v}^\sharp}\boldsymbol{\rho}, \quad e_\rho = \dfrac{1}{2}\langle \boldsymbol{v}^\sharp, \boldsymbol{v}^\sharp \rangle + w(*\boldsymbol{\rho}), \\ g_1 = -(*\boldsymbol{\rho})^{-1}*(*d\boldsymbol{v} \wedge *e_v), \quad g_2 = (*\boldsymbol{\rho})^{-1}*(*\boldsymbol{J} \wedge *\boldsymbol{B}), \\ \boldsymbol{B} = \mu*\boldsymbol{H}, \quad *(\boldsymbol{E} + \boldsymbol{i}_{\boldsymbol{v}^\sharp}\boldsymbol{B}) = \eta\boldsymbol{J}, \end{cases} \tag{19}$$

having set $p = 3, q = 1$, and $r = 3 \cdot 1 + 1$ for the fluid, and $p = 2, q = 2$, and $r = 2 \cdot 2 + 1$ for the electromagnetic field. The DPH system satisfies the following power balance equations:

$$\int_Z (-e_\rho \wedge \boldsymbol{\rho}_t - e_v \wedge \boldsymbol{v}_t) - \int_Z e_v \wedge g_2 - \int_{\partial Z} e_v \wedge e_\rho = 0, \tag{20}$$

$$\int_Z (-\boldsymbol{E} \wedge \boldsymbol{D}_t - \boldsymbol{H} \wedge \boldsymbol{B}_t) - \int_Z \boldsymbol{E} \wedge \boldsymbol{J} + \int_{\partial Z} \boldsymbol{H} \wedge \boldsymbol{E} = 0, \tag{21}$$

where $e_v \wedge g_1 = -(*\boldsymbol{\rho})^{-1} e_v \wedge *(*d\boldsymbol{v} \wedge *e_v) = -(*\boldsymbol{\rho})^{-1} *e_v \wedge *d\boldsymbol{v} \wedge *e_v = 0$ and (21) which corresponds to Poynting's theorem. Note that the definition of the boundary energy flow in (21) is invariant even if $\boldsymbol{D}_t$ is assumed to be zero, as is done in the standard theory of MHD. The first integrals of (20) and (21) correspond to the total change in energy of the system defined on Z, and the third integral is equal to the energy flowing across ∂Z.

2.6 Passivity-based boundary controls

The basic strategy of passivity-based controls is to connect controllers through pairs of port variables, e.g., new port-Hamiltonian systems for changing the total Hamiltonians, or dissipative elements for stabilizing the system to the global minimum of the shaped Hamiltonian. The passivity-based boundary controls for DPH systems are applied to the boundary port variables f_j^b and e_j^b for $j \in \{f, e\}$. The product $e_j^b \wedge f_j^b$ has the dimension of power; therefore, f_j^b and e_j^b can be considered to be a generalized velocity and a generalized force in analogy to mechanical systems (the correspondence might be the inverse in some cases).

Applying the output f_j^b magnified by a negative gain to the input e_j^b means velocity feedback. This is one of most important passivity-based controls, i.e, damping assignment. Moreover, the boundary energy flow $e_j^b \wedge f_j^b$ balances the internal energy of DPH systems; therefore, the total energy of the controlled system decreases, and the system becomes stable in the sense of passivity (16).

On the other hand, the Hamiltonian of the original DPH system can be changed by connecting other DPH systems to the original. The connection by means of port variables is expressed by bond graph theory (Karnopp et al., 2006), which is a generalized circuit theory for describing physical systems from the viewpoint of energy flows. For instance, the following diagram is the bound graph representation of the DPH system of MHD:

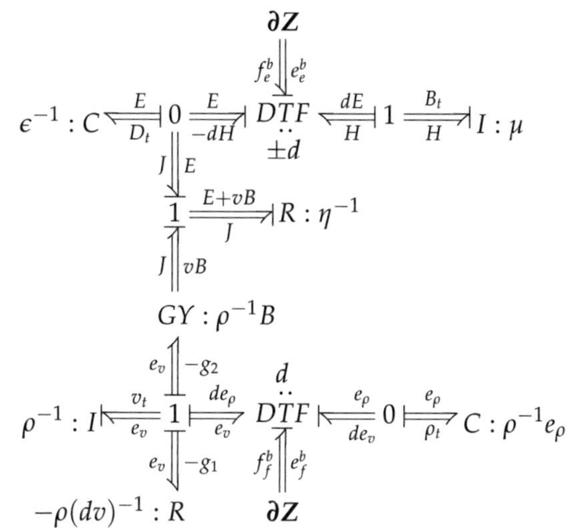

(22)

where $\partial \boldsymbol{Z}$ is the boundary of the systems, and we have defined the following bond graph elements:

- The arrow with the pair of variables e and f means the energy flow $e \wedge f$.
- The direction arrow indicates the sign of the energy flow.
- The causal stroke $|$ at the edge of the arrows indicates the direction in which the effort signal is directed.
- The n pairs of variables e_i and f_i around the 0-junction satisfy $e_1 = e_2 = \cdots = e_n$ and $\sum_{i=1}^{n} s_i f_i = 0$, where $s_i = 1$ if the arrow is directed towards the junction and $s_i = -1$ otherwise.
- The n pairs of variables e_i and f_i around the 1-junction satisfy $f_1 = f_2 = \cdots = f_n$ and $\sum_{i=1}^{n} s_i e_i = 0$.
- The C element with a parameter K means the capacitor satisfies $e = K \int_{-\infty}^{t} f dt$.
- The I element with a parameter K means the inductor satisfies $f = K^{-1} \int_{-\infty}^{t} e dt$.
- The R element with a parameter K means the resister satisfies $e = Kf$.
- The GY element with a parameter M means the gyrator satisfies $e_2 = Mf_1$ and $e_1 = Mf_2$.
- The DTF element means the differential transformer that has a Stokes-Dirac structure. In the case with the symbol d, $e_2 = de_1$, $f_1 = df_2$, $f^b = e_1|_{\partial Z}$ and $e^b = -f_2|_{\partial Z}$. In the case with the symbol $\pm d$, $e_2 = de_1$, $f_1 = -df_2$, $f^b = e_1|_{\partial Z}$ and $e^b = f_2|_{\partial Z}$.

The Hamiltonian is shaped by connecting new systems to it through the pairs of boundary port variables (f_f^b, e_f^b) and (f_e^b, e_e^b). For example, we can connect an electromagnetic system as a controller on the boundary ∂Z of the upper part of (22) as follows:

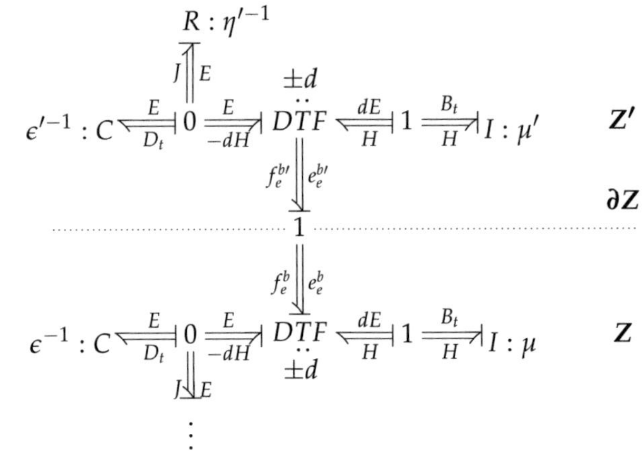

(23)

where Z' is the domain of the new electromagnetic system and each system is connected through the common boundary $\partial Z = \partial Z'$. In this case, the original Hamiltonian $\mathcal{H}_f + \mathcal{H}_e$ is changed into the controlled Hamiltonian $\mathcal{H}_f + \mathcal{H}_e + \mathcal{H}'_e$, where $\mathcal{H}'_e$ is the Hamiltonian of the new electromagnetic system. Note that the Hamiltonians can only be shaped to control the energy flows of boundary port variables or energy levels of the original system, not to control the distributed states in the sense of boundary value problems.

The energy flow through the boundary $\partial Z = \partial Z'$ can be described as

$$\mathcal{H}_{\delta t} = \int_{\partial Z} e^b \wedge f^b - e^{b\prime} \wedge f^{b\prime}, \tag{24}$$

where $e^{b\prime}$ and $f^{b\prime}$ are the pair of the boundary port variables defined on $\partial Z'$. In general, when the port variable e^b is regarded as an input, the power balance (15) is changed into

$$\int_Z (e^p \wedge f^p + e^q \wedge f^q) + \int_Z \left\{ e_d^p \wedge (f_d^p + u_d^p) + e_d^q \wedge (f_d^q + u_d^q) \right\} + \int_{\partial Z} u^b \wedge f^b = 0, \tag{25}$$

where $e^b = u^b$ is the boundary control, and u_d^p and u_d^q are the distributed controls. If f^b is regarded as an input, then the boundary control is replaced by $f^b = u^b$.

Damping terms are assigned by connecting of resisters to the pair on the system domain; they are illustrated as R elements in the bond graph. If systems with dissipative elements are connected to the boundary of a controlled system, it corresponds to a boundary damping assignment that absorbs the energy of the original system through the boundary. For example, in (25), the controls

$$u^b = -K^b f^b, \quad u_d^p = -K_d^p \alpha^p, \quad u_d^q = -K_d^q \alpha^q \tag{26}$$

are equivalent to connecting an R element to the port variables, where K^b is the gain function defined on ∂Z, K_d^p and K_d^p are the gain functions defined on Z, and $f^i = -(\partial \alpha^i / \partial t)$. For eliminating distributed energy flows f_d^p and f_d^q that are exactly known, we can use the controls

$$u_d^p = -f_d^p, \quad u_d^q = -f_d^q, \tag{27}$$

where the inputs u_d^p and u_d^q distributed on Z. Moreover, in (23), R: η'^{-1} distributed on Z' is considered as an element to create energy flowing across the boundary of the original MHD system.

A practical problem is whether the boundary port variables e_i^b and f_i^b can actually be used as inputs and outputs. In this section, we show all possible boundary port variables of MHD regardless of whether they are actually usable or not. The input/output pairs for the passivity-based boundary control of MHD are the boundary port variables

$$(e_f^b, f_f^b) = (-e_v|_{\partial Z}, e_\rho|_{\partial Z}), \quad (e_e^b, f_e^b) = (\boldsymbol{H}|_{\partial Z}, \boldsymbol{E}|_{\partial Z}). \tag{28}$$

(e_f^b, f_f^b) can be transformed as follows:

$$\begin{aligned}\int_{\partial Z} e_v \wedge e_\rho &= \int_{\partial Z} \boldsymbol{i}_{\boldsymbol{v}^\sharp}\boldsymbol{\rho} \wedge \left(\frac{1}{2}\langle \boldsymbol{v}^\sharp, \boldsymbol{v}^\sharp\rangle + w(*\boldsymbol{\rho})\right) \\ &= \int_{\partial Z} \boldsymbol{i}_{\boldsymbol{v}^\sharp}\left(\frac{1}{2}\langle \boldsymbol{v}^\sharp, \boldsymbol{v}^\sharp\rangle\boldsymbol{\rho} + U(*\boldsymbol{\rho})\boldsymbol{\rho}\right) + \int_{\partial Z} \boldsymbol{i}_{\boldsymbol{v}^\sharp}(*p),\end{aligned} \tag{29}$$

where the first term corresponds to the boundary energy flow of convections and the second term means external work. Hence, the altered port variables are

$$(e_{f1}^b, f_{f1}^b) = (\mathcal{H}_f|_{\partial Z}, \boldsymbol{v}|_{\partial Z}), \quad (e_{f2}^b, f_{f2}^b) = (p|_{\partial Z}, \boldsymbol{v}|_{\partial Z}). \tag{30}$$

2.7 Port representation of balanced MHD

This section discusses the stability of the DPH systems of MHD (18) with (19) in a balanced state. If the change in the potential energy of MHD caused by physically admissible perturbation is positive, then the equilibrium of MHD is stable. This fact is called *the energy principle of MHD* (Wesson, 2004). We derive the basic equation of the energy principle from the DPH system.

If the 2-form $d\boldsymbol{v}$ is zero at a certain time $t = t_0$, it continues to be zero after t_0. Accordingly, (5) can be reduced as follows:

$$\frac{\partial \boldsymbol{\rho}}{\partial t} = 0, \quad \frac{\partial \boldsymbol{v}}{\partial t} = (*\boldsymbol{\rho})^{-1}\left\{-dp(*\boldsymbol{\rho}) + *(*\boldsymbol{J} \wedge *\boldsymbol{B})\right\} = 0. \tag{31}$$

Now, let us consider the variation in energy with respect to an infinitesimal variation in displacement:

$$\mathcal{W}_{\delta t} = \int_Z \delta\boldsymbol{x}\frac{\delta}{\delta t}\left\{-dp(*\boldsymbol{\rho}) + *(*\boldsymbol{J} \wedge *\boldsymbol{B})\right\}, \tag{32}$$

where the subscript δt means the variational derivative with respect to the time, and δ means an infinitesimal variation. From (9), we obtain

$$\frac{\delta \boldsymbol{J}}{\delta t} = d\frac{\delta \boldsymbol{H}}{\delta t}, \tag{33}$$

where we have assumed that $D_t = 0$ and $\eta = 0$; therefore,

$$dD_t = \varrho_t = 0, \quad \varrho_t = dJ = 0, \quad E = -i_{v^\sharp} B. \tag{34}$$

The DPH system of balanced MHD can be constructed as follows:

$$\begin{cases} \begin{bmatrix} -\rho_t \\ 0 \end{bmatrix} = \begin{bmatrix} 0 & d \\ d & 0 \end{bmatrix} \begin{bmatrix} w_{\delta t}(*\rho) \\ i_{v^\sharp}\rho \end{bmatrix} - \begin{bmatrix} 0 \\ (*\rho)^{-1} dp_{\delta t} \end{bmatrix}, & \begin{bmatrix} f^b_{fs} \\ e^b_{fs} \end{bmatrix} = \begin{bmatrix} w_{\delta t}(*\rho)|_{\partial Z} \\ -i_{v^\sharp}\rho|_{\partial Z} \end{bmatrix}, \\ \begin{bmatrix} 0 \\ -B_t \end{bmatrix} = \begin{bmatrix} 0 & -d \\ d & 0 \end{bmatrix} \begin{bmatrix} -i_{v^\sharp} B \\ H_{\delta t} \end{bmatrix} + \begin{bmatrix} J_{\delta t} \\ 0 \end{bmatrix}, & \begin{bmatrix} f^b_{es} \\ e^b_{es} \end{bmatrix} = \begin{bmatrix} -i_{v^\sharp} B|_{\partial Z} \\ H_{\delta t}|_{\partial Z} \end{bmatrix}, \end{cases} \tag{35}$$

where $\delta x = v$. The DPH system (35) satisfies the power balance equations,

$$-\int_Z w_{\delta t}(*\rho) \wedge \rho_t + \int_Z i_{v^\sharp}\rho \wedge (*\rho)^{-1} dp_{\delta t} - \int_{\partial Z} i_{v^\sharp}\rho \wedge w_{\delta t}(*\rho) = 0, \tag{36}$$

$$-\int_Z H_{\delta t} \wedge B_t + \int_Z i_{v^\sharp} B \wedge J_{\delta t} - \int_{\partial Z} H_{\delta t} \wedge i_{v^\sharp} B = 0. \tag{37}$$

As a result, we obtain the boundary port variables

$$(f^b_{fs}, e^b_{fs}) = (w_{\delta t}(*\rho)|_{\partial Z}, -i_{v^\sharp}\rho|_{\partial Z}), \quad (f^b_{es}, e^b_{es}) = (-i_{v^\sharp} B|_{\partial Z}, H_{\delta t}|_{\partial Z}) \tag{38}$$

from (35).

The energy principle is frequently used to analyze the stability of MHD. The DPH system of MHD generates the power balance equation (37) for an analysis. The boundary port variables of (35) correspond to those of the DPH system of dynamical MHD (18) except for the term depending on v. Hence, (18) can be considered to be a generalized system following the energy principle of MHD. If active controls are used in MHD systems, e.g., in Tokamaks, the control side of the DPH system able to be used, e.g., as a boundary control for subdivided MHD systems.

3. Construction pseudo potentials for non-Hamiltonian subsystems

3.1 DPH systems of MHD with perturbations

Section 2 discussed the energy structure of the DPH system of MHD on the basis of its physical meaning. However, model perturbations caused by, for instance, disturbances, additional terms derived by using system identification methods for model refinements, or controllers designed by a control theory do not always have physical interpretations. In this section, we show a method of determining the energy structure of such perturbations. Precisely speaking, we decompose a given perturbation into a Hamiltonian subsystem and a non-Hamiltonian subsystem that can be regarded as an external force in terms of Euler-Lagrange equations (Nishida et al., 2007a).

In this section, we consider an n-dimensional smooth Riemannian manifold Y that is homeomorphic to an n-dimensional Euclidian space (i.e., topologically same, and one can be deformed into the other). Let Z be an n-dimensional smooth Riemannian submanifold of Y with a smooth boundary ∂Z. The DPH system (18) of MHD defined on a domain Z is extended so as to have perturbations as follows:

$$\begin{cases} \begin{bmatrix} -\rho_t \\ -v_t \end{bmatrix} = \begin{bmatrix} 0 & d \\ d & 0 \end{bmatrix} \begin{bmatrix} e_\rho \\ e_v \end{bmatrix} - \begin{bmatrix} 0 \\ g_1 + g_2 \end{bmatrix} + \begin{bmatrix} \Delta_f^p \\ \Delta_f^q \end{bmatrix}, & \begin{bmatrix} f_f^b \\ e_f^b \end{bmatrix} = \begin{bmatrix} e_\rho|_{\partial Z} \\ -e_v|_{\partial Z} \end{bmatrix}, \\ \begin{bmatrix} -\boldsymbol{D}_t \\ -\boldsymbol{B}_t \end{bmatrix} = \begin{bmatrix} 0 & -d \\ d & 0 \end{bmatrix} \begin{bmatrix} \boldsymbol{E} \\ \boldsymbol{H} \end{bmatrix} + \begin{bmatrix} \boldsymbol{J} \\ 0 \end{bmatrix} + \begin{bmatrix} \Delta_e^p \\ \Delta_e^q \end{bmatrix}, & \begin{bmatrix} f_e^b \\ e_e^b \end{bmatrix} = \begin{bmatrix} \boldsymbol{E}|_{\partial Z} \\ \boldsymbol{H}|_{\partial Z} \end{bmatrix}, \end{cases} \tag{39}$$

where each Δ_j^i for $i \in \{p,q\}$ and $j \in \{f,e\}$ means a perturbation. Now, let us consider the subsystem of DPH systems, $\Delta_j^i(u_I^a)$, where $i \in \{p,q\}$, $j \in \{f,e\}$, u^a for $1 \le a \le l$ is the function defined by the local coordinates x^k of Y for $1 \le k \le n$, and we denote all possible derivatives up to the order r of u^a by u_I^a and denote the order by $0 \le |I| \le r$. For example, u_I^a for $r = 2$ means $\{u^a, u_t^a, u_y^a, u_z^a, u_{tt}^a, u_{ty}^a, u_{tz}^a, u_{yy}^a, u_{yz}^a, u_{zz}^a\}$ for $(x^1, x^2, x^3) = (t, y, z)$, and the subscript means the partial derivative.

3.2 Decomposition of model perturbations of DPH systems

Consider the DPH system (39) of MHD with perturbations. We assume that the DPH system includes up to second-order derivatives: $r = 2$. Accordingly, Δ_j^i can be uniquely decomposed into

$$\Delta_j^i = d\varphi_j^i + \gamma_j^i, \tag{40}$$

where φ_j^i is a *pseudo potential* derived from

$$\gamma_j^i\, du = \Delta\, du - d\tilde{\varphi}_j^i, \quad \varphi_j^i\, du = \Delta\, du - \gamma_j^i\, du, \tag{41}$$

the temporal variable $\tilde{\varphi}_j^i$ is calculated as

$$\tilde{\varphi}_j^i = h_v(\Delta_j^i\, du^a) = \int_0^1 u^a \cdot \Delta_j^i(x^k, \lambda u_I^a)\, d\lambda, \tag{42}$$

h_v is *the homotopy operator* for $\omega \in \Omega^k(Z)$ with respect to an equilibrium point u_{cI}^a, called a *homotopy center*, defined by

$$h_v(\omega) = \int_0^1 \boldsymbol{i}_{\bar{v}}\omega(x, \lambda \bar{u}_I)\, \lambda^{-1} d\lambda, \quad \bar{v} = \sum_{a,I} (u_I^a - u_{cI}^a) \frac{\partial}{\partial u_I^a}, \tag{43}$$

where $\bar{u}_I^a = u_{cI}^a + \lambda(u_I^a - u_{cI}^a)$, and usually $u_{cI}^a = 0$. In (40), we call $d\varphi_j^i$ an *exact system* and call γ_j^i a *dual exact system*, which corresponds to a distributed energy variable.

For example, let us consider $\Delta_j^i = 1 + w_t + w_{tt}$ for some i and j, where $u^1 = w$ and $x^0 = t$. The temporal variable

$$\tilde{\varphi}_j^i = w + \frac{1}{2} w w_t + \frac{1}{2} w w_{tt} \tag{44}$$

is derived from $h_v(\Delta^i_j\, dw)$. Hence,

$$d\tilde{\varphi}^i_j = (1 + w_{tt})\, dw, \quad \gamma^i_j = \Delta^i_j\, dw - d\tilde{\varphi}^i_j = w_t\, dw. \tag{45}$$

On the other hand, from the relation

$$\begin{aligned}\Delta^i_j\, dw &= (1 + w_t + w_{tt})\, dw \\ &= \left(1 + \frac{1}{2}w_t + w_{tt}\right) dw + \left(-\frac{1}{2}w - w_t\right) dw_t\end{aligned} \tag{46}$$

that is transformed in terms of an integration by parts, we obtain

$$\tilde{\varphi}^i_j = w - \frac{1}{2}w_t^2. \tag{47}$$

This result yields the same relation $d\tilde{\varphi}^i_j = (1 + w_{tt})\, dw$. Thus, the expression $\tilde{\varphi}^i_j$ has variations generated by an integration by parts; therefore, we should recalculate φ^i_j as in (41).

3.3 Necessary and sufficient condition of decomposition

We can check whether a given Δ^i_j is an exact system or a dual exact system from the self-adjointness of the differential operator $\mathcal{D}_{\Delta^i_j}$ defining Δ^i_j: $\mathcal{D}^*_{\Delta^i_j} = \mathcal{D}_{\Delta^i_j}$ (Olver, 1993, pp. 109, 307, 329 and 364). Here, *the Fréchet derivative* $\mathcal{D}_{\mathcal{F}}$ of a second-order subsystem $\mathcal{F}(u_I)$ is an $(l \times k)$-matrix with elements

$$(\mathcal{D}_{\mathcal{F}})_{ab}(h) = \left(\frac{\partial \mathcal{F}_a}{\partial u^b} + \sum_{i=0}^{n} \frac{\partial \mathcal{F}_a}{\partial u^b_{x^i}} \frac{\partial}{\partial x^i} + \sum_{i=0}^{n}\sum_{j=0}^{n} \frac{\partial \mathcal{F}_a}{\partial u^b_{x^i x^j}} \frac{\partial}{\partial x^i}\frac{\partial}{\partial x^j}\right) h \tag{48}$$

and the adjoint operator $\mathcal{D}^*_{\mathcal{F}}$ of $\mathcal{D}_{\mathcal{F}}$ is a $(k \times l)$-matrix with elements

$$(\mathcal{D}^*_{\mathcal{F}})_{ba}(h) = \frac{\partial \mathcal{F}_a}{\partial u^b} h - \sum_{i=0}^{n} \frac{\partial}{\partial x^i}\left(\frac{\partial \mathcal{F}_a}{\partial u^b_{x^i}} h\right) + \sum_{i=0}^{n}\sum_{j=0}^{n} \frac{\partial}{\partial x^i}\frac{\partial}{\partial x^j}\left(\frac{\partial \mathcal{F}_a}{\partial u^b_{x^i x^j}} h\right) \tag{49}$$

for $a = 1, \cdots, k$ and $b = 1, \cdots, l$, where $h = h(u_I)$ is any function and we assume $k = l$.

For example, consider $\Delta^q_f = 1 + \nu v + v_t$ in (39), where $u^1 = w$, $w_t = v$ and $x^0 = t$. Then, $\varphi^q_f = 1 + v_t$ and $\gamma^q_f = \nu v$, because $g = \nu v$ is non-self-adjoint: $\mathcal{D}^*_g \neq \mathcal{D}_g$, and we have used (48) and (49) with $a = b = 1$, i.e.,

$$\mathcal{D}_g(h) = \frac{\partial g}{\partial u_{x^0}}\frac{\partial}{\partial x^0}(h) = \nu\frac{\partial h}{\partial t}, \quad \mathcal{D}^*_g(h) = -\frac{\partial}{\partial x^0}\left(\frac{\partial g}{\partial u_{x^0}} h\right) = -\nu\frac{\partial h}{\partial t}. \tag{50}$$

3.4 Elimination of decomposed perturbations

The uniqueness of the decomposition is determined by the topology of Y. That is, differential k-forms for $k \geq 1$ defined on such a domain can be always described as in (40). If a pseudo potential can be defined for a perturbation, the perturbation can be included in the variables e^p or e^q of the Stokes-Dirac structure. Hence, such a perturbation can be detected in terms of the following boundary power balances:

$$\int_Z \left(-(e_\rho + \varphi_f^q) \wedge \rho_t - (e_v + \varphi_f^p) \wedge v_t\right) - \int_Z e_\rho \wedge \gamma_f^p - \int_Z e_v \wedge (g_2 + \gamma_f^q) - \int_{\partial Z} (e_v + \varphi_f^p) \wedge (e_\rho + \varphi_f^q) = 0, \quad (51)$$

$$\int_Z \left\{-(E + \varphi_e^q) \wedge D_t - (H - \varphi_e^p) \wedge B_t\right\} - \int_Z E \wedge (J + \gamma_e^p) - \int_Z H \wedge \gamma_e^q + \int_{\partial Z} (H - \varphi_e^p) \wedge (E + \varphi_e^q) = 0. \quad (52)$$

Moreover, from these relations, we can see that the exact subsystem of perturbations can be controlled by boundary port variables. Indeed, we can construct the boundary controls in the fourth integrals of the power balance equations (51) and (52) as follows:

$$\int_{\partial Z} (e_v + \varphi_f^p + u_f^q) \wedge (e_\rho + \varphi_f^q + u_f^p), \quad (53)$$

$$\int_{\partial Z} (H - \varphi_e^p + u_e^q) \wedge (E + \varphi_e^q + u_e^p), \quad (54)$$

where u_j^i is the boundary input for compensating pseudo potentials such that

$$u_f^q = -\varphi_f^p, \quad u_f^p = -\varphi_f^q, \quad u_e^q = \varphi_e^p, \quad u_e^p = -\varphi_e^q. \quad (55)$$

On the other hand, the decomposed perturbations corresponding dual exact subsystems cannot be eliminated by boundary controls. Hence, we should introduce the distributed controls in the second and third integrals of the power balance equations (51) and (52) as follows:

$$-\int_Z e_\rho \wedge (\gamma_f^p + u_{df}^p) - \int_Z e_v \wedge (g_2 + \gamma_f^q + u_{df}^q), \quad (56)$$

$$-\int_Z E \wedge (J + \gamma_e^p + u_{de}^p) - \int_Z H \wedge (\gamma_e^q + u_{de}^q), \quad (57)$$

where u_{dj}^i is the distributed input for eliminating dual exact subsystems such that

$$u_{df}^p = -\gamma_f^p, \quad u_{df}^q = -\gamma_f^q, \quad u_{de}^p = -\gamma_e^p, \quad u_{de}^q = -\gamma_e^q. \quad (58)$$

4. Boundary observer for detecting topological symmetry breaking

4.1 Symmetry and power balance equations

In this section, we first discuss the influence of topological variations in the system domains on the power balance equation of DPH systems. We can detect such changes by checking the boundary power balance of the original system; if there is an imbalance. According to Noether's theorem (Olver, 1993), conservation laws are associated with symmetries present in systems. That is, our purpose is to construct a boundary observer for detecting symmetry breaking (Nishida et al., 2009). Finally, we derive a boundary control for creating desired energy flows from topological properties of manifolds.

We shall clarify the first problem by means of the following example. Consider a DPH system defined on a 2-dimensional domain Z. We assume that the energy flow of the system can be split along the x- and y-axis. Next, we divide the domain Z into subdomains, i.e., $Z = \bigcup_i Z^i$, where Z^i is the i-th subdomain of Z. We denote the common boundary between Z^i and Z^j by ∂Z^{ij}. The following power balance holds:

$$\mathcal{H}_{\delta t} = \int_{\partial Z^{ij}} \sum_{i,j} \left(e^{bi} \wedge f^{bi} - e^{bj} \wedge f^{bj} \right) = 0, \tag{59}$$

where e^{bi} and f^{bi} are the boundary port variables defined on ∂Z^i. The DPH system can be regarded as a connected structure of DPH systems defined on Z^i in terms of boundary port variables of ∂Z^{ij}. We shall simplify the shapes of Z and each Z^i to be squares as in the left diagram below:

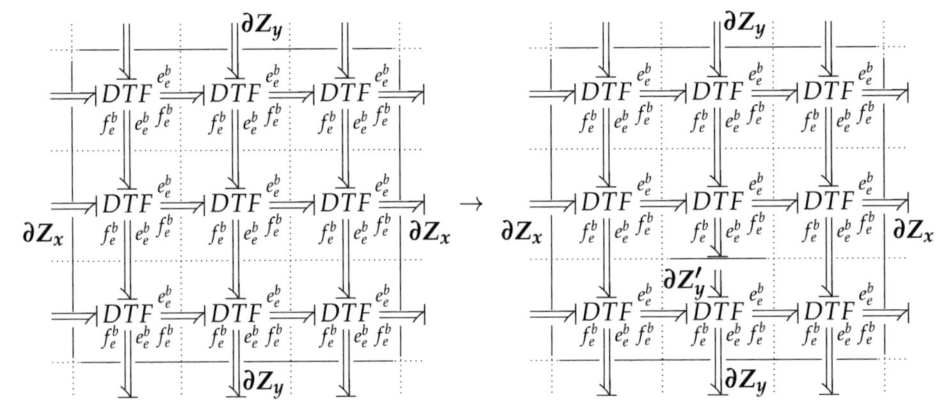

(60)

Accordingly, we can split the original boundary ∂Z and denote the boundaries with respect to the x- and y-axis by ∂Z_x and ∂Z_y, respectively. Hence, the following power balance holds:

$$\mathcal{H}_{\delta t} = \mathcal{H}_{\delta t}|_{\partial Z_x} + \mathcal{H}_{\delta t}|_{\partial Z_y} = 0. \tag{61}$$

Now, let us assume that a structural change occurs in the inner part of Z on a segment along x-axis that we denote as $\partial Z'_y$ in the right diagram of (60). Such changes are caused by, for instance, energy dissipations, or energy transformations to other physical systems, and they can be illustrated as a new element connected to $\partial Z'_y$ in the bond graph. This means the energy preserving symmetry is broken along the x-axis. In this case, (61) should be revised to

$$\mathcal{H}'_{\delta t} = \mathcal{H}_{\delta t}|_{\partial Z_x} + \mathcal{H}_{\delta t}|_{\partial Z'_y} + \mathcal{H}_{\delta t}|_{\partial Z_y} = 0. \tag{62}$$

Hence, we can detect that the power on ∂Z_y: $\mathcal{H}_{\delta t}|_{\partial Z_y} = 0$ becomes imbalanced if the port variables in (61) are observable. In other words, this change can be regarded as a change in the topology of the system domain, i.e., a deformation from $Z \simeq \mathbb{R}^n$ to $Z \setminus \partial Z'_y \simeq \mathbb{R}^n \setminus \{0\}$, where $\simeq$ means topological equivalence (i.e., homeomorphic), $\setminus$ means subtraction of sets, and $\{0\}$ is a point.

4.2 Topological decomposition of differential forms and vector fields

This section discusses the relation between the topology of the domain Z of DPH systems and the decomposable components of vector fields on Z. After this discussion, the symmetry breaking explained in the previous section will be extended to a change in energy flows of DPH systems defined on compact manifolds.

In *Section 2*, we assumed that the system domain Z is a subdomain of a manifold that is topologically the same as a Euclidian space. Actually, this assumption restricted the form of diffrential forms. In this case, differential k-forms for $k \geq 1$ can be decomposed into two types, i.e., an exact form and a dual exact form as in (40). That is, differential forms $\omega_e \in \Omega^k(Z)$ are called *exact forms* if there exists some $\eta \in \Omega^{k-1}(Z)$ such that $\omega_e = d\eta$, i.e., $d\omega_e = d(d\eta) = 0$ because of the nature of exterior differentiation. The forms $\omega_d \in \Omega^k(Z)$ such that $d\omega_d \neq 0$ are called *dual exact forms*. In general, there might also exist *harmonic forms* $\omega_h \in \Omega^k(Z)$ satisfying $\triangle\omega_h = 0$, where $\triangle = dd^\dagger + d^\dagger d$ is the Laplacian and $d^\dagger = (-1)^{n(k+1)+1} * d *$ is the adjoint operator of exterior differentiation. The components of differential forms depend on the topology of domains. All classifications of differential forms defined on a compact domain with a smooth boundary are given by *the Hodge decomposition theorem* (Morita, 2001); i.e., an arbitrary differential form on an oriented compact Riemannian manifold can be uniquely decomposed into an exact form, a dual exact form, and a harmonic form:

$$\omega = \omega_e + \omega_d + \omega_h \in \Omega^k(Z). \tag{63}$$

Moreover, a unique harmonic form on an oriented compact Riemannian manifold corresponds to a topological quantity of the manifolds called a *homology*. Precisely speaking, from *Hodge theorem*, *Poincaré duality thorem* and the duality between homology and (de Rham) cohomology, we obtain the isomorphism $H_k(Z, \partial Z) \cong \Omega_h^{n-k}(Z)$ (Morita, 2001; Gross and Kotiuga, 2004, pp. 102), where $H_k(Z)$ is the vector space with real coefficients of the k-th homology of Z, and $\Omega_h^k(Z)$ is the space of harmonic forms.

If $n = 3$, the homology of Z consists of the following vector spaces:

- $H_0(Z) \cdots$ The vector space is generated by such equivalence classes of points in Z as two points are equivalent if they can be connected by a path in Z. $\dim H_0(Z)$ is the number of components of Z. Note that $H_0(Z) \cong \mathbb{R}$ for a connected Z and the element of $H_0(Z)$ is a constant function.
- $H_1(Z) \cdots$ The vector space is generated by such equivalence classes of oriented loops in Z as two loops are equivalent if their difference is the boundary of an oriented surface in Z. The number of holes of closed surfaces is called a *genus*. $\dim H_1(Z)$ is the number of total genus of Z.
- $H_2(Z) \cdots$ The vector space is generated by such equivalence surfaces of points in Z as two surfaces are equivalent if their difference is the boundary of some oriented subregion of Z. $\dim H_2(Z)$ is the number of the difference between components of ∂Z and those of Z.
- $H_3(Z) \cdots$ $\dim H_3(Z)$ is always 0.

On the other hand, the dual space of $H_k(Z)$ is $H_{n-k}(Z, \partial Z)$, where $H_k(Z, \partial Z)$ is called *the k-th relative homology of Z modulo* ∂Z. In $n = 3$, the relative homology of Z modulo ∂Z consists of the following vector spaces with real coefficients:

- $H_0(Z, \partial Z) \cdots \dim H_0(Z)$ is always 0.
- $H_1(Z, \partial Z) \cdots$ The vector space is generated by such equivalence classes of oriented paths whose endpoints lie on ∂Z as two such paths are equivalent if their difference (possibly paths on ∂Z) is the boundary of an oriented surface in Z.
- $H_2(Z, \partial Z) \cdots$ The vector space is generated by such equivalence classes of oriented surface whose boundaries lie on ∂Z as two such surfaces are equivalent if their difference (possibly portions of ∂Z) is the boundary of some oriented subregion of Z.
- $H_3(Z, \partial Z) \cdots$ The vector space has the oriented components of Z as a basis. Thus, $\dim H_3(Z, \partial Z)$ is the number of components of the subregions of Z whose boundaries lie on ∂Z. Note that $H_3(Z, \partial Z) \cong \mathbb{R}$ for a connected Z and the element of $H_3(Z, \partial Z)$ is a constant function.

Hence, $H_k(Z, \partial Z) \cong H_{3-k}(Z)$ for $0 \leq k \leq 3$.

As we mentioned before, the space of vector fields $\mathfrak{X}$ can be identified with that of 1-forms Ω^1 in the sense of a Riemannian metric. Thus, vector fields are affected by the decomposition of differential forms. Indeed, the space of vector fields on a compact domain Z in three-dimensional space with a smooth boundary can be decomposed as follows.

Theorem 4.1 (Cantarella et al. (2002)). *Consider vector fields $v^\sharp \in \mathfrak{X}(Z)$ on a compact domain Z with a smooth boundary ∂Z in three-dimensional space. Let W denote any smooth orientable surface in Z whose boundary ∂W lies on the boundary ∂Z: $W \subset Z$ and $\partial W \subset \partial Z$, and called it a* cross-sectional surface. *The space $\mathfrak{X}(Z)$ is the direct sum of five mutually orthogonal subspaces:*

$$\mathfrak{X}(Z) = \mathfrak{X}_K(Z) \oplus \mathfrak{X}_G(Z), \tag{64}$$

where $v \in \Omega^1(Z)$, $v^\sharp \in \mathfrak{X}(Z)$, $\varphi \in \Omega^0(Z)$,

$$\mathfrak{X}_K(Z) = \left\{ v^\sharp \in \mathfrak{X}(Z) \colon *d{*v} = 0,\ v^\sharp \cdot \boldsymbol{n}^\sharp = 0 \right\}, \quad \mathfrak{X}_G(Z) = \left\{ v^\sharp \in \mathfrak{X}(Z) \colon v = d\varphi \right\}, \tag{65}$$

which are called knots *and* gradients, *respectively, and $\boldsymbol{n}^\sharp$ means all unit vector fields normal to ∂Z. Furthermore,*

$$\mathfrak{X}_K(Z) = \mathfrak{X}_{FK}(Z) \oplus \mathfrak{X}_{HK}(Z), \quad \mathfrak{X}_G(Z) = \mathfrak{X}_{CG}(Z) \oplus \mathfrak{X}_{HG}(Z) \oplus \mathfrak{X}_{GG}(Z), \tag{66}$$

where

$$\mathfrak{X}_{FK}(Z) = \left\{ v^\sharp \in \mathfrak{X}(Z) \colon *d{*v} = 0,\ \langle v, \boldsymbol{n} \rangle_{\partial Z} = 0,\ \langle v, \boldsymbol{m} \rangle_W = 0 \right\}, \tag{67}$$

$$\mathfrak{X}_{HK}(Z) = \left\{ v^\sharp \in \mathfrak{X}(Z) \colon *d{*v} = 0,\ \langle v, \boldsymbol{n} \rangle_{\partial Z} = 0,\ dv = 0 \right\}, \tag{68}$$

$$\mathfrak{X}_{CG}(Z) = \left\{ v^\sharp \in \mathfrak{X}(Z) \colon v = d\varphi,\ *d{*v} = 0,\ \langle v, \boldsymbol{n} \rangle_{\partial Z} = 0 \right\}, \tag{69}$$

$$\mathfrak{X}_{HG}(Z) = \left\{ v^\sharp \in \mathfrak{X}(Z) \colon v = d\varphi,\ *d*v = 0,\ \varphi = C \right\}, \tag{70}$$

$$\mathfrak{X}_{GG}(Z) = \left\{ v^\sharp \in \mathfrak{X}(Z) \colon v = d\varphi,\ \varphi|_{\partial Z} = 0 \right\} \tag{71}$$

$$\dim H_1(Z) = \dim \mathfrak{X}_{HK}(Z), \quad \dim H_2(Z) = \dim \mathfrak{X}_{HG}(Z). \tag{72}$$

which are respectively called fluxless knots, harmonic knots, curly gradients, harmonic gradients *and* grounded gradients, *and* $\boldsymbol{m}^\sharp$ *means all unit vector fields normal to W, and C is a function on* ∂Z *that is locally constant.*

For example, consider a vector field defined on a three-dimensional disc. There is no $v^\sharp \in \mathfrak{X}_{HK}(Z)$ on the disc, because the genus is 0 and $\dim H_1(Z) = \dim \mathfrak{X}_{HK}(Z) = 0$. Thus, all rotation vector fields on the disc are $v^\sharp \in \mathfrak{X}_{FK}(Z)$ that is the rotating vector field whose axis is an inner point of the disc. $v^\sharp \in \mathfrak{X}_{CG}(Z)$ is a constant vector field flowing across the disc; therefore, it is divergence-free and zero flux through the one and only component of ∂Z. $v^\sharp \in \mathfrak{X}_{GG}(Z)$ is a radiational vector field flowing from an inner point of the disc, where the potential φ is constant on ∂Z. There is no $v^\sharp \in \mathfrak{X}_{HG}(Z)$ on the disc, because the numbers of components of ∂Z and Z are each 1, i.e., $\dim \mathfrak{X}_{HG}(Z) = 0$. However, a three-dimensional solid torus has a hole; therefore, $\dim \mathfrak{X}_{HK}(Z) = 1$, but $\dim \mathfrak{X}_{HG}(Z) = 0$. $v^\sharp \in \mathfrak{X}_{HK}(Z)$ is a circulative vector field flowing around the hole. Moreover, for a region between two concentric round spheres, $\dim \mathfrak{X}_{HG}(Z) = 1$. $v^\sharp \in \mathfrak{X}_{HK}(Z)$ is a radiational vector field flowing from a common center in the small sphere.

4.3 DPH systems with harmonic energy flows

In this section, we extend the DPH system of MHD to include the global energy flows originating from topological shapes of manifolds.

Let Z be a three-dimensional smooth Riemannian submanifold of Y with a smooth boundary ∂Z. The DPH system (18) of MHD defined on a domain Z is extended to have energy flows regarding harmonic knots and harmonic gradients that we call *harmonic energy flows* as follows:

$$\begin{cases} \begin{bmatrix} -\rho_t \\ -v_t \end{bmatrix} = \begin{bmatrix} 0 & d \\ d & 0 \end{bmatrix} \begin{bmatrix} e_\rho + e^p_{hf} \\ e_v + e^q_{hf} \end{bmatrix} - \begin{bmatrix} 0 \\ g_1 + g_2 \end{bmatrix} + \begin{bmatrix} f^p_{hf} \\ f^q_{hf} \end{bmatrix}, & \begin{bmatrix} f^b_f \\ e^b_f \end{bmatrix} = \begin{bmatrix} (e_\rho + e^p_{hf})|_{\partial Z} \\ -(e_v + e^q_{hf})|_{\partial Z} \end{bmatrix}, \\ \begin{bmatrix} -\boldsymbol{D}_t \\ -\boldsymbol{B}_t \end{bmatrix} = \begin{bmatrix} 0 & -d \\ d & 0 \end{bmatrix} \begin{bmatrix} \boldsymbol{E} + e^p_{he} \\ \boldsymbol{H} + e^q_{he} \end{bmatrix} + \begin{bmatrix} \boldsymbol{J} \\ 0 \end{bmatrix} + \begin{bmatrix} f^p_{he} \\ f^q_{he} \end{bmatrix}, & \begin{bmatrix} f^b_e \\ e^b_e \end{bmatrix} = \begin{bmatrix} (\boldsymbol{E} + e^p_{he})|_{\partial Z} \\ (\boldsymbol{H} + e^q_{he})|_{\partial Z} \end{bmatrix}, \end{cases} \tag{73}$$

where we defined the following harmonic forms yielding harmonic energy flows:

$$\begin{cases} (f^p_{hf}, e^p_{hf}) \in \Omega^3(Z) \times \Omega^0(Z), & (f^q_{hf}, e^q_{hf}) \in \Omega^2(Z) \times \Omega^1(Z), \\ (f^p_{he}, e^p_{he}) \in \Omega^2(Z) \times \Omega^1(Z), & (f^q_{he}, e^q_{he}) \in \Omega^2(Z) \times \Omega^1(Z). \end{cases} \tag{74}$$

Note that $H_k(Z) \cong H_{n-k}(Z, \partial Z) \cong \Omega^k_h(Z)$, there is the dual from of ω_h with respect to $\langle\ ,\ \rangle_Z$, called a *Poincaré dual*: $\Omega^k_h(Z) \cong \Omega^{n-k}_h(Z)$, and f^p_{hf} and e^p_{hf} are constant functions. The system

(73) satisfies the power balances

$$\int_Z \left(-(e_\rho + e^p_{hf}) \wedge \rho_t - (e_v + e^q_{hf}) \wedge v_t\right) - \int_Z (e_\rho + e^p_{hf}) \wedge f^p_{hf}$$
$$- \int_Z (e_v + e^q_{hf}) \wedge (g_2 + f^q_{hf}) - \int_{\partial Z} (e_v + e^q_{hf}) \wedge (e_\rho + e^p_{hf}) = 0, \tag{75}$$
$$\int_Z \left\{-(E + e^p_{he}) \wedge D_t - (H + e^q_{he}) \wedge B_t\right\} - \int_Z (E + e^p_{he}) \wedge (J + f^p_{he})$$
$$- \int_Z (H + e^q_{he}) \wedge f^q_{he} + \int_{\partial Z} (H + e^q_{he}) \wedge (E + e^p_{he}) = 0. \tag{76}$$

4.4 Boundary detection and control of topological transitions

In fact, it is difficult to determine specific harmonic forms in (74). Hence, let us apply the classification of vector fields to the power balance equation for detecting topological transitions of systems and controlling energy flows.

Consider the cross-sectional surface W of Z such that $W \subset Z$ and $\partial W \subset \partial Z$. Let $\partial Z = \cup_i \partial Z^i$ be a set of subdivided domains of ∂Z or W in which each ∂Z^i is homeomorphic to Euclidian spaces (e.g., each component of ∂Z_x and ∂Z_y in (60)). In this setting, we can approximate port variables distributed on ∂Z^i, for instance, by using those on the boundary of each subdivided domain $\partial(\partial Z^i)$ if the subdivision is sufficiently fine. Let

$$(v^\sharp_1, v^\sharp_2, v^\sharp_3, v^\sharp_4, v^\sharp_5) \in \mathfrak{X}_{FK}(Z) \oplus \mathfrak{X}_{HK}(Z) \oplus \mathfrak{X}_{CG}(Z) \oplus \mathfrak{X}_{HG}(Z) \oplus \mathfrak{X}_{GG}(Z). \tag{77}$$

Then, we can rewrite (61) as follows:

$$\mathcal{H}_{\delta t} = \sum_i \left\{\mathcal{H}_{\delta t}(v^\sharp_1) + \mathcal{H}_{\delta t}(v^\sharp_2) + \mathcal{H}_{\delta t}(v^\sharp_3) + \mathcal{H}_{\delta t}(v^\sharp_4) + \mathcal{H}_{\delta t}(v^\sharp_5)\right\}\Big|_{\partial Z^i} = 0, \tag{78}$$

where $\mathcal{H}_{\delta t}(v^\sharp_r)$ means the split energy flow generated by $v^\sharp_r$ for $1 \leq r \leq 5$. If all boundary port variables are available as inputs and outputs, the balance of each decomposed energy flows can be confirmed from (78).

On the other hand, desired energy flows depending on the topology of system domain can be reinforced by servo feedback in terms of boundary port variables. If the cause of a change is a known structural perturbation and the boundary surrounds all energy flows generated by the perturbation, we can use the power balance defined on such appropriate boundaries to realize an energy flow control. Indeed, the control law is

$$\int_{\partial Z^j} \sum_{r=1}^{5} (e^b_j(v^\sharp_r) - u^q_{jr}) \wedge (f^b_j(v^\sharp_r) - u^p_{jr}), \tag{79}$$
$$u^i_{jr} = g^{ij}(e^b_j(v^\sharp_r) - \bar{e}^b_j(v^\sharp_r)) \wedge (f^b_j(v^\sharp_r) - \bar{f}^b_j(v^\sharp_r))|_{\partial Z^j}, \tag{80}$$

where e^{bi} is the boundary control input or output, f^{bi} is the boundary output or input, $\bar{e}^{bi}$ and $\bar{f}^{bi}$ are the desired energy flows, and g^{ij} is the feedback gain.

5. Conclusion

This chapter derived the boundary controls based on passivity for ideal magnetohydrodynamics (MHD) systems in terms of distributed port-Hamiltonian (DPH) representations. In *Section 2*, We first rewrote the geometric formulation of MHD as a DPH system. Next, we explained the passivity-based controls for the DPH system of MHD by using collocated input/output pairs, i.e., port variables for stabilizing and assigning a global stable point. The boundary power balance equation of the DPH system could be considered as an extended energy principle of MHD in the sense of dynamical systems and boundary controls. In *Section 3*, we considered the DPH model of MHD with model perturbations. The perturbation can be uniquely decomposed into a Hamiltonian subsystem, called an exact subsystem, and a non-Hamiltonian subsystem, called a dual-exact subsystem. We presented the method of creating a pseudo potential for an exact subsystem of the DPH model. In *Section 4*, we explained a symmetry breaking of conservation laws associated with the DPH system. The breaking can be detected by checking quantities with the boundary port variables of the DPH system. Finally, we showed that the boundary port variables can detect the topological change of the domain of DPH systems and can create desired topological energy flows.

These results open the way to active disturbance rejections or plasma shape controls. If an actual MHD system is not ideal or includes modeling errors, the power balance equations should be revised. In this case, the pseudo potential construction might be used for improving the model. The boundary control using the boundary port variables might be approximated by the discretization of port-Hamiltonian systems (Golo et al., 2004).

6. Acknowledgement

The authors would like to thank Professor Bernhard Maschke for fruitful discussions with us.

7. References

A.J. van der Schaft and B.M. Maschke, (2002). "Hamiltonian formulation of distributed-parameter systems with boundary energy flow", *Journal of Geometry and Physics*, Vol. 42, pp. 166–194.

S. Arimoto, (1996). *Control Theory of Non-linear Mechanical Systems: A Passivity-based and Circuit-Theoretic Approach*, Oxford Univ. Press.

R. Ortega, J.A.L. Perez, P.J. Nichlasson and H.J. Sira-Ramirez, (1998). *Passivity-based Control of Euler-Lagrange Systems: Mechanical, Electrical and Electromechanical Applications*, Springer.

A.J. van der Schaft, (2000). *L_2-Gain and Passivity Techniques in Nonlinear Control*, 2nd revised and enlarged edition, Springer Communications and Control Engineering series, Springer-Verlag, London.

V. Duindam, A. Macchelli, S. Stramigioli and H. Bruyninckx (Eds.), (2009). *Modeling and Control of Complex Physical Systems - The Port-Hamiltonian Approach*, Springer.

J. Wesson, (2004). *Tokamaks*, 3rd ed., Oxford Univ. Press.

A. Pironti and M. Walker, (2005). *Control System Magazine*, Vol. 25, No. 5.

M. Ariola and A. Pironti, (2008). *Magnetic Control of Tokamak Plasmas*, Springer.

K. Zhou, J.C. Doyle and K. Glover, (1996). *Robust and Optimal Control*, Prentice Hall.

H. K. Khalil, (2001). *Nonlinear Systems*, 3rd Ed., Prentice Hall.

A. Isidori, (1995). *Nonlinear Control systems*, third ed., Springer.

B. Brogliato, R. Lozano, B. Maschke and O. Egeland, (2006). *Dissipative Systems Analysis and Control: Theory and Applications*, 2nd ed., Springer.

T. Courant, (1990). "Dirac manifolds", *Trans. American Math. Soc.*, 319, pp.631-661.

V.I. Arnold, (1989). *Mathematical Methods of Classical Mechanics*, Second Edition, Springer.

Y. Le Gorrec, H. Zwart and B. Maschke, (2005). "Dirac structures and Boundary Control Systems associated with Skew-Symmetric Differential Operators", *SIAM Journal on Control and Optimization*, Vol. 44, No. 5, pp. 1864–1892.

G. Nishida and M. Yamakita, (2004). "A Higher Order Stokes-Dirac Structure for Distributed-Parameter Port-Hamiltonian Systems", in *Proc. of 2004 American Control Conference*, Boston, pp. 5004–5009.

D. Eberard, B.M. Maschke and A.J. van der Schaft, (2007). "An extension of Hamiltonian systems to the thermodynamic phase space: Towards a geometry of nonreversible processes", *Reports on Mathematical Physics*, Vol. 60, No. 2, pp. 175–198.

G. Golo, V. Talasila, A. van der Schaft, B. Maschke, (2004). "Hamiltonian discretization of boundary control systems", *Automatica*, Vol. 40, pp.757–771.

T. Voss and J.M.A. Scherpen (2011)., "Structure Preserving Spatial Discretization of a 1-D Piezoelectric Timoshenko Beam", *SIAM Multiscale Model. Simul.*, Vol. 9, No. 1, pp. 129–154.

G. Nishida, M. Yamakita and Z. Luo, (2007a). "Virtual Lagrangian Construction Method for Infinite Dimensional Systems with Homotopy Operators", in F. Bullo et al. (Eds.): *Lagrangian and Hamiltonian methods for nonlinear control 2006*, LNCIS 366, pp. 75–86, Springer.

G. Nishida, M. Sugiura, M. Yamakita, B. Maschke and R. Ikeura, (2009). "Boundary Detection of Variational Symmetry Breaking using Port-Representation of Conservation Laws", in *Proc. of the 48th IEEE Conf. on Decision and Control*, pp. 2861–2868, Shanghai.

H. Flanders, (1963), *Differential Forms with Applications to the Physical Sciences*, Academic Press, New York.

S. Morita, (2001). *Geometry of Differential Forms*, American Mathematical Society.

D. Karnopp, D. Margolis and R. Rosenberg, (2006). *System Dynamics, Modeling and Simulation of Mechatronic Systems*, 4th Edition. Wiley.

P.J. Olver, (1993). *Applications of Lie Groups to Differential Equations: Second Edition*, Springer-Verlag, New York.

R. Bott and L.W.Tu, (1982). *Differential Forms in Algebraic Topology*, Springer-Verlag.

P.W. Gross and P.R. Kotiuga, (2004). *Electromagnetic Theory and Computation: A Topological Approach*, Cambridge Univ. Press.

J. Cantarella, D. DeTurck and H. Gluck, (2002). "Vector Calculus and the Topology of Domains in 3-Space", American Math. Monthly, 109, 5, pp. 409–442.

8

Magnetohydrodynamic Rotating Flow of a Fourth Grade Fluid Between Two Parallel Infinite Plates

M.A. Rana[1*], Akhlaq Ahmed[2] and Rashid Qamar[3]
[1]*Department of Basic Sciences, Riphah International University, Sector I-14, Islamabad,*
[2]*Department of Mathematics, Quaid-i-Azam University, Islamabad,*
[3]*Directorate of Management Information System, PAEC HQ, Islamabad,*
Pakistan

1. Introduction

Mechanics of non-linear fluids present a special challenge to physicists, mathematician and engineers. The non-linearity can manifest itself in a variety of ways. Materials such as clay coatings and other suspensions, polymer melts, drilling muds, certain oils and greases, elastomers and many emulsions have been treated as non-Newtonian fluids. There is no single model which clearly exhibits all properties of non-Newtonian fluids and there has been much confusion over the classification of non-Newtonian fluids. However, non-Newtonian fluid may be classified as (1) fluids for which the shear stress depends only on the rate of shear; (2) fluids for which the relation between shear stress and shear rate depends on time; (3) the visco-elastoic fluids, which possess both elastic and viscous properties.

It is not possible to recommend a single constitutive equation which exhibits all properties of non-Newtonian fluids due to the great diversity in the physical structure of non-Newtonian fluids. For this reason, several non-Newtonian models or constitutive equations have been proposed and most of them are empirical or semi empirical. One of the simplest ways in which the visco-elastic fluids have been classified is the methodology given in [1,2]. They present constitutive relations for the stress tensor as a function of the symmetric part of the velocity gradient and its higher derivatives. Another class of models is the rate-type fluid models such as the Oldroyd model [3]. Although many constitutive equations have been suggested, many questions are still unsolved. Some of the continuum models do not give satisfactory results in accordance with the available experimental data. For this reason, in many practical applications, empirical or semi empirical equations have been used. A complete and thorough discussion of various models can be found in [4-7]. Various authors [8-12] investigated non-Newtonian fluids.

---

*Corresponding author

The study of an electrically conducting fluid flows in channels under the action of a transversely applied magnetic field has important applications in many devices such as magnetohydrodynamic (MHD) pumps, aerodynamics heating, MHD power generators, accelerators, centrifugal separation of matter from fluid, flow meters, electrostatic precipitation, fluid droplets sprays, purification of crude oil, petroleum industries and polymer technology. Hartmann [13] first studied an incompressible viscous electrically conducting fluid under the action of a transverse magnetic field. Under different physical conditions it was considered by Sutton and Shermann [14], Hughes and Young [15], Cowling [16] and Pai [17]. Rajagopal and Na [18] studied the flow of a third grade fluid due to an oscillation of plate, Mollica and Rajagopal [19] examined secondary flows due to axial shearing of a third grade fluid between two eccentrically placed cylinders, Siddiqui and Kaloni [20] investigated plane flow of a third grade fluid. Rotating disk flows of conducting fluids have practical applications in many areas such as computer storage devices, lubrication, crystal growth processes, viscometry and rotating machinery. The effect of an external uniform magnetic field on the flow due to a rotating disk was studied [21-25], and eccentric rotation of disks was studied [26-29]. In many process of industries, the cooling of threads or sheets of some polymer materials is of great importance in the production line. Magneto convection plays an important role in various industrial applications including magnetic control of molten iron flow in the steel industry and liquid metal cooling in nuclear reactors. Palani and Abbas [30] investigated the combined effects of magnetohydrodynamic and radiation on free convection flow past an impulsively started isothermal vertical plate with Rosseland diffusion approximation, Farzaneh-Gord et al. [31] studied two-dimensional steady-state incompressible viscoelastic boundary layer magnetohydrodynamics flow and heat transfer over a stretching sheet in the presence of electric and magnetic fields. The highly non-linear momentum and heat transfer equations are solved analytically.

The MHD fluid flow as lubricant is of interest in industrial applications, because it prevents the unexpected variation of lubricant viscosity with temperature under certain extreme operating conditions. The MHD lubrication in an externally pressurized thrust bearing has been investigated both theoretically and experimentally by Maki et al. [32]. Hughes and Elco [33] and Kuzma et al. [34] have investigated the effects of a magnetic field in lubrication. These authors had neglected the inertial terms in the Navier–Stokes equations. Hamza [35] considered the squeezing flow between two discs in the presence of a magnetic field. The problem of squeezing flow between rotating discs has been studied by Hamza [36] and Bhattacharyya and Pal [37]. Considering two-dimensional unsteady MHD flow of a viscous fluid between two moving parallel plates, Sweet et al. [38] have shown that the flow is strongly influenced by the strength of the magnetic field and the density of the fluid. Abbas et al. [39] have investigated the unsteady MHD boundary layer flow and heat transfer in an incompressible rotating viscous fluid over a stretching continuous sheet. The resulting system of partial differential equations is solved numerically using Keller-box method. Turkylimazoglu [40] has analyzed the MHD time-dependent von Karman swirling electrically conducting viscous fluid flow having a temperature-dependent viscosity due to a rotating disk impulsively set into motion.

Hayat et al. [41] have considered the unsteady rotating MHD flow of an incompressible second grade fluid in a porous half space. The flow is induced by a suddenly moved plate in its own plane. Both the fluid and plate rotate in unison with the same angular velocity. Assuming the velocity field of the form $\mathbf{V}=[u(z,t),v(z,t),w(z,t)]$, analytical solutions are presented using Fourier sine transforms and it is shown that with an increase in MHD parameter the real and imaginary parts of velocity as well as the boundary layer thickness decreases.

The classical theories of continuum mechanics are inadequate to explicate the microscopic manifestations of microscopic events. The fluids with microstructure belonging to a class of fluid with non-symmetrical stress tensor referred to as polar fluids are called Micropolar fluids. Physically they represent fluids consisting of randomly oriented particles suspended in a viscous medium. Eringen [42] presented the earliest formulation of a general theory of fluid microcontinua taking into account the inertial characteristics of the substructure particles which are allowed to undergo rotation in 1964. This theory has been extended by Eringen [43] to take into account thermal effects. The theory of micropolar fluids and its extension thermomicropolar fluids [44] may form suitable non-Newtonian fluid models which can be used to explain the flow of colloidal fluids, polymeric suspensions, liquid crystals, animal blood, etc. Eldabe et al. [45] have discussed the problem of heat transfer to MHD flow of a micropolar fluid from a stretching sheet with suction and blowing through a porous medium. The numerical results indicate that the velocity and the angular velocity increase as the permeability parameter increases but they decrease as the magnetic field increases. On the other hand, the temperature decreases as the permeability parameter increases but it increases as the magnetic field increases.

The study of laminar boundary layer flow of non-Newtonian fluids over continuous moving surfaces is very important because of its practical importance in a number of engineering processes. For example, cooling of an infinite metallic plate in a cooling bath, the boundary layer along a liquid film in condensation processes, aerodynamic extrusion of plastic sheets and a polymer sheet or filament extruded continuously from a dye. Furthermore, it has several practical applications in the field of metallurgy and chemical engineering such as material manufactured by extrusion process and heat-treated materials traveling between a feed roll and a wind-up roll or on conveyor belt possess, the feature of a moving continuous surface. Also, glass blowing, continuous casting, and spinning of fibers involve the flow due to a stretching surface. Sarpakaya [46] studied the MHD flow of a non-Newtonian fluid, Char [47] studied the MHD flow of a viscoelastic fluid over a stretching sheet by considering the thermal diffusion in the energy equation. However, the effects of thermal radiation on the viscoelastic boundary layer flow and heat transfer can be quite significant at high operating temperatures. In view of this, Raptis [48], Raptis and Perdikis [49] and Raptis et al. [50] studied the viscoelastic flow and heat transfer over a flat plate with constant suction, thermal radiation and viscous dissipation. Recently, the effects of viscous dissipation, radiation, in presence of temperature dependent heat sources/sinks on heat transfer characteristics of a viscoelastic fluid is considered by Siddheshwar and Mahabaleswar [51]. Khan [52] extended the problem by including the effects of suction/injection, heat source/sink and radiation effects. Abel et al. [53] investigated the effects of viscous dissipation and non-uniform heat source on viscoelastic boundary layer flow over a linear stretching sheet. Abel and Nandeppanavar [54] studied the effect of non-uniform heat source/sink on MHD viscoelastic boundary layer flow, further Nandeppanavar et al. [55] studied the effects of elastic deformation and non-uniform heat source on viscoelastic boundary layer flow. Motivated by these studies, Mahantesh et al. [56] extended the results of researchers [53,54,55] for MHD viscoelastic boundary layer flow with combined effects of viscous dissipation, thermal radiation and non-uniform heat source which was ignored by [53,54,55]. Furthermore, they analyzed the effects of radiation, viscous dissipation, viscoelasticity, magnetic field on the heat transfer characteristics in the presence of non-uniform heat source with variable PST and PHF temperature boundary conditions. Kayvan Sadeghy et al. [57] have investigated theoretically the applicability of magnetic fields for controlling hydrodynamic separation in Jeffrey-Hamel flows of viscoelastic fluids. It is shown that for viscoelastic fluids, it is possible to delay flow separation in a diverging channel provided that the magnetic field is sufficiently strong. It is also shown that the effect of magnetic field on flow separation becomes more pronounced the higher the fluid's elasticity.

In the present paper we have modeled the unsteady flow equations of a fourth grade fluid bounded between two non-conducting rigid plates in a rotating frame of reference with imposed uniform transverse magnetic field. It is interesting to note that we are able to couple the equations arising for the velocity field. The steady rotating flow of the non-Newtonian fluid subject to a uniform transverse magnetic field is studied. The non-linear differential equations resulting from the balance of momentum and mass are solved numerically. The effects of exerted magnetic field, Ekman number and material parameter on the velocity distribution are presented graphically. The results for Newtonian and non-Newtonian fluids are compared.

2. Mathematical model of the problem

We introduce a Cartesian coordinate system with z-axis normal to the plane of the parallel plates. The plates are located at $z=0$ and $z=L$ and the plates and the fluid bounded between them are in a rigid body rotation with constant angular velocity Ω about the z-axis. The fluid is electrically conducting and assumed to be permeated by an imposed magnetic field B_0 perpendicular to the parallel plates. The disturbance in the fluid is produced by small amplitude non-torsional oscillations of the lower plate. For the present model we take the velocity field of the form.

$$\mathbf{V}=\left[u(z,t),\ v(z,t),\ 0\right], \tag{1}$$

where u and v are the x and y components of the velocity field. The Cauchy stress tensor for the fourth grade fluid can be obtained by the model introduced by Coleman and Noll [58]

$$\mathbf{T}=-p\mathbf{I}+\sum_{j=1}^{n}\mathbf{S}_j. \tag{2}$$

For the fourth grade fluid we have $n=4$ and the first four tensors S_j are given by

$$\mathbf{S}_1 = \mu \mathbf{A}_1, \tag{3}$$

$$\mathbf{S}_2 = \alpha_1 \mathbf{A}_2 + \alpha_2 \mathbf{A}_1^2, \tag{4}$$

$$\mathbf{S}_3 = \beta_1 \mathbf{A}_3 + \beta_2 (\mathbf{A}_2 \mathbf{A}_1 + \mathbf{A}_1 \mathbf{A}_2) + \beta_3 (tr\mathbf{A}_1^2)\mathbf{A}_1, \tag{5}$$

$$\begin{aligned}\mathbf{S}_4 &= \gamma_1 \mathbf{A}_4 + \gamma_2 (\mathbf{A}_3 \mathbf{A}_1 + \mathbf{A}_1 \mathbf{A}_3) + \gamma_3 \mathbf{A}_2^2 \\ &+ \gamma_4 (\mathbf{A}_2 \mathbf{A}_1^2 + \mathbf{A}_1^2 \mathbf{A}_2) + \gamma_5 (tr\mathbf{A}_2)\mathbf{A}_2 + \gamma_6 (tr\mathbf{A}_2)\mathbf{A}_1^2 \\ &+ \{\gamma_7 (tr\mathbf{A}_3) + \gamma_8 (tr\mathbf{A}_2 \mathbf{A}_1)\}\mathbf{A}_1,\end{aligned} \tag{6}$$

where μ is the co-efficient of shear viscosity; and

$$\alpha_i \ (i = 1,2),\ \beta_j \ (j = 1,2,3),\ \gamma_k \ (k = 1,2,...,8)$$

are material constants. The Rivlin- Ericken tensors $\mathbf{A}_n$ are defined by the recursion relation

$$\mathbf{A}_n = \frac{d\mathbf{A}_{n-1}}{dt} + \mathbf{A}_{n-1}(grad\mathbf{V}) + (grad\mathbf{V})^T \mathbf{A}_{n-1},\ n > 1, \tag{7}$$

$$\mathbf{A}_1 = (grad\mathbf{V}) + (grad\mathbf{V})^T, \tag{8}$$

where

$$\frac{d}{dt}(.) = \left(\frac{\partial}{\partial t} + \mathbf{V}.\nabla\right)(.). \tag{9}$$

When $\gamma_k = 0\ (k = 1,2,...,8)$, the fourth grade model reduces to third grade model, when $\beta_j = 0\ (j = 1,2,3)$ and $\gamma_k = 0\ (k = 1,2,...,8)$ then above model reduces to second grade model and if $\alpha_i = 0\ (i = 1,2),\ \beta_j = 0\ (j = 1,2,3),\ \gamma_k = 0\ (k = 1,2,...,8)$ the flow model reduces to classical Navier-Stokes viscous fluid model.

The hydromagnetic flow is generated in the uniformly rotating fluid by small amplitude non-torsional oscillations of the plate located at $z = 0$. With the Cartesian coordinate system O_{xyz} the unsteady motion of the incompressible fourth grade conducting fluid in the presence of magnetic field **B** is governed by the law of balance of linear momentum and balance of mass i.e.

$$\frac{d\mathbf{V}}{dt} + 2(\mathbf{\Omega} \times \mathbf{V}) + \mathbf{\Omega} \times (\mathbf{\Omega} \times \mathbf{r}) = \frac{1}{\rho}\text{div}\mathbf{T} + \frac{1}{\rho}(\mathbf{J} \times \mathbf{B}), \tag{10}$$

$$\text{div}\,\mathbf{V} = 0, \tag{11}$$

where ρ is the density, $\mathbf{J}$ is the current density and $\mathbf{B}(=\mathbf{B}_0+\mathbf{b}$, $\mathbf{b}$ being the induced magnetic field) is the total magnetic field.

In the absence of displacement currents, the Maxwell equations and the generalized Ohm's law can be written as

$$\nabla\cdot\mathbf{B}=0,\ \nabla\times\mathbf{B}=\mu_m\mathbf{J},\ \nabla\times\mathbf{E}=-\frac{\partial\mathbf{B}}{\partial t}, \tag{12}$$

$$\mathbf{J}=\sigma(\mathbf{E}+\nabla\times\mathbf{B}), \tag{13}$$

where μ_m is the magnetic permeability, $\mathbf{E}$ is the electric field and σ is the electrical conductivity of the fluid.

The magnetic Reynolds number is assumed to be very small so that the induced magnetic field is negligible [14]. This assumption is reasonable for the flow of liquid metals, e.g. mercury or liquid sodium (which are electrically conducting under laboratory conditions). The electron–atom collision frequency is assumed to be relatively high so that the Hall effect can be included [14]. The Lorentz force per unit volume is given by

$$\mathbf{J}\times\mathbf{B}=-\sigma B_0^2\mathbf{V}. \tag{14}$$

For the velocity field defined in Eq. (1), the equation of continuity (11) is identically satisfied and Eq. (10) in component form can be written as

$$\begin{aligned}
\frac{\partial u}{\partial t}-2\Omega v-\Omega^2 x={}&-\frac{1}{\rho}\frac{\partial p}{\partial x}+\frac{\mu}{\rho}\frac{\partial^2 u}{\partial z^2}+\frac{\alpha_1}{\rho}\frac{\partial^3 u}{\partial z^2\partial t}+\frac{\beta_1}{\rho}\frac{\partial^4 u}{\partial z^2\partial t^2}\\
&+\frac{2(\beta_2+\beta_3)}{\rho}\frac{\partial}{\partial z}\left[2\frac{\partial u}{\partial z}\left\{\left(\frac{\partial u}{\partial z}\right)^2+\left(\frac{\partial v}{\partial z}\right)^2\right\}\right]\\
&+\frac{\gamma_2}{\rho}\frac{\partial}{\partial z}\left[\frac{\partial u}{\partial z}\left\{\begin{array}{l}2\frac{\partial}{\partial t}\left(\left(\frac{\partial u}{\partial z}\right)^2+\left(\frac{\partial v}{\partial z}\right)^2\right)\\+2\frac{\partial u}{\partial z}\frac{\partial^2 u}{\partial z\partial t}+2\frac{\partial v}{\partial z}\frac{\partial^2 v}{\partial z\partial t}\end{array}\right\}\right]\\
&+\frac{(\gamma_3+\gamma_5)}{\rho}\frac{\partial}{\partial z}\left[2\frac{\partial^2 u}{\partial z\partial t}\left\{\left(\frac{\partial u}{\partial z}\right)^2+\left(\frac{\partial v}{\partial z}\right)^2\right\}\right]\\
&+\frac{\partial}{\partial z}\left[\frac{\partial u}{\partial z}\left\{\begin{array}{l}2\frac{\gamma_7}{\rho}\frac{\partial}{\partial t}\left(\left(\frac{\partial u}{\partial z}\right)^2+\left(\frac{\partial v}{\partial z}\right)^2\right)\\+\frac{(\gamma_7+\gamma_8)}{\rho}\left(\frac{\partial u}{\partial z}\frac{\partial^2 u}{\partial z\partial t}+\frac{\partial v}{\partial z}\frac{\partial^2 v}{\partial z\partial t}\right)\end{array}\right\}\right]+\frac{\gamma_1}{\rho}\frac{\partial^5 u}{\partial z^2\partial t^3}-\frac{\sigma}{\rho}B_0^2 u,
\end{aligned} \tag{15}$$

$$\frac{\partial v}{\partial t}+2\Omega u-\Omega^2 y=-\frac{1}{\rho}\frac{\partial p}{\partial y}+\frac{\mu}{\rho}\frac{\partial^2 v}{\partial z^2}+\frac{\alpha_1}{\rho}\frac{\partial^3 v}{\partial z^2\partial t}+\frac{\beta_1}{\rho}\frac{\partial^4 v}{\partial z^2\partial t^2}$$
$$+\frac{2(\beta_2+\beta_3)}{\rho}\frac{\partial}{\partial z}\left[2\frac{\partial v}{\partial z}\left\{\left(\frac{\partial u}{\partial z}\right)^2+\left(\frac{\partial v}{\partial z}\right)^2\right\}\right]$$
$$+\frac{\gamma_2}{\rho}\frac{\partial}{\partial z}\left[\frac{\partial v}{\partial z}\left\{\begin{array}{l}2\frac{\partial}{\partial t}\left(\left(\frac{\partial u}{\partial z}\right)^2+\left(\frac{\partial v}{\partial z}\right)^2\right)\\+2\frac{\partial u}{\partial z}\frac{\partial^2 u}{\partial z\partial t}+2\frac{\partial v}{\partial z}\frac{\partial^2 v}{\partial z\partial t}\end{array}\right\}\right] \tag{16}$$
$$+\frac{(\gamma_3+\gamma_5)}{\rho}\frac{\partial}{\partial z}\left[2\frac{\partial^2 v}{\partial z\partial t}\left\{\left(\frac{\partial u}{\partial z}\right)^2+\left(\frac{\partial v}{\partial z}\right)^2\right\}\right]$$
$$+\frac{\partial}{\partial z}\left[\frac{\partial v}{\partial z}\left\{\begin{array}{l}2\frac{\gamma_7}{\rho}\frac{\partial}{\partial t}\left(\left(\frac{\partial u}{\partial z}\right)^2+\left(\frac{\partial v}{\partial z}\right)^2\right)\\+\frac{(\gamma_7+\gamma_8)}{\rho}\left(\frac{\partial u}{\partial z}\frac{\partial^2 u}{\partial z\partial t}+\frac{\partial v}{\partial z}\frac{\partial^2 v}{\partial z\partial t}\right)\end{array}\right\}\right]+\frac{\gamma_1}{\rho}\frac{\partial^5 v}{\partial z^2\partial t^3}-\frac{\sigma}{\rho}B_0^2 v,$$

$$0=-\frac{1}{\rho}\frac{\partial p}{\partial z}+\frac{(2\alpha_1+\alpha_2)}{\rho}\frac{\partial}{\partial z}\left\{\left(\frac{\partial u}{\partial z}\right)^2+\left(\frac{\partial v}{\partial z}\right)^2\right\}+\frac{\beta_1}{\rho}\frac{\partial}{\partial z}\left\{\begin{array}{l}2\frac{\partial}{\partial t}\left(\left(\frac{\partial u}{\partial z}\right)^2+\left(\frac{\partial v}{\partial z}\right)^2\right)\\+2\frac{\partial u}{\partial z}\frac{\partial^2 u}{\partial z\partial t}+2\frac{\partial v}{\partial z}\frac{\partial^2 v}{\partial z\partial t}\end{array}\right\}$$
$$+\frac{\beta_2}{\rho}\frac{\partial}{\partial z}\left\{2\frac{\partial u}{\partial z}\frac{\partial^2 u}{\partial z\partial t}+2\frac{\partial v}{\partial z}\frac{\partial^2 v}{\partial z\partial t}\right\}+\frac{\gamma_1}{\rho}\frac{\partial}{\partial z}\left\{\begin{array}{l}2\frac{\partial^2}{\partial t^2}\left(\left(\frac{\partial u}{\partial z}\right)^2+\left(\frac{\partial v}{\partial z}\right)^2\right)\\+2\frac{\partial}{\partial t}\left(\frac{\partial u}{\partial z}\frac{\partial^2 u}{\partial z\partial t}+\frac{\partial v}{\partial z}\frac{\partial^2 v}{\partial z\partial t}\right)\\+2\left(\frac{\partial u}{\partial z}\frac{\partial^3 u}{\partial z\partial t^2}+\frac{\partial v}{\partial z}\frac{\partial^3 v}{\partial z\partial t^2}\right)\end{array}\right\} \tag{17}$$
$$+\frac{\gamma_3}{\rho}\frac{\partial}{\partial z}\left\{\begin{array}{l}\left(\frac{\partial^2 u}{\partial z\partial t}\right)^2+\left(\frac{\partial^2 v}{\partial z\partial t}\right)^2\\+4\left(\left(\frac{\partial u}{\partial z}\right)^2+\left(\frac{\partial v}{\partial z}\right)^2\right)^2\end{array}\right\}+\frac{(4\gamma_4+4\gamma_5+2\gamma_6}{\rho}\frac{\partial}{\partial z}\left(\left(\frac{\partial u}{\partial z}\right)^2+\left(\frac{\partial v}{\partial z}\right)^2\right)^2.$$

Defining the modified pressure

$$\hat{p} = \frac{p}{\rho} - \frac{(2\alpha_1+\alpha_2)}{\rho}\left\{\left(\frac{\partial u}{\partial z}\right)^2 + \left(\frac{\partial v}{\partial z}\right)^2\right\} - \frac{\beta_1}{\rho}\left\{\begin{array}{l} 2\frac{\partial}{\partial t}\left(\left(\frac{\partial u}{\partial z}\right)^2 + \left(\frac{\partial v}{\partial z}\right)^2\right) \\ +2\frac{\partial u}{\partial z}\frac{\partial^2 u}{\partial z\partial t} + 2\frac{\partial v}{\partial z}\frac{\partial^2 v}{\partial z\partial t} \end{array}\right\}$$

$$- \frac{\beta_2}{\rho}\left\{2\frac{\partial u}{\partial z}\frac{\partial^2 u}{\partial z\partial t} + 2\frac{\partial v}{\partial z}\frac{\partial^2 v}{\partial z\partial t}\right\} - \frac{\gamma_1}{\rho}\left\{\begin{array}{l} 2\frac{\partial^2}{\partial t^2}\left(\left(\frac{\partial u}{\partial z}\right)^2 + \left(\frac{\partial v}{\partial z}\right)^2\right) \\ +2\frac{\partial}{\partial t}\left(\frac{\partial u}{\partial z}\frac{\partial^2 u}{\partial z\partial t} + \frac{\partial v}{\partial z}\frac{\partial^2 v}{\partial z\partial t}\right) \\ +2\left(\frac{\partial u}{\partial z}\frac{\partial^3 u}{\partial z\partial t^2} + \frac{\partial v}{\partial z}\frac{\partial^3 v}{\partial z\partial t^2}\right) \end{array}\right\} \tag{18}$$

$$- \frac{\gamma_3}{\rho}\left\{\begin{array}{l} \left(\frac{\partial^2 u}{\partial z\partial t}\right)^2 + \left(\frac{\partial^2 v}{\partial z\partial t}\right)^2 \\ +4\left(\left(\frac{\partial u}{\partial z}\right)^2 + \left(\frac{\partial v}{\partial z}\right)^2\right)^2 \end{array}\right\} - \frac{(4\gamma_4+4\gamma_5+2\gamma_6)}{\rho}\left(\left(\frac{\partial u}{\partial z}\right)^2 + \left(\frac{\partial v}{\partial z}\right)^2\right)^2,$$

then Eqs. (15)-(17) become

$$\begin{aligned}
\frac{\partial u}{\partial t} - 2\Omega v - \Omega^2 x = & -\frac{\partial \hat{p}}{\partial x} + \frac{\mu}{\rho}\frac{\partial^2 u}{\partial z^2} + \frac{\alpha_1}{\rho}\frac{\partial^3 u}{\partial z^2\partial t} + \frac{\beta_1}{\rho}\frac{\partial^4 u}{\partial z^2\partial t^2} \\
& + \frac{2(\beta_2+\beta_3)}{\rho}\frac{\partial}{\partial z}\left[2\frac{\partial u}{\partial z}\left\{\left(\frac{\partial u}{\partial z}\right)^2 + \left(\frac{\partial v}{\partial z}\right)^2\right\}\right] \\
& + \frac{\gamma_2}{\rho}\frac{\partial}{\partial z}\left[\frac{\partial u}{\partial z}\left\{2\frac{\partial}{\partial t}\left(\left(\frac{\partial u}{\partial z}\right)^2 + \left(\frac{\partial v}{\partial z}\right)^2\right) + 2\frac{\partial u}{\partial z}\frac{\partial^2 u}{\partial z\partial t} + 2\frac{\partial v}{\partial z}\frac{\partial^2 v}{\partial z\partial t}\right\}\right] \\
& + \frac{(\gamma_3+\gamma_5)}{\rho}\frac{\partial}{\partial z}\left[2\frac{\partial^2 u}{\partial z\partial t}\left\{\left(\frac{\partial u}{\partial z}\right)^2 + \left(\frac{\partial v}{\partial z}\right)^2\right\}\right] \\
& + \frac{\partial}{\partial z}\left[\frac{\partial u}{\partial z}\left\{2\frac{\gamma_7}{\rho}\frac{\partial}{\partial t}\left(\left(\frac{\partial u}{\partial z}\right)^2 + \left(\frac{\partial v}{\partial z}\right)^2\right) + \frac{(\gamma_7+\gamma_8)}{\rho}\left(\frac{\partial u}{\partial z}\frac{\partial^2 u}{\partial z\partial t} + \frac{\partial v}{\partial z}\frac{\partial^2 v}{\partial z\partial t}\right)\right\}\right] \\
& + \frac{\gamma_1}{\rho}\frac{\partial^5 u}{\partial z^2\partial t^3} - \frac{\sigma}{\rho}B_0^2 u,
\end{aligned} \tag{19}$$

$$\frac{\partial v}{\partial t}+2\Omega u-\Omega^2 y=-\frac{\partial \hat{p}}{\partial y}+\frac{\mu}{\rho}\frac{\partial^2 v}{\partial z^2}+\frac{\alpha_1}{\rho}\frac{\partial^3 v}{\partial z^2\partial t}+\frac{\beta_1}{\rho}\frac{\partial^4 v}{\partial z^2\partial t^2}$$
$$+\frac{2(\beta_2+\beta_3)}{\rho}\frac{\partial}{\partial z}\left[2\frac{\partial v}{\partial z}\left\{\left(\frac{\partial u}{\partial z}\right)^2+\left(\frac{\partial v}{\partial z}\right)^2\right\}\right]$$
$$+\frac{\gamma_2}{\rho}\frac{\partial}{\partial z}\left[\frac{\partial v}{\partial z}\left\{2\frac{\partial}{\partial t}\left(\left(\frac{\partial u}{\partial z}\right)^2+\left(\frac{\partial v}{\partial z}\right)^2\right)+2\frac{\partial u}{\partial z}\frac{\partial^2 u}{\partial z\partial t}+2\frac{\partial v}{\partial z}\frac{\partial^2 v}{\partial z\partial t}\right\}\right]$$
$$+\frac{(\gamma_3+\gamma_5)}{\rho}\frac{\partial}{\partial z}\left[2\frac{\partial^2 v}{\partial z\partial t}\left\{\left(\frac{\partial u}{\partial z}\right)^2+\left(\frac{\partial v}{\partial z}\right)^2\right\}\right] \quad (20)$$
$$+\frac{\partial}{\partial z}\left[\frac{\partial v}{\partial z}\left\{2\frac{\gamma_7}{\rho}\frac{\partial}{\partial t}\left(\left(\frac{\partial u}{\partial z}\right)^2+\left(\frac{\partial v}{\partial z}\right)^2\right)+\frac{(\gamma_7+\gamma_8)}{\rho}\left(\frac{\partial u}{\partial z}\frac{\partial^2 u}{\partial z\partial t}+\frac{\partial v}{\partial z}\frac{\partial^2 v}{\partial z\partial t}\right)\right\}\right]$$
$$+\frac{\gamma_1}{\rho}\frac{\partial^5 v}{\partial z^2\partial t^3}-\frac{\sigma}{\rho}B_0^2 v,$$

$$0=-\frac{\partial \hat{p}}{\partial z}. \quad (21)$$

Since $r^2=x^2+y^2$, therefore $x=\frac{\partial}{\partial x}\left(\frac{1}{2}r^2\right)$ *and* $y=\frac{\partial}{\partial y}\left(\frac{1}{2}r^2\right)$. In view these substitutions we can write Eqs. (19)-(21) in the following manner:

$$\frac{\partial u}{\partial t}-2\Omega v=-\frac{\partial}{\partial x}\left(\hat{p}-\frac{1}{2}\Omega^2 r^2\right)+\frac{\mu}{\rho}\frac{\partial^2 u}{\partial z^2}+\frac{\alpha_1}{\rho}\frac{\partial^3 u}{\partial z^2\partial t}+\frac{\beta_1}{\rho}\frac{\partial^4 u}{\partial z^2\partial t^2}$$
$$+\frac{2(\beta_2+\beta_3)}{\rho}\frac{\partial}{\partial z}\left[2\frac{\partial u}{\partial z}\left\{\left(\frac{\partial u}{\partial z}\right)^2+\left(\frac{\partial v}{\partial z}\right)^2\right\}\right]$$
$$+\frac{\gamma_2}{\rho}\frac{\partial}{\partial z}\left[\frac{\partial u}{\partial z}\left\{2\frac{\partial}{\partial t}\left(\left(\frac{\partial u}{\partial z}\right)^2+\left(\frac{\partial v}{\partial z}\right)^2\right)+2\frac{\partial u}{\partial z}\frac{\partial^2 u}{\partial z\partial t}+2\frac{\partial v}{\partial z}\frac{\partial^2 v}{\partial z\partial t}\right\}\right]$$
$$+\frac{(\gamma_3+\gamma_5)}{\rho}\frac{\partial}{\partial z}\left[2\frac{\partial^2 u}{\partial z\partial t}\left\{\left(\frac{\partial u}{\partial z}\right)^2+\left(\frac{\partial v}{\partial z}\right)^2\right\}\right] \quad (22)$$
$$+\frac{\partial}{\partial z}\left[\frac{\partial u}{\partial z}\left\{2\frac{\gamma_7}{\rho}\frac{\partial}{\partial t}\left(\left(\frac{\partial u}{\partial z}\right)^2+\left(\frac{\partial v}{\partial z}\right)^2\right)+\frac{(\gamma_7+\gamma_8)}{\rho}\left(\frac{\partial u}{\partial z}\frac{\partial^2 u}{\partial z\partial t}+\frac{\partial v}{\partial z}\frac{\partial^2 v}{\partial z\partial t}\right)\right\}\right]$$
$$+\frac{\gamma_1}{\rho}\frac{\partial^5 u}{\partial z^2\partial t^3}-\frac{\sigma}{\rho}B_0^2 u,$$

$$\begin{aligned}\frac{\partial v}{\partial t}+2\Omega u=&-\frac{\partial}{\partial y}\left(\hat{p}-\frac{1}{2}\Omega^2r^2\right)+\frac{\mu}{\rho}\frac{\partial^2 v}{\partial z^2}+\frac{\alpha_1}{\rho}\frac{\partial^3 v}{\partial z^2\partial t}+\frac{\beta_1}{\rho}\frac{\partial^4 v}{\partial z^2\partial t^2}\\&+\frac{2(\beta_2+\beta_3)}{\rho}\frac{\partial}{\partial z}\left[2\frac{\partial v}{\partial z}\left\{\left(\frac{\partial u}{\partial z}\right)^2+\left(\frac{\partial v}{\partial z}\right)^2\right\}\right]\\&+\frac{\gamma_2}{\rho}\frac{\partial}{\partial z}\left[\frac{\partial v}{\partial z}\left\{2\frac{\partial}{\partial t}\left(\left(\frac{\partial u}{\partial z}\right)^2+\left(\frac{\partial v}{\partial z}\right)^2\right)+2\frac{\partial u}{\partial z}\frac{\partial^2 u}{\partial z\partial t}+2\frac{\partial v}{\partial z}\frac{\partial^2 v}{\partial z\partial t}\right\}\right]\\&+\frac{(\gamma_3+\gamma_5)}{\rho}\frac{\partial}{\partial z}\left[2\frac{\partial^2 v}{\partial z\partial t}\left\{\left(\frac{\partial u}{\partial z}\right)^2+\left(\frac{\partial v}{\partial z}\right)^2\right\}\right]\\&+\frac{\partial}{\partial z}\left[\frac{\partial v}{\partial z}\left\{2\frac{\gamma_7}{\rho}\frac{\partial}{\partial t}\left(\left(\frac{\partial u}{\partial z}\right)^2+\left(\frac{\partial v}{\partial z}\right)^2\right)+\frac{(\gamma_7+\gamma_8)}{\rho}\left(\frac{\partial u}{\partial z}\frac{\partial^2 u}{\partial z\partial t}+\frac{\partial v}{\partial z}\frac{\partial^2 v}{\partial z\partial t}\right)\right\}\right]\\&+\frac{\gamma_1}{\rho}\frac{\partial^5 v}{\partial z^2\partial t^3}-\frac{\sigma}{\rho}B_0^2 v,\end{aligned}\tag{23}$$

$$0=-\frac{\partial}{\partial z}\left(\hat{p}-\frac{1}{2}\Omega^2r^2\right).\tag{24}$$

Redefining the modified pressure

$$\tilde{p}=\hat{p}-\frac{1}{2}\Omega^2r^2,\tag{25}$$

Eqs. (22)-(24) become

$$\begin{aligned}\frac{\partial u}{\partial t}-2\Omega v=&-\frac{\partial\tilde{p}}{\partial x}+\frac{\mu}{\rho}\frac{\partial^2 u}{\partial z^2}+\frac{\alpha_1}{\rho}\frac{\partial^3 u}{\partial z^2\partial t}+\frac{\beta_1}{\rho}\frac{\partial^4 u}{\partial z^2\partial t^2}\\&+\frac{2(\beta_2+\beta_3)}{\rho}\frac{\partial}{\partial z}\left[2\frac{\partial u}{\partial z}\left\{\left(\frac{\partial u}{\partial z}\right)^2+\left(\frac{\partial v}{\partial z}\right)^2\right\}\right]\\&+\frac{\gamma_2}{\rho}\frac{\partial}{\partial z}\left[\frac{\partial u}{\partial z}\left\{2\frac{\partial}{\partial t}\left(\left(\frac{\partial u}{\partial z}\right)^2+\left(\frac{\partial v}{\partial z}\right)^2\right)+2\frac{\partial u}{\partial z}\frac{\partial^2 u}{\partial z\partial t}+2\frac{\partial v}{\partial z}\frac{\partial^2 v}{\partial z\partial t}\right\}\right]\\&+\frac{(\gamma_3+\gamma_5)}{\rho}\frac{\partial}{\partial z}\left[2\frac{\partial^2 u}{\partial z\partial t}\left\{\left(\frac{\partial u}{\partial z}\right)^2+\left(\frac{\partial v}{\partial z}\right)^2\right\}\right]\\&+\frac{\partial}{\partial z}\left[\frac{\partial u}{\partial z}\left\{2\frac{\gamma_7}{\rho}\frac{\partial}{\partial t}\left(\left(\frac{\partial u}{\partial z}\right)^2+\left(\frac{\partial v}{\partial z}\right)^2\right)+\frac{(\gamma_7+\gamma_8)}{\rho}\left(\frac{\partial u}{\partial z}\frac{\partial^2 u}{\partial z\partial t}+\frac{\partial v}{\partial z}\frac{\partial^2 v}{\partial z\partial t}\right)\right\}\right]\\&+\frac{\gamma_1}{\rho}\frac{\partial^5 u}{\partial z^2\partial t^3}-\frac{\sigma}{\rho}B_0^2 u,\end{aligned}\tag{26}$$

$$\begin{aligned}\frac{\partial v}{\partial t}+2\Omega u=&-\frac{\partial\tilde{p}}{\partial y}+\frac{\mu}{\rho}\frac{\partial^2 v}{\partial z^2}+\frac{\alpha_1}{\rho}\frac{\partial^3 v}{\partial z^2\partial t}+\frac{\beta_1}{\rho}\frac{\partial^4 v}{\partial z^2\partial t^2}\\&+\frac{2(\beta_2+\beta_3)}{\rho}\frac{\partial}{\partial z}\left[2\frac{\partial v}{\partial z}\left\{\left(\frac{\partial u}{\partial z}\right)^2+\left(\frac{\partial v}{\partial z}\right)^2\right\}\right]\\&+\frac{\gamma_2}{\rho}\frac{\partial}{\partial z}\left[\frac{\partial v}{\partial z}\left\{2\frac{\partial}{\partial t}\left(\left(\frac{\partial u}{\partial z}\right)^2+\left(\frac{\partial v}{\partial z}\right)^2\right)+2\frac{\partial u}{\partial z}\frac{\partial^2 u}{\partial z\partial t}+2\frac{\partial v}{\partial z}\frac{\partial^2 v}{\partial z\partial t}\right\}\right]\\&+\frac{(\gamma_3+\gamma_5)}{\rho}\frac{\partial}{\partial z}\left[2\frac{\partial^2 v}{\partial z\partial t}\left\{\left(\frac{\partial u}{\partial z}\right)^2+\left(\frac{\partial v}{\partial z}\right)^2\right\}\right]\\&+\frac{\partial}{\partial z}\left[\frac{\partial v}{\partial z}\left\{2\frac{\gamma_7}{\rho}\frac{\partial}{\partial t}\left(\left(\frac{\partial u}{\partial z}\right)^2+\left(\frac{\partial v}{\partial z}\right)^2\right)+\frac{(\gamma_7+\gamma_8)}{\rho}\left(\frac{\partial u}{\partial z}\frac{\partial^2 u}{\partial z\partial t}+\frac{\partial v}{\partial z}\frac{\partial^2 v}{\partial z\partial t}\right)\right\}\right]\\&+\frac{\gamma_1}{\rho}\frac{\partial^5 v}{\partial z^2\partial t^3}-\frac{\sigma}{\rho}B_0^2 v,\end{aligned}\tag{27}$$

$$0=-\frac{\partial\tilde{p}}{\partial z}.\tag{28}$$

Differentiating Eqs. (26) and (27) with respect to z and making use of Eq. (28), and then integrating with respect to z to obtain

$$\begin{aligned}\frac{\partial u}{\partial t}-2\Omega v=&\frac{\mu}{\rho}\frac{\partial^2 u}{\partial z^2}+\frac{\alpha_1}{\rho}\frac{\partial^3 u}{\partial z^2\partial t}+\frac{\beta_1}{\rho}\frac{\partial^4 u}{\partial z^2\partial t^2}\\&+\frac{2(\beta_2+\beta_3)}{\rho}\frac{\partial}{\partial z}\left[2\frac{\partial u}{\partial z}\left\{\left(\frac{\partial u}{\partial z}\right)^2+\left(\frac{\partial v}{\partial z}\right)^2\right\}\right]\\&+\frac{\gamma_2}{\rho}\frac{\partial}{\partial z}\left[\frac{\partial u}{\partial z}\left\{2\frac{\partial}{\partial t}\left(\left(\frac{\partial u}{\partial z}\right)^2+\left(\frac{\partial v}{\partial z}\right)^2\right)+2\frac{\partial u}{\partial z}\frac{\partial^2 u}{\partial z\partial t}+2\frac{\partial v}{\partial z}\frac{\partial^2 v}{\partial z\partial t}\right\}\right]\\&+\frac{(\gamma_3+\gamma_5)}{\rho}\frac{\partial}{\partial z}\left[2\frac{\partial^2 u}{\partial z\partial t}\left\{\left(\frac{\partial u}{\partial z}\right)^2+\left(\frac{\partial v}{\partial z}\right)^2\right\}\right]\\&+\frac{\partial}{\partial z}\left[\frac{\partial u}{\partial z}\left\{2\frac{\gamma_7}{\rho}\frac{\partial}{\partial t}\left(\left(\frac{\partial u}{\partial z}\right)^2+\left(\frac{\partial v}{\partial z}\right)^2\right)+\frac{(\gamma_7+\gamma_8)}{\rho}\left(\frac{\partial u}{\partial z}\frac{\partial^2 u}{\partial z\partial t}+\frac{\partial v}{\partial z}\frac{\partial^2 v}{\partial z\partial t}\right)\right\}\right]\\&+\frac{\gamma_1}{\rho}\frac{\partial^5 u}{\partial z^2\partial t^3}-\frac{\sigma}{\rho}B_0^2 u+\lambda(t),\end{aligned}\tag{29}$$

$$\begin{aligned}
\frac{\partial v}{\partial t}+2\Omega u=&\frac{\mu}{\rho}\frac{\partial^2 v}{\partial z^2}+\frac{\alpha_1}{\rho}\frac{\partial^3 v}{\partial z^2\partial t}+\frac{\beta_1}{\rho}\frac{\partial^4 v}{\partial z^2\partial t^2}\\
&+\frac{2(\beta_2+\beta_3)}{\rho}\frac{\partial}{\partial z}\left[2\frac{\partial v}{\partial z}\left\{\left(\frac{\partial u}{\partial z}\right)^2+\left(\frac{\partial v}{\partial z}\right)^2\right\}\right]\\
&+\frac{\gamma_2}{\rho}\frac{\partial}{\partial z}\left[\frac{\partial v}{\partial z}\left\{2\frac{\partial}{\partial t}\left(\left(\frac{\partial u}{\partial z}\right)^2+\left(\frac{\partial v}{\partial z}\right)^2\right)+2\frac{\partial u}{\partial z}\frac{\partial^2 u}{\partial z\partial t}+2\frac{\partial v}{\partial z}\frac{\partial^2 v}{\partial z\partial t}\right\}\right]\\
&+\frac{(\gamma_3+\gamma_5)}{\rho}\frac{\partial}{\partial z}\left[2\frac{\partial^2 v}{\partial z\partial t}\left\{\left(\frac{\partial u}{\partial z}\right)^2+\left(\frac{\partial v}{\partial z}\right)^2\right\}\right]\\
&+\frac{\partial}{\partial z}\left[\frac{\partial v}{\partial z}\left\{2\frac{\gamma_7}{\rho}\frac{\partial}{\partial t}\left(\left(\frac{\partial u}{\partial z}\right)^2+\left(\frac{\partial v}{\partial z}\right)^2\right)+\frac{(\gamma_7+\gamma_8)}{\rho}\left(\frac{\partial u}{\partial z}\frac{\partial^2 u}{\partial z\partial t}+\frac{\partial v}{\partial z}\frac{\partial^2 v}{\partial z\partial t}\right)\right\}\right]\\
&+\frac{\gamma_1}{\rho}\frac{\partial^5 v}{\partial z^2\partial t^3}-\frac{\sigma}{\rho}B_0^2 v+\delta(t).
\end{aligned}\tag{30}$$

On multiplying Eq. (30) by i and then adding the resulting equation in Eq. (29) we get

$$\begin{aligned}
\frac{\partial q}{\partial t}+2i\Omega q=&\frac{\mu}{\rho}\frac{\partial^2 q}{\partial z^2}+\frac{\alpha_1}{\rho}\frac{\partial^3 q}{\partial z^2\partial t}+\frac{\beta_1}{\rho}\frac{\partial^4 q}{\partial z^2\partial t^2}\\
&+\frac{4(\beta_2+\beta_3)}{\rho}\frac{\partial}{\partial z}\left[\left\{\left(\frac{\partial q}{\partial z}\right)^2\left(\frac{\partial \bar{q}}{\partial z}\right)\right\}\right]\\
&+\frac{\gamma_2}{\rho}\frac{\partial}{\partial z}\left[\left\{3\left(\frac{\partial q}{\partial z}\right)\frac{\partial}{\partial t}\left(\frac{\partial q}{\partial z}\frac{\partial \bar{q}}{\partial z}\right)\right\}\right]+\frac{\gamma_1}{\rho}\frac{\partial^5 q}{\partial z^3\partial t^2}\\
&+\frac{(\gamma_3+\gamma_5)}{\rho}\frac{\partial}{\partial z}\left[\left\{2\left(\frac{\partial^2 q}{\partial z\partial t}\right)\left(\frac{\partial q}{\partial z}\frac{\partial \bar{q}}{\partial z}\right)\right\}\right]\\
&+\frac{\partial}{\partial z}\left[\begin{array}{c}\frac{2\gamma_7}{\rho}\left(\frac{\partial q}{\partial z}\right)\frac{\partial}{\partial t}\left(\frac{\partial q}{\partial z}\frac{\partial \bar{q}}{\partial z}\right)\\ +\frac{(\gamma_7+\gamma_8)}{2\rho}\left(\frac{\partial q}{\partial z}\right)\frac{\partial}{\partial t}\left(\frac{\partial q}{\partial z}\frac{\partial \bar{q}}{\partial z}\right)\end{array}\right]-\frac{\sigma}{\rho}B_0^2 q+\psi(t),
\end{aligned}\tag{31}$$

where

$$q = u + iv, \quad \bar{q} = u - iv, \quad \psi(t) = \lambda(t) + i\delta(t) \tag{32}$$

For steady state the Equations (29) and (30) reduce to

$$-2\Omega v = \frac{\mu}{\rho}\frac{\partial^2 u}{\partial z^2} + \frac{2(\beta_2+\beta_3)}{\rho}\frac{\partial}{\partial z}\left[2\frac{\partial u}{\partial z}\left\{\left(\frac{\partial u}{\partial z}\right)^2 + \left(\frac{\partial v}{\partial z}\right)^2\right\}\right] - \frac{\sigma}{\rho}B_0^2 u, \tag{33}$$

$$2\Omega u = \frac{\mu}{\rho}\frac{\partial^2 v}{\partial z^2} + \frac{2(\beta_2+\beta_3)}{\rho}\frac{\partial}{\partial z}\left[2\frac{\partial v}{\partial z}\left\{\left(\frac{\partial u}{\partial z}\right)^2 + \left(\frac{\partial v}{\partial z}\right)^2\right\}\right] - \frac{\sigma}{\rho}B_0^2 v, \tag{34}$$

Introducing the dimensionless variables

$$z^* = \frac{z}{L}, \; u^* = \frac{uL}{\nu}, \text{ and } v^* = \frac{vL}{\nu} \tag{35}$$

in above equations and simplifying the resulting equations and dropping '*' to obtain

$$-2vE^{-1} = \frac{\partial^2 u}{\partial z^2} + 4\beta\left[3\left(\frac{\partial^2 u}{\partial z^2}\right)\left(\frac{\partial u}{\partial z}\right)^2 + \left(\frac{\partial^2 u}{\partial z^2}\right)\left(\frac{\partial v}{\partial z}\right)^2 + 2\left(\frac{\partial u}{\partial z}\right)\left(\frac{\partial v}{\partial z}\right)\left(\frac{\partial^2 v}{\partial z^2}\right)\right] - H^2 u \tag{36}$$

$$2E^{-1}u = \frac{\partial^2 v}{\partial z^2} + 4\beta\left[3\left(\frac{\partial^2 v}{\partial z^2}\right)\left(\frac{\partial v}{\partial z}\right)^2 + \left(\frac{\partial^2 v}{\partial z^2}\right)\left(\frac{\partial u}{\partial z}\right)^2 + 2\left(\frac{\partial u}{\partial z}\right)\left(\frac{\partial v}{\partial z}\right)\left(\frac{\partial^2 u}{\partial z^2}\right)\right] - H^2 v \tag{37}$$

where

$$E^{-1} = \frac{\Omega L^2}{\nu}, \; \beta = \frac{4(\beta_2+\beta_3)\nu}{\rho L^4}, \; H^2 = \frac{nL^2}{\nu}, \; n = \frac{\sigma B_0^2}{\rho} \tag{38}$$

and E is the Ekman number while H is the Hartmann number.

3. Numerical procedure

Consider a simplest boundary value problem

$$F(u'', u', u, z) = 0, \tag{39}$$

$$u(a) = A \text{ and } u(b) = B. \tag{40}$$

To solve the boundary value problem the derivatives u' and u'' involved in the problem are approximated by finite differences of appropriate order. If we employ second order central difference formulation, then we can write

$$u'(z) = \frac{u(z+h) - u(z-h)}{2h} + O(h^2), \tag{41}$$

$$u''(z) = \frac{u(z+h) - 2u(z) + u(z-h)}{h^2} + O(h^2). \quad (42)$$

This converts the given boundary value problem into a linear system of equations involving values of the function u at a, $a+h$, $a+2h,\cdots$,b. For higher accuracy, one should choose h small. However, this increases the number of equations in the system which in turn increases the computational time.

Depending upon the size of this resulting system of linear equations, it can either be solved by exact methods or approximate methods.

In the present problem the governing differential equations (36) and (37) are highly non-linear which cannot be solved analytically. These equations are discretized using second order central finite difference approximations defined in Eqs. (41) and (42). The resulting system of algebraic equations is solved using successive under relaxation scheme. The difference equations are linearized employing a procedure known as lagging the coefficients [59]. The iterative procedure is repeated until convergence is obtained according to the following criterion

$$\max\left|u^{(n+1)} - u^{(n)}\right| < \varepsilon,$$

where superscript $'n'$ represents the number of iteration and $'\varepsilon'$ is the order of accuracy. In the present case ε is taken as 10^{-8}.

4. Numerical results and discussion

The steady velocity components u and v are plotted against independent variable z for different values of Ekman number E, Hartmann number H and material parameter β and results are compared for two types of fluids: the Newtonian fluid, for which $\beta_i = 0$ $(i = 1,2,3)$, and the non-Newtonian fluid, in which we choose $\beta = 1$. Fig. 1 shows the effect of Hartmann number H on the velocity components u and v. We fixed $E = 1$ and varied H= $0,1,3,5$. It is observed that an increase in the Hartmann number reduces the velocity components u and v due to the effects of the magnetic force against the flow direction. Figs. 1a and 1b show that with an increase of Hartmann number H, the curvature of the velocity component u profile increases for both a Newtonian fluid and non-Newtonian fluid. Quite contrary, increasing Hartmann number H causes the velocity component v profile to become less parabolic, see Figs. 1c and 1d. It is also noted that decrease in u and v in the Newtonian fluid is larger as compared with non-Newtonian fluid. Furthermore, the boundary layer thickness is drastically decreased by increasing H. It means that the magnetic field provides some mechanism to control the boundary layer thickness.

The dependence of the velocity components u and v on the Ekman number is shown in Fig. 2. In Fig. 2 we fixed $H = 1$ and varied $E = 0.1$, 0.2, 0.3. It is observed that velocity component u increases with an increase in Ekman number E for the Newtonian fluid while it remains almost unaffected for non-Newtonian fluid (Figs. 2a, 2b). Moreover, a

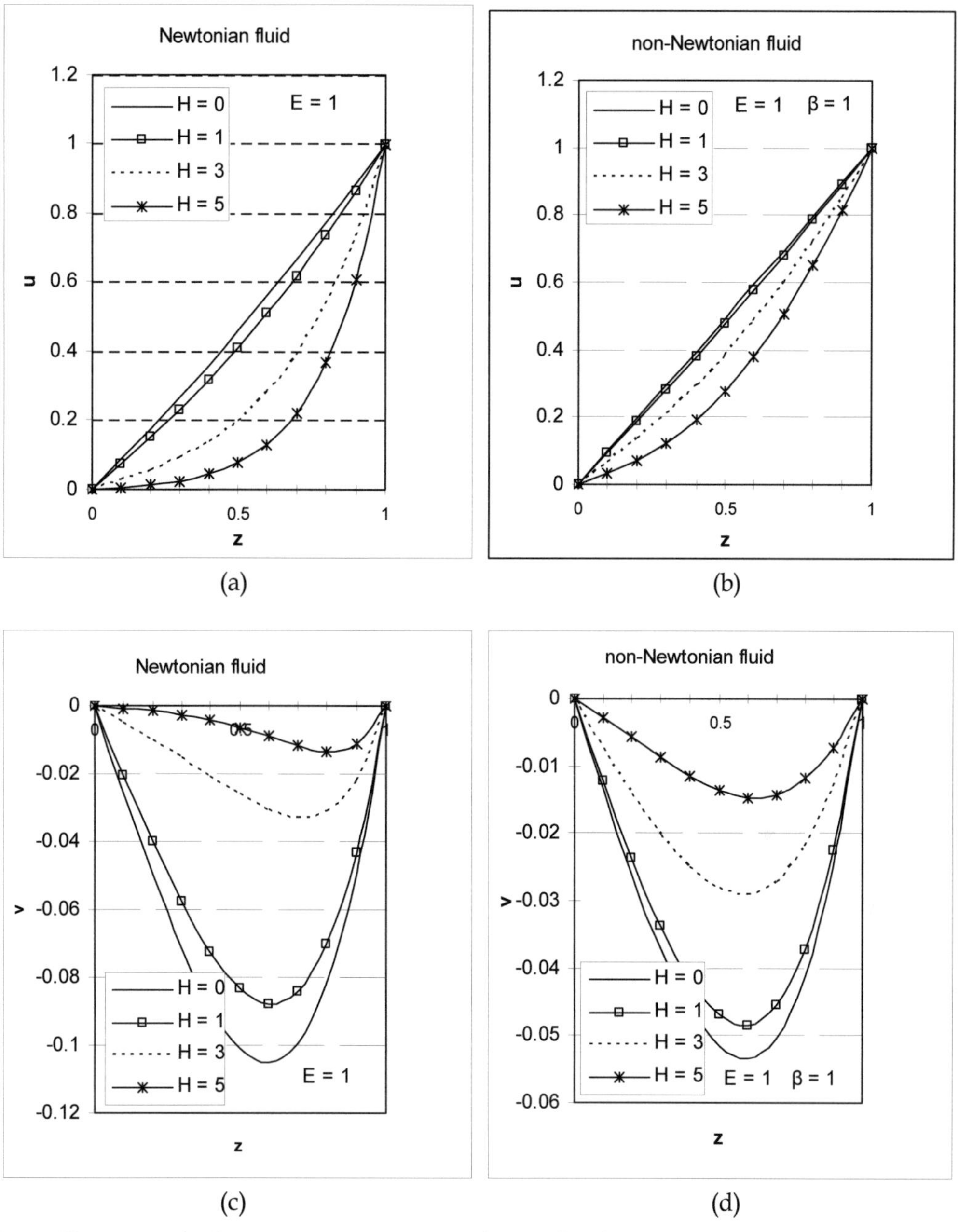

Fig. 1. Variation of velocity components u and v with z for $H = 0,1,3,5;\ E = 1, \beta = 0$ in (a), (c); $E = 1, \beta = 1$ in (b), (d).

backflow is observed near the boundary $z = 0$ for $E = 0.1$. On the contrary, the magnitude of velocity component v decreases with an increase in Ekman number E for the both types of the fluids (Figs. 2c, 2d). This velocity component has larger magnitude in Newtonian fluid as compared with non-Newtonian fluid.

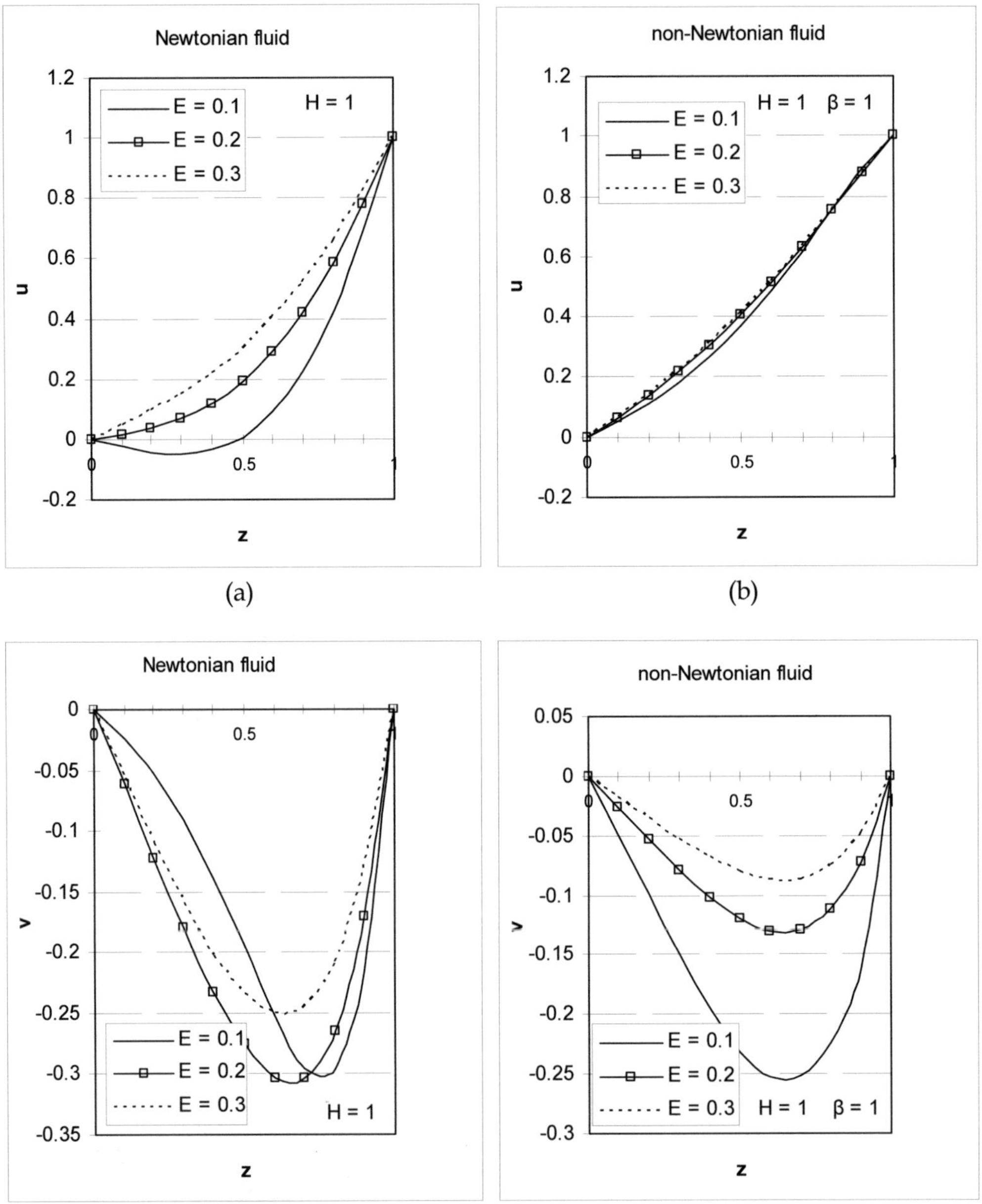

Fig. 2. Variation of velocity components u and v with z for $E = 0.1,\ 0.2,\ 0.3;\ H = 1, \beta = 0$ in (a), (c); $H = 1, \beta = 1$ in (b), (d).

Figs. 3,4 depict the variation of the velocity components u and v with z for various values of material parameter β fixing $E = 1$ and taking $H = 1$ in Figs. 3a and 3c, while $H = 5$ in 3b, 3d, and in Fig. 4. It is observed from Fig. 3b and 4a that when the material parameter β increases from $\beta = 1$ to a large value of 20, the velocity component u tend to approach the

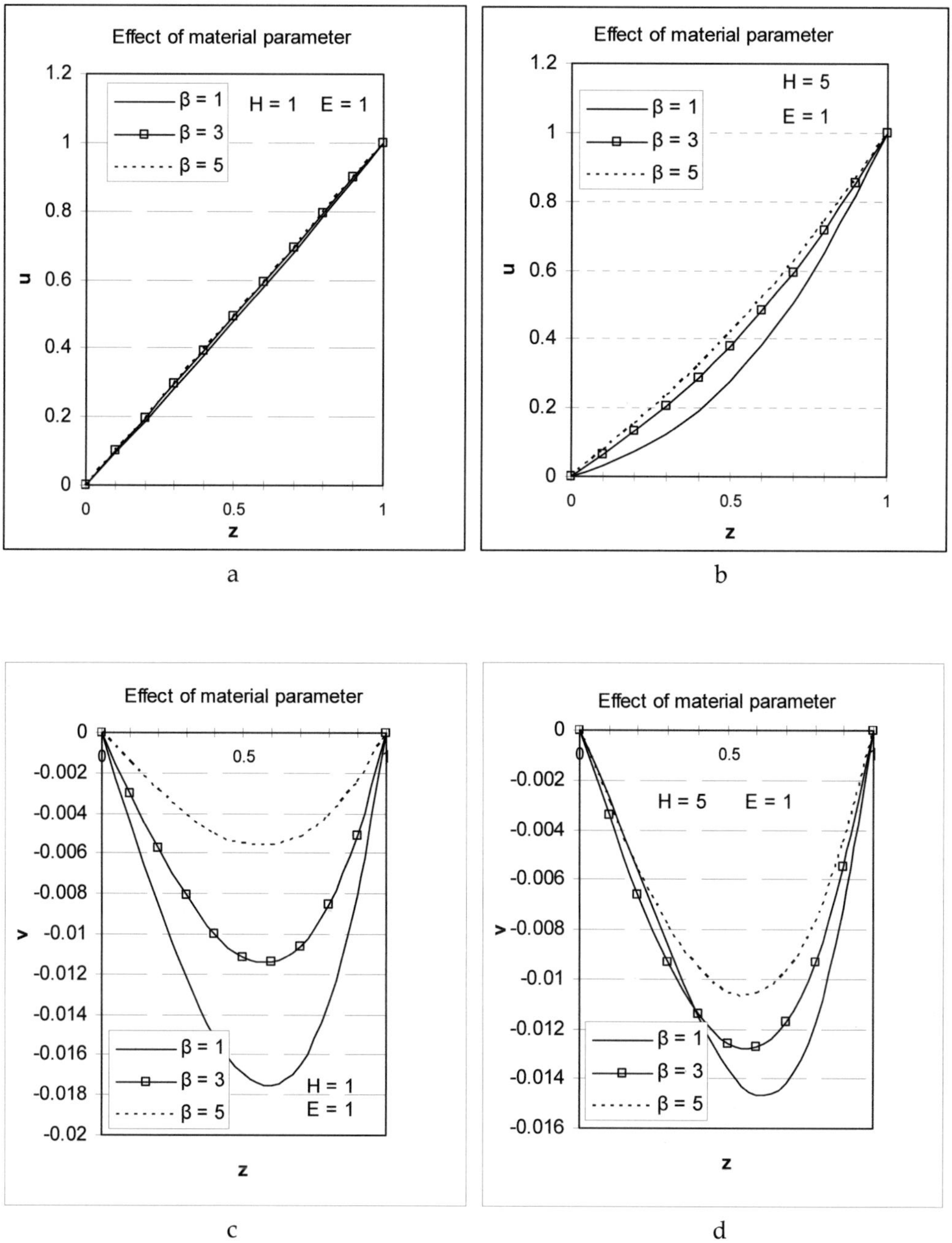

Fig. 3. Variation of velocity components u and v with z for $\beta = 1,\ 3,\ 5$; fixing $E = 1$, $H = 5$.

linear distribution; thus, the shearing can unattenuately extend to the whole flow domain from the boundaries, corresponding to a shear-thickening phenomenon. A further increase of β will not effect this velocity component further. The magnitude of velocity component v decreases when β increases and the curvature of the velocity profile decreases with an increase in material parameter β (see Figs. 3c, 3d and 4b). It is also found that the flow behaviour depends strongly on the choice of the parameters, for example, for large H, u increases with an increase of material parameter β, whereas this velocity component is independent of β for small H. On the contrary, the magnitude of velocity component v decreases with β for both small and large values of H.

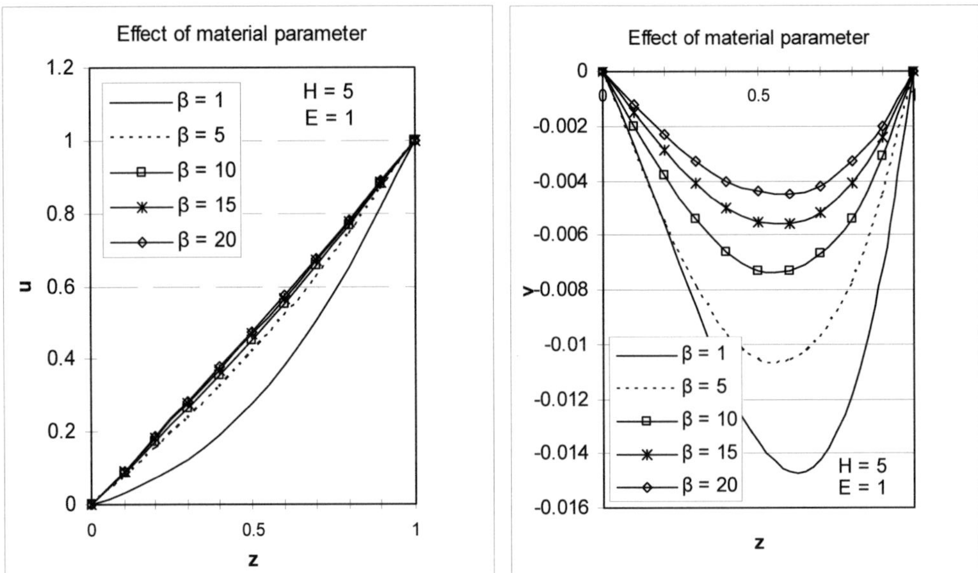

(a) (b)

Fig. 4. Variation of velocity components u and v with z for large values of $\beta = 1, 5,\ 10,\ 15, 20;$ fixing $E = 1, H = 5$.

5. Conclusions

The unsteady rotating flow of a uniformly conducting incompressible fourth-grade fluid between two parallel infinite plates in the presence of a magnetic field is modeled. The steady rotating flow of the non-Newtonian fluid subject to a uniform transverse magnetic field is studied. The governing non-linear equations are solved numerically. The numerical results of the non-Newtonian fluid are compared with those of a Newtonian fluid. The major findings of the present works can be summarized as follows:

- The transverse magnetic field decelerates the fluid motion. When the strength of the magnetic field increases, the flow velocity decreases.
- It is observed that the boundary layer thickness decreases drastically by increasing H. It means that the magnetic field provides some mechanism to control the boundary layer thickness.

- It is noted that the flow behaviour depends strongly on the choice of the parameters.

6. References

[1] R.S. Rivlin, J.L. Ericksen, Stress deformation relations for isotropic material, J. Rational Mech. Anal. 4 (1955) 323-425.

[2] C. Truesdell and W. Noll, The Nonlinear Field Theories of Mechanics, 2nd ed. Springer-Verlog, Brelin, 1992.

[3] J. G. Oldroyd, On the formulation of rheological equations of state, Proc. Roy. Soc. Lond. A 200 (1950) 523-541.

[4] K.R. Rajagopal, Mechanics of non-Newtonian fluids in Recent Developments in Theoretical Fluid Mechanics, Pitman Res. Notes Math. Ser. 291, eds. G.P. Galdi and J. Necas (Springer), 1993.

[5] J.E. Dunn, K.R. Rajagopal, Fluids of differential type: critical review and thermodynamic analysis, Int. J. Engng. Sci. 33 (1995) 689-729.

[6] R.R. Huilgol, Continuum mechanics of viscoelastic liquids, Hindushan Publishing Corporation, 1975.

[7] W.R. Schowalter, Mechanics of non-Newtonian fluids, Pergamon, 1978.

[8] M.A. Rana, A.M. Siddiqui, Rashid Qamar, Hall effects on hydromagnetic flow of an Oldroyd 6-constant fluid between concentric cylinders, Chaos, Solitons and Fractals 39 (2009) 204–213.

[9] R. Bandelli, K. R. Rajagopal and G.P. Galdi, On some unsteady motions of fluids of second grade. Arch. Mech., 47 (1995) 661 -676.

[10] R.L. Fosdick, K.R. Rajagopal, Anomalous features in the model of "second order fluids", Arch. Ration. Mech. Anal. 70 (1979) 145-152.

[11] R.L. Fosdick, K.R. Rajagopal, Thermaodynamics and stability of fluids of third grade, Proc. Roy. Soc. London Ser. A 369 (1980), no. 1738, 351-377.

[12] J. Malek, K.R. Rajagopal, and M. Ruzicka, Existence and regularity of solutions and stability of the rest state for the fluids with shear dependent viscosity, Math. Models Methods Appl. Sci. 5 (1995), no. 6, 789-812.

[13] Hartmann Hydrodynamics. 1. theory of the lamina flow of an electrically conductive liquid in a homogeneous magnetic field. Kgi Danske Videnslab, Sleskuh, Mat. Fys. Medd. 15 (1937).

[14] G.W. Sutton and A. Shermann, Engineering magneto-hydrodynamics, McGraw-Hill Co. Inc. 1965.

[15] W.F. Hughes and Y.J. Young, The electromagnetinodynamics of fluids, John Wiley, New York, 1966.

[16] T.G. Cowling, Magnetohydrodynamics, Interscience Publishing, Inc., New York, 1957.

[17] S.I. Pai, Magneto-gas dynamics and plasma dynamics, Springer-Verlag, 1962.

[18] K. R. Rajagopal, T.Y. Na, On Stokes problem for non-Newtonian fluid. Acta Mech. 48 (1983) 233 -239.

[19] F. Mollica, K.R. Rajagopal, Secondary flows due to axial shearing of a third grade fluid between two eccentrically placed cylinders. Int. J. Engng. Sci. 37 (1999) 411 - 429.

[20] A.M. Siddiqui, P.N. Kaloni, Plane steady flows of a third grade fluid. Int. J. Engng. Sci. 25 (2) (1987) 171 -188.

[21] A.M. Sidiqui, M.A. Rana, Naseer Ahmed, Magnetohydrodynamic flow of a Burgers' fluid in an orthogonal rheometer, Applied Mathematical Modelling, 34(2010) 2881-2892.

[22] A.M. Sidiqui, M.A. Rana, Naseer Ahmed, Hall effects on flow and heat transfer in the hydromagnetic Burgers Ekman layer, Int. J. Moder. Math. 2(2) (2007) 255-268.

[23] S.R. Kasiviswanathan and A.R. Rao, On exact solutions of unsteady MHD flow between eccentrically rotating disks, Arch, Mech., 39(4) (1987) 411-418.

[24] H.K. Mohanty, Hydromagnetic flow between two rotating disk with non-coincident parallel axes of rotation, Phys., Fluids 15 (1972) 1456-1458.

[25] C.S. Erkman, Lett. Appl. Engng. Sci. 3(1975) 51.

[26] S.N. Murthy, R.K.P. Ram, MHD flow and heat transfer due to eccentric rotation of a porous disk and a fluid at infinity, Int. J. Engng. Sci. 16(1978) 943-949.

[27] P.N. Kaloni and A.M. Siddiqui, A note on the flow of viscoelastic fluid between eccentric disks, J. Non-Newtonian Fluid Mech. 26 (1987). 125–133.

[28] A.M. Siddiqui, T. Haroon and S. Asghar, Unsteady MHD flow of non-Newtonian fluid due to eccentric rotations of a porous disk and a fluid at infinity, Acta Mech. 147 (2000), 99-109.

[29] A.M. Sidiqui, M.A. Rana, Naseer Ahmed, Effects of Hall current and heat transfer on MHD flow of a Burgers' fluid due to a pull of eccentric rotating disks, Commun. Nonlinear Sci. Numer. Simul. 13(2008) 1554-1570.

[30] G. Palani, I.A. Abbas, Free Convection MHD Flow with Thermal Radiation from an Impulsively-Started Vertical Plate, Nonlinear Analysis: Modelling and Control, 14(1) (2009) 73–84.

[31] M. Farzaneh-Gord, A. A. Joneidi, and B. Haghighi, Investigating the effects of the important parameters on magnetohydrodynamics flow and heat transfer over a stretching sheet, Proceedings of the Institution of Mechanical Engineers, Part E: Journal of Process Mechanical Engineering 2010 224: 1DOI:10.1243/09544089JPME258.

[32] E.R. Maki, D .Kuzma, R.L. Donnelly, B. Kim, Magnetohydrodynamic lubrication flow between parallel plates. J. Fluid Mech. 26 (1966) 537–543.

[33] W.F. Hughes, R.A. Elco, Magnetohydrodynamic lubrication flow between parallel rotating disks. J. Fluid Mech. 13 (1962) 21–32.

[34] D.C. Kuzma, E.R. Maki, R.J. Donnelly, The magnetohydrodynamic squeeze film, J. Tribol 110 (1988) 375–377.

[35] E.A. Hamza, The magnetohydrodynamic squeeze film, J. Fluid Mech. 19 (1964) 395–400.

[36] E.A. Hamza, The magnetohydrodynamic effects on a fluid film squeezed between two rotating surfaces, J. Phys. D: Appl. Phys. 24 (1991) 547–554.

[37] S. Bhattacharyya, A. Pal, Unsteady MHD squeezing flow between two parallel rotating discs, Mech. Res. Commun. 24 (1997) 615–623.

[38] Erik Sweet, K. Vajravelu, A. Robert. Van Gorder, I. Pop, Analytical solution for the unsteady MHD flow of a viscous fluid between moving parallel plates, Commun. Nonlinear Sci. Numer. Simulat. 16 (2011) 266–273.

[39] Z. Abbas, T. Javed, M. Sajid, N. Ali, Unsteady MHD flow and heat transfer on a stretching sheet in a rotating fluid, Journal of the Taiwan Institute of Chemical Engineers 41 (2010) 644–650.

[40] M. Turkylimazoglu, Unsteady MHD flow with variable viscosity: Applications of spectral scheme, Int. J. Thermal Sci. 49 (2010) 563-570.

[41] T. Hayat, C. Fetecaua, M. Sajid, Analytic solution for MHD Transient rotating flow of a second grade fluid in a porous space, Nonlinear Analysis: Real World Applications 9 (2008) 1619 - 1627.

[42] A.C. Eringen, Simple microfluids, Int. J. Eng. Sci. 2 (1964) 203–217.

[43] A.C. Eringen, Theory of micropolar fluids, Math. Mech. 16 (1966) 1–18.

[44] A.C. Eringen, Theory of thermomicropolar fluids, Math. Anal. Appl. 38 (1972) 480– 496.

[45] N.T. Eldabe, E.F. Elshehawey, Elsayed M.E. Elbarbary, Nasser S. Elgazery, Chebyshev finite difference method for MHD flow of a micropolar fluid past a stretching sheet with heat transfer, Applied Mathematics and Computation 160 (2005) 437–450.

[46] T. Sarpakaya, Flow of non-Newtonian fluids in magnetic field, AIChE J 7 (1961) 324–328.

[47] M.I. Char, Heat and mass transfer in a hydromagnetic flow of viscoelastic fluid over a stretching sheet, J. Math. Annal Appl. 186 (1994) 674–689.

[48] A. Raptis, Flow of a micropolar fluid past a continuously moving plate by the presence of radiation, Int. J. Heat Mass transfer 41 (1998) 2865–2866.

[49] A. Raptis, C. Perdikis, Viscoelastic flow by the presence of radiation. ZAMM 78 (1998) 277–279.

[50] A. Raptis, C. Pardikis, H.S. Takhar, Effect of thermal radiation on MHD flow, Appl. Math. Comput. 153 (2004) 645–649.

[51] P. G. Siddheshwar, U.S. Mahabaleswar, Effects of radiation and heat source on MHD flow of a visco-elastic liquid and heat transfer over a stretching sheet, Int. J. Non-linear Mech. 40 (2005) 807–821.

[52] S.K. Khan, Heat transfer in a viscoelastic fluid flow over a stretching surface with heat source/sink, suction/blowing and radiation, Int. J. Heat Mass transfer 49 (2006) 628–639.

[53] M.S. Abel, P.G. Siddheshwar, M. Nandeppanavar Mahantesh, Heat transfer in a viscoelastic boundary layer flow over a stretching sheet with viscous dissipation and non-uniform heat source, Int. J. Heat Mass Transfer 50 (2007) 960–966.

[54] M.S. Abel, M.M. Nandeppanavar, Heat transfer in MHD viscoelastic boundary layer flow over a stretching sheet with non-uniform heat source/sink, Commun. Nonlinear Sci. Numer. Simul. 14 (2009) 2120–2131.

[55] M.M. Nandeppanavar, M.S. Abel, J. Tawade, Heat transfer in a Walter's liquid B fluid over an impermeable stretching sheet with non-uniform heat source/sink and elastic deformation, Commun. Nonlinear Sci. Numer. Simul. 15 (2010) 1791–1802.

[56] M. Mahantesh, Nandeppanavar, K. Vajravelu, M. Subhas Abel, Heat transfer in MHD viscoelastic boundary layer flow over a stretching sheet with thermal radiation and non-uniform heat source/sink, Commun. Nonlinear Sci. Numer. Simulat. 16 (2011) 3578–3590.

[57] Kayvan Sadeghy, Navid Khabazi, Seyed-Mohammad Taghavi, Magnetohydrodynamic (MHD) flows of viscoelastic fluids in converging/diverging Channels, Int. J. Engng. Sci. 45 (2007) 923–938.

[58] B. D. Coleman and W. Noll, An approximation theorem for functionals with applications continuum mechanics, Arch. Rational Mech. Anal. 6 (1960) 355-370.

[59] K.A Hoffmann, S. Chiang, Computational fluid dynamics for engineers, I, 1995, P. 92.